THE PHYSICOCHEMICAL BASIS OF
PHARMACEUTICALS

THE PHYSICOCHEMICAL BASIS OF
PHARMACEUTICALS

HUMPHREY A. MOYNIHAN
ABINA M. CREAN

University College Cork

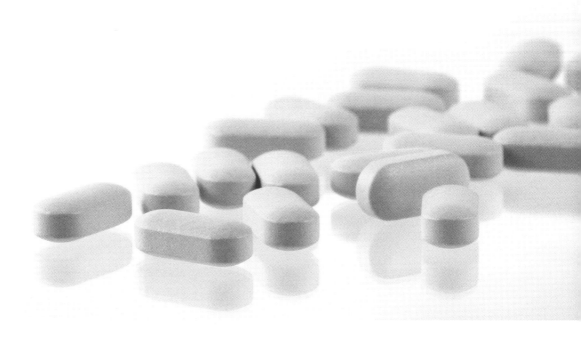

OXFORD
UNIVERSITY PRESS

OXFORD

UNIVERSITY PRESS

Great Clarendon Street, Oxford OX2 6DP

Oxford University Press is a department of the University of Oxford.
It furthers the University's objective of excellence in research, scholarship,
and education by publishing worldwide in

Oxford New York

Auckland Cape Town Dar es Salaam Hong Kong Karachi
Kuala Lumpur Madrid Melbourne Mexico City Nairobi
New Delhi Shanghai Taipei Toronto

With offices in

Argentina Austria Brazil Chile Czech Republic France Greece
Guatemala Hungary Italy Japan Poland Portugal Singapore
South Korea Switzerland Thailand Turkey Ukraine Vietnam

Oxford is a registered trade mark of Oxford University Press
in the UK and in certain other countries

Published in the United States
by Oxford University Press Inc., New York

British Library Cataloguing in Publication Data

Data available

Library of Congress Cataloging in Publication Data

Data available

Typeset by Graphicraft Limited, Hong Kong
Printed in Italy by L.E.G.O. S.p.A

ISBN: 978–0–19–923284–0

1 3 5 7 9 10 8 6 4 2

Dedicated to Nuala, Helen & Martha and William & Helen

PREFACE

Recent years have seen a tremendous growth in education provision in the pharmaceutical sciences. Many new courses in pharmacy, medicinal chemistry, pharmacology and related disciplines have come on stream, and new textbooks covering many of the underpinning sciences and technologies have appeared.

One issue that affects a range of pharmaceutical disciplines is that of the physical properties of pharmaceutical chemicals, better known as their physicochemical properties. Whether studying pharmaceutical issues within courses in pharmacy, chemistry, chemical engineering, biochemistry or other disciplines, engagement with physicochemical properties becomes necessary, affecting as they do issues as diverse as bioavailability, formulation, patents and manufacturing.

This text aims to provide an introduction to the physicochemical issues that underpin the pharmaceutical sciences. The book is particularly aimed at students of pharmacy, pharmaceutical chemistry, medicinal chemistry, chemical engineering, biochemistry, biotechnology, or of pharmaceutically related interdisciplinary courses.

The book briefly explains what pharmaceuticals actually are, and describes the environments in which pharmaceutical compounds exist and act. The book covers the fundamental theories required to deal with physicochemical issues in a quantitative way, which is essential if we wish to engage with 'real-world' pharmaceutical issues. Pharmaceutical materials existing in liquid, solid or gaseous states are considered. A central theme of the book concerns the different states and environments in which a pharmaceutical compound may exist between its formulation in a medicine and the exertion of its therapeutic action in the body. Special features of the book include chapters on pharmaceutical solids and on disperse pharmaceutical systems that present the underlying science of these topics in an accessible way.

USING THE BOOK

The book includes a number of features and approaches that we hope will enhance the learning experience.

In certain topics covered in the book, exposure to mathematics is unavoidable. We have strived in the text itself to identify key equations, and explain how they are applicable to the topic under discussion. Any key derivations, which delve a little deeper into the underlying mathematics, are presented in **Boxes**.

Examples are included. These are usually a multistep problem related to a pharmaceutical scenario that requires the application of concepts covered in the preceding section(s) of the chapter.

A small number of **Appendices** are included at the end of the book. These cover a number of topics that lie somewhat outside of the main subject of the book, but an understanding of which would help in grasping some central physicochemical ideas as applied to pharmaceuticals.

Each chapter ends with several features:

Summary: an overview of the key concepts covered in the chapter. The summary is an ideal revision tool: read through the summary to check you have picked up on, and understood, the key points covered in the chapter.

References: detailed treatments of the topics covered in the chapter, from which key themes and concepts have been extracted for presentation in the chapter.

Further Reading: a small selection of materials, related to the topics covered in each chapter, which present an ideal step up from this text if you're looking to take your understanding one stage further.

There are also a number of **Exercises**; selected answers to these are given in Appendix 5.

 **ONLINE RESOURCE CENTRE**

This book is supported by the Online Resource Centre at
www.oxfordtextbooks.co.uk/orc/moynihan/
which features:

For registered adopters of the book:
Figures from the book in electronic format, ready to download for use in lectures.

For students:
Hyperlinked bibliography: links to reference materials cited in the text (institutional subscriptions to relevant journals required for full-text access).

With any first edition, there are always things we could have approached differently, or presented in slightly different ways. We would welcome any comments and suggestions for improvement. You can share your feedback by visiting **www.oxfordtextbooks.co.uk/orc/ moynihan/** and clicking on the 'contact us' link on the right-hand side.

H. A. M. & A. M. C.

ACKNOWLEDGEMENTS

The authors are very grateful to Jonathan Crowe, Professor Paul Brint and Dr. Simon Lawrence for much advice and input during the preparation of the manuscript. We are also grateful to OUP's reviewers for their many helpful and constructive comments: Alison Marks, University of Bradford; Ken Rutt, University of Brighton; and Peter Wyatt, Queen Mary, University of London.

CONTENTS

LIST OF ABBREVIATIONS

ACE	Angiotensin Converting Enzyme
API	Active Pharmaceutical Ingredient
AUC	Area Under Curve
BBB	Blood/Brain Barrier
CMC	Critical Micelle Concentration
CNS	Central Nervous System
CYP	Cytochrome P450
DNA	Deoxyribonucleic Acid
DPI	Dry Powder Inhaler
DSC	Differential Scanning Calorimetry
FDA	Food and Drug Administration
GI	Gastrointestinal
HPLC	High-performance Liquid Chromatography
HLB	Hydrophilic–Lipophilic Balance
L-DOPA	L-Dihydroxy Phenylacetic Acid
LUV	Large Unilamellar Vesicles
LCST	Lower Critical Solution Temperature
MAO	Monoamine Oxygenase
MDI	Metered Dose Inhaler
MLV	Multilamellar Vesicles
MSZW	Metastable Zone Width
NADP(H)	Nicotinamide Adenine Dinucleotide Phosphate (Reduced form)
OLV	Oligolamellar Vesicles
PABA	*para*-Aminobenzoic Acid
PEG	Polyethylene Glycol
PLA	Polylactic Acid
PLGA	Poly (lactic-co-glycolic acid)
PXRD	Powder X-Ray Diffraction
RES	Reticuloendothelial System
RNA	Ribonucleic Acid
ROF	Rule of Five
SEDDS	Self-emulsifying Drug-delivery System
SUV	Small Unilamellar Vesicles
TGA	Thermogravimetric Analysis
TPN	Total Parenteral Nutrition
UDP	Uridine Diphosphate
XRD	X-ray Diffraction

PHARMACEUTICALS AND MEDICINES

01

Learning objectives

Having studied this chapter, you should be able to:

- understand the terms pharmaceutical, medicine, active pharmaceutical ingredient, drug substance and excipient
- appreciate the chemical nature of pharmaceuticals
- broadly understand the role of pharmaceutical excipients
- understand the terms drug delivery and dosage form
- describe the various pharmaceutical dosage forms and routes of administration
- appreciate the factors that must be considered when selecting an appropriate dosage form and route of administration for a drug substance
- understand what the term physicochemical properties means and why these properties are important.

1.1 INTRODUCTION TO THE ESSENTIAL PROPERTIES OF PHARMACEUTICALS

The overall goal of pharmaceutical research is to identify molecules with pharmacological activity to treat, prevent or alleviate disease and illness. To successfully progress through the highly regulated drug development process (Fig. 1.1), these molecules must possess certain attributes. Firstly, they must have a proven pharmacological efficacy. Secondly, they must demonstrate an acceptable safety profile. Numerous compounds fail during development or are withdrawn from the market post-approval due to adverse effects. Reliability of pharmacological activity is a third essential criterion. A pharmaceutical product will only be approved by national and international regulatory bodies when these three quality attributes have been demonstrated.

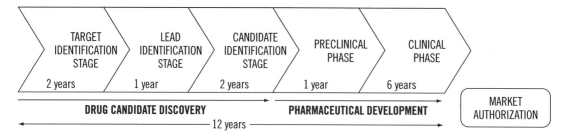

FIGURE 1.1 Schematic diagram of the drug development process (adapted from Food and Drug Administration (FDA), (2004)).

1.1.1 Some key concepts

We begin with an overview of some of the key terms and concepts that will be used in this book. **Physicochemical** is the adjective used to describe the properties relating to both the physical and chemical behaviour of a substance. A good understanding of physical chemistry is required to understand these properties. Physical chemistry is the application of physical concepts to particulate, macroscopic, microscopic, atomic and subatomic phenomena in chemical systems. Table 1.1 lists some key molecular and crystalline physicochemical properties of particular interest to the efficacy of pharmaceutical compounds.

The components of medicines

Pharmaceutical compounds can be divided into **drug substances** [also referred to as **active pharmaceutical ingredients (APIs)**] and **excipients**. Excipients are pharmacologically inert materials that are combined with APIs to aid their processing into dosage forms (e.g. tablets, injections and ointments) and to facilitate API administration to patients. As well as being pharmacologically inert, excipients are also required to be chemically and physically inert. A more detailed classification of APIs is given in Section 1.2. The physicochemical properties of these materials can influence their performance during processing, storage and subsequent administration to the patient. Alterations in the performance of these compounds can undermine the required quality attributes of **efficacy, safety and reliability**. (The **efficacy** of a drug is the maximum possible pharmacological effect it can achieve.) It is an essential pre-requisite for medicinal chemists and pharmaceutical scientists to possess a fundamental understanding of these physicochemical properties in relation to pharmaceutical compounds. The aim of this text is to explain the fundamental theories behind key physicochemical properties and how they impact upon the biological behaviour of pharmaceutical compounds.

APIs exert their desired pharmacological effect by interacting with specific **biological targets**. In modern pharmacy, APIs are rarely administrated to patients in their pure form. Normally they are combined with excipients via one or more processing steps to produce a **dosage form**. The primary goal in transforming an API into a dosage form is to facilitate delivery of that API to the biological target. The delivery of an API to a biological target is commonly referred to as **drug delivery.**

TABLE 1.1 Molecular and crystalline physicochemical properties related to pharmaceutical compounds.

Scale of scrutiny	Physicochemical properties
Molecular	Structure
	Molecular weight
	Dissociation constant
	Equilibrium solubility
	Partition coefficient
Crystalline	Polymorphism/solvates
	Crystalline or amorphous state
	Thermal behaviour
	Surface energy
	Dissolution rate
	Crystal strength
	Particle-size distribution
	Morphology
	Compressibility
	Density

Physicochemical properties of pharmaceutical compounds, both APIs and excipients, can have a significant influence on the drug-delivery process. A basic knowledge of the classes of pharmaceutical compounds, the drug-delivery process, **routes of drug administration** and types of dosage forms is required. The remaining sections of this chapter provide this basic knowledge.

1.2 CLASSES OF PHARMACEUTICAL COMPOUNDS

By definition, pharmaceuticals are the chemicals used in pharmacy – that is, they are the chemical constituents of medicines. These chemicals can be further divided into APIs, and excipients. These groupings can be further subclassified in a number of ways, for instance on the basis of chemical structure. However, it is most common to classify pharmaceuticals on the basis of activity or function, and that basis will be used here.

1.2.1 Active pharmaceutical ingredients

APIs can be categorized into four major classes, as follows.

Psychopharmacological agents, or compounds acting on the central nervous system

The central nervous system, or CNS, consists of the brain and the spinal cord. The CNS controls thought processes, emotions, senses and motor functions. Drugs that act on the CNS include antidepressants, antipsychotics, anxiolytics, psychomimetics, anticonvulsants, sedative-hypnotics, analgesics, and anti-Parkinsonian agents.

Pharmacodynamic agents

These are drugs that interact with the normal dynamic process of the body, for example circulation of blood. Included are antiarrhythmics, antianginals, vasodilators, anti-hypertensives, antithrombotics, antiallergy agents, and drugs that interact with the gastrointestinal system.

Chemotherapeutic agents

These are agents that are selectively more toxic for disease-causing microorganisms than for the host. As such, this category includes antibiotics, antiparasitics, antivirals and antifungals. Cancer-controlling compounds are also usually included in this category.

Agents acting on metabolic diseases and endocrine function

This group includes drugs for the treatment of diabetes, inflammation, atherosclerosis and arthritis. Also included are drugs not falling under any of the previous categories, for example hormones.

An appreciation of molecular structure and function is required to understand the physicochemical properties of these compounds. A thorough review of drug structure is beyond the scope of this book. However, Fig. 1.2 gives the molecular structures of a representative selection from each of the above categories. The following general points should be noticed.

- The great majority of the examples are molecular organic compounds. The molecular masses of these compounds usually lie within the approximate range of 300 to 1200 atomic mass units. These compounds are often referred to as 'low to medium weight' compounds, or even as 'small molecule' drugs. Simple inorganic compounds are represented, for example lithium carbonate. Organometallic drugs are also known, for example oxaliplatin. Larger molecular weight biomolecules are also represented, for example salmon calcitonin, a peptide with 32 amino acid residues.

FIGURE 1.2 (*opposite*) Molecular structures of a representative selection of (a) psychopharmacological agents, (b) pharmacodynamic agents, (c) chemotherapeutic agents, and (d) other agents. In (d), the structure of salmon calcitonin uses the following abbreviations for amino acids: Cys (cysteine), Ser (serine), Asn (aspargine), Leu (leucine), Thr (threonine), Val (valine), Gly (glycine), Lys (lysine), Gln (glutamine), Glu (glutamic acid), His (histidine), Tyr (tyrosine), Pro (proline), Arg (arginine); for further information on amino acids and proteins, see Patrick (2005) under Further Reading at the end of the chapter.

(a) Psychopharmacological agents

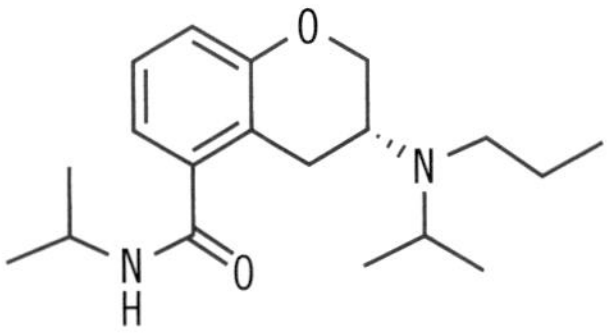

Ebalzotan
(antidepressant)

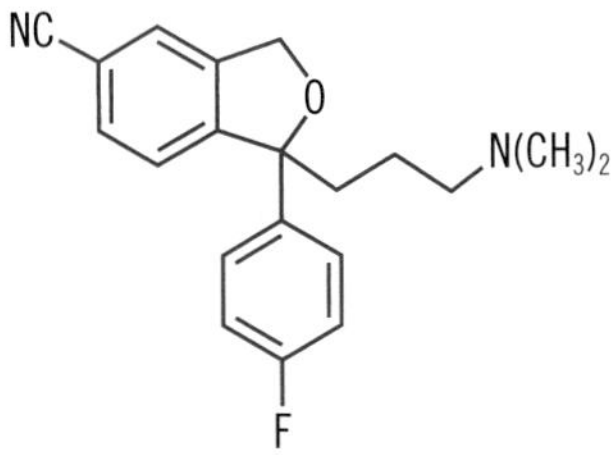

Citalopram
(antidepressant)

Li_2CO_3

Lithium carbonate
(mood stabilizer)

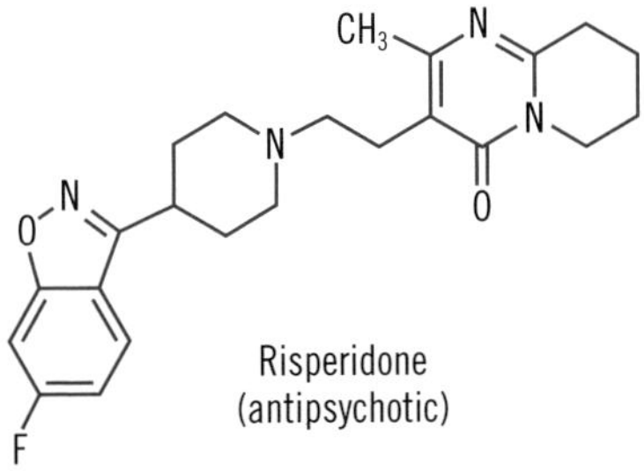

Risperidone
(antipsychotic)

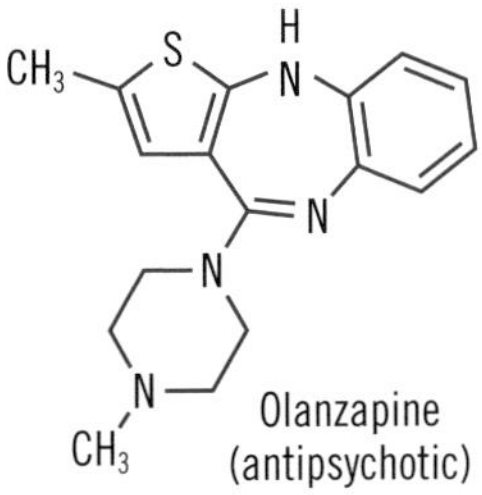

Olanzapine
(antipsychotic)

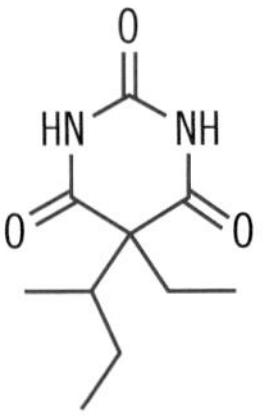

Pentobarbital
(sedative-hypnotic)

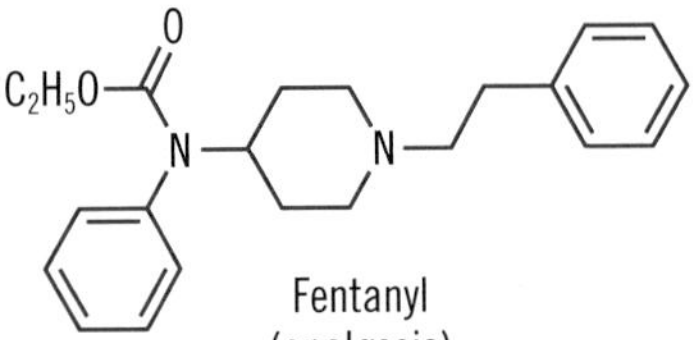

Fentanyl
(analgesic)

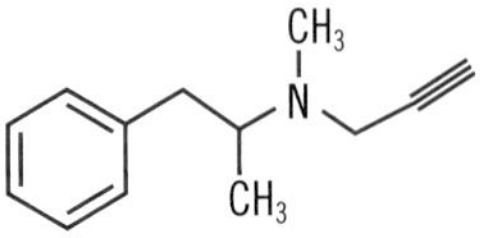

Selegiline
(antiParkinsonian)

(b) Pharmacodynamic agents

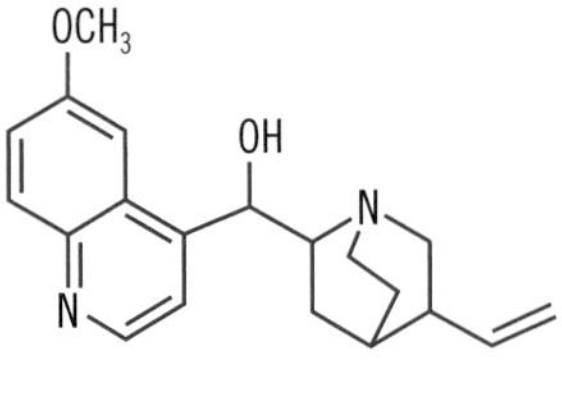

Quinidine
(antiarrhythmic)

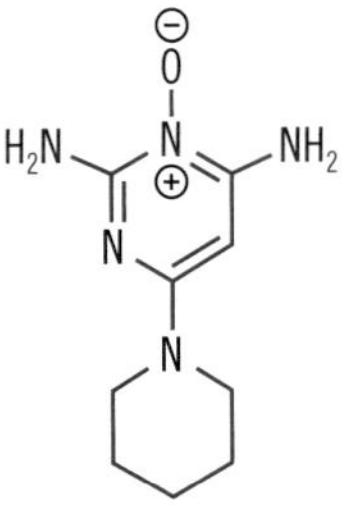

Minoxidil
(vasodilator)

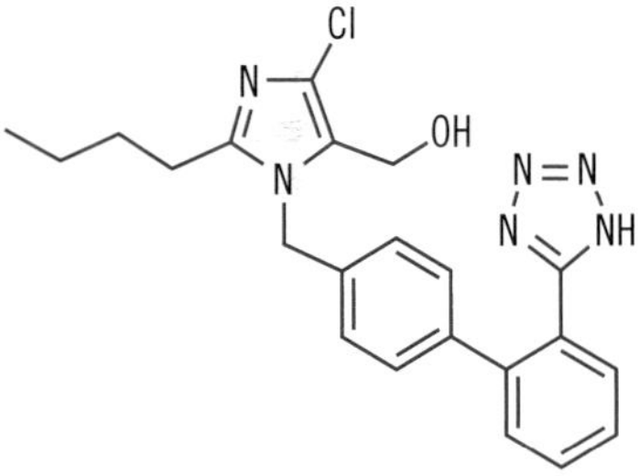

Losartan
(antihypertensive)

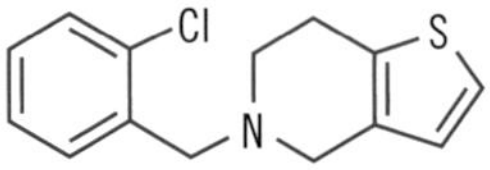

Ticlopidine
(antithrombotic)

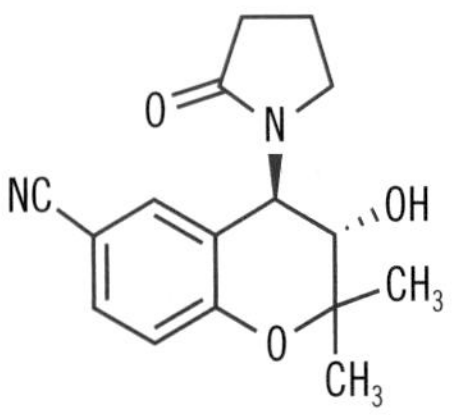

Cromakalim
(antihypertensive)

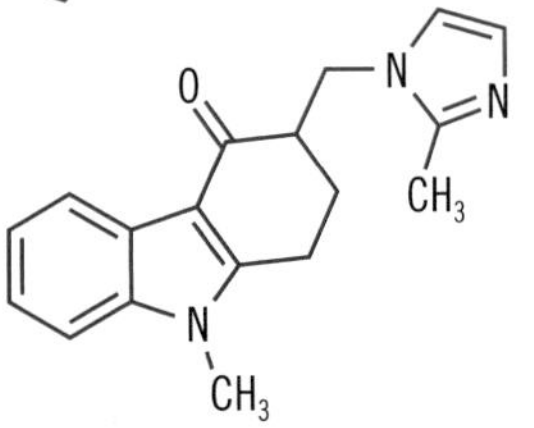

Ondansetran
(antiemetic)

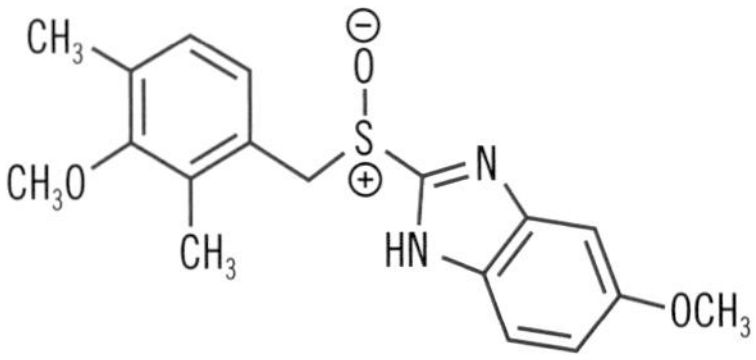

Omeprazole
(gastrointestinal agent)

(c) Chemotherapeutic agents

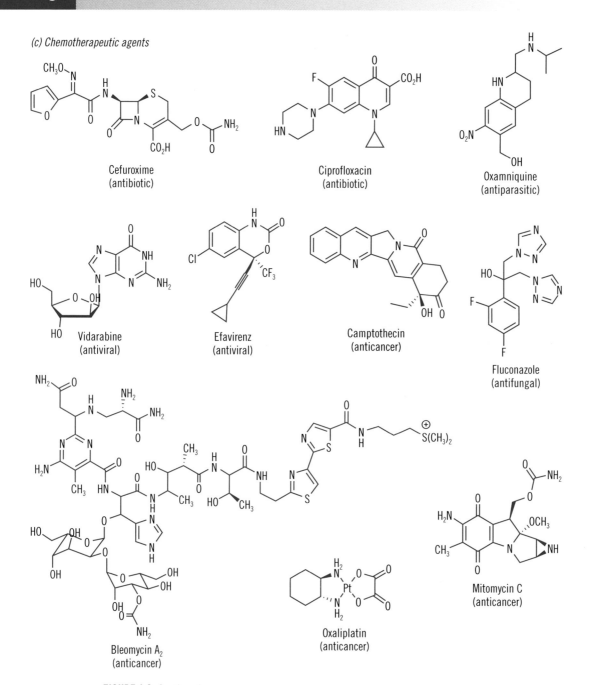

Cefuroxime
(antibiotic)

Ciprofloxacin
(antibiotic)

Oxamniquine
(antiparasitic)

Vidarabine
(antiviral)

Efavirenz
(antiviral)

Camptothecin
(anticancer)

Fluconazole
(antifungal)

Bleomycin A$_2$
(anticancer)

Oxaliplatin
(anticancer)

Mitomycin C
(anticancer)

FIGURE 1.2 Continued

(d) Other agents

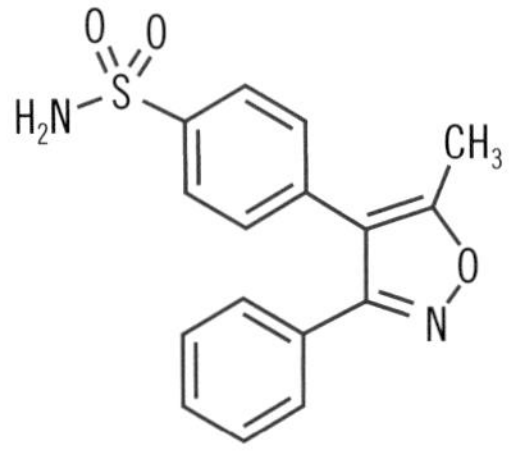

| Valdecoxib (anti-inflammatory) | Indomethacin (anti-inflammatory) | Mestranol (estrogenic) |

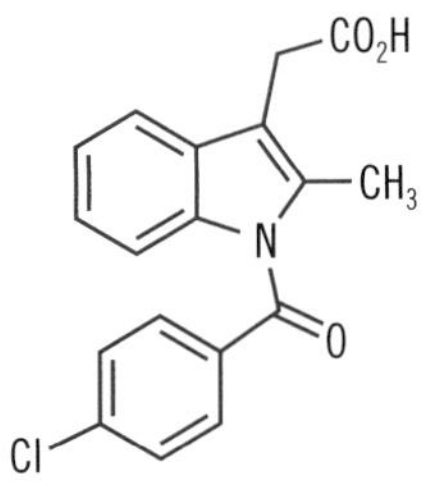

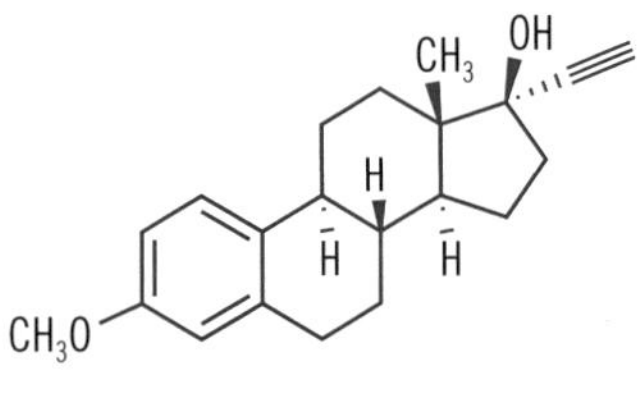

Cys-Ser-Asn-Leu-Ser-Thr-Cys-Val-Leu-Gly-Lys-Leu-Ser-Gln-Glu-Leu-His-
Lys-Leu-Gln-Thr-Tyr-Pro-Arg-Thr-Asn-Thr-Asn-Thr-Gly-Ser-Gly-Thr-Pro-NH$_2$
Salmon Calcitonin
(osteoarthritis treatment)

FIGURE 1.2 Continued

- Cyclic systems (for example benzene rings) are very well represented. Carbocyclic and heterocyclic systems are ubiquitous in drug structure. Five- and six-membered rings are the most common, but small ring systems occur with reasonable frequency, (for example, the cyclopropane ring in ciprofloxacin and the aziridine ring in mitomycin C). Of the five- and six-membered systems, the majority are aromatic or pseudo-aromatic. Hence, substituted benzene rings are very common, and heterocycles such as pyridines, furans, thiophenes, imidazoles, isoxazoles and others occur commonly in drug structure.

- Polar functional groups are very common. These include ethers, amines, amides, esters, carboxylic acids, nitriles, halides, alcohols, thiols, N-oxides, sulfoxides, sulfonamides and others. Many of these functional groups are also acidic or basic to some extent. This especially applies to carboxylic acids, phenols, thiols, sulfonamides, amines and other nitrogen-containing functional groups such as amidines. Such groups are capable of existing in both un-ionized or ionized forms. This possibility is often of great pharmaceutical importance, as it allows drugs to travel through different types of environment in the body (see Chapter 7). For example, many of the drug compounds shown in Fig. 1.2 are used as salt forms, rather than as the un-ionized forms shown.

1.2.2 Excipients

Excipients are the chemical components of medicines other than the active ingredients. The most convenient classification of excipients is in terms of the functions they perform in dosage formulations.

Diluents (or fillers)

These are used to increase the bulk volume of material, especially for tablet and capsule formation. Examples include carbohydrates such as glucose, sucrose, lactose, sorbitol, mannitol, cellulose and starch. Calcium phosphates and carbonates are also used.

Surfactants

These are used to aid formation of suspensions, emulsions or solutions. Surfactants are amphipathic compounds, that is, they contain both hydrophobic and hydrophilic regions. The hydrophobic regions are generally long alkyl chains. The hydrophilic regions can be either 'ionic' or 'non-ionic'. The ionic groups are positively or negatively charged functional groups. The non-ionic groups are most commonly polyethers, in particular, polyethylene glycol.

Lubricants

These are used to reduce friction between powders and metal surfaces during tablet manufacture. Examples include paraffin, magnesium stearate, polyethylene glycol and sodium dodecyl sulfate.

Disintegrants

Disintegrants are added to aid break up of solid dosage forms such as tablets. Examples include cellulose and cellulose derivatives, and cros-povidone.

Viscosity enhancers

These can be added to liquid formulations to control properties such as ease of pouring and sedimentation of suspensions. An example would be xanthan gum.

Binders (or adhesives)

These are added to give adherence to powder mixtures for manufacture of solid-state dosage forms such as tablets. Examples include cellulose derivatives, polyvinylpyrrolidone and polyethylene glycol.

Figure 1.3 gives the molecular structures of a selection of excipients. While some of these are small to medium molecular weight molecules, many are high molecular weight polymers. Most of the structures shown have hydrophilic groups, for example alcohol groups, polyether segments or ionized groups. Some, such as the surfactants, also have hydrophobic groups. It should be noted that the cellulose derivatives often do not have very uniform repeating structures. The structures shown in Fig. 1.3 at best illustrate representative portions of these derivatives. Excipients such as calcium phosphate are not molecular in nature and possess complex mineral structures.

1.3 DRUG DELIVERY: GETTING THE ACTIVE PHARMACEUTICAL INGREDIENT TO THE SITE OF ACTION

Dosage forms are also referred to as '**drug-delivery systems**' and '**finished drug product**'. An ideal dosage form should reliably deliver the specified level of drug substance, to the specified biological target, for the specified duration. It should minimize exposure of the drug substance to other receptors that might result in the patient experiencing adverse effects. The inconvenience or discomfort associated with administering a dosage form should not outweigh its therapeutic benefits. For example, a patient may be willing to have an analgesic

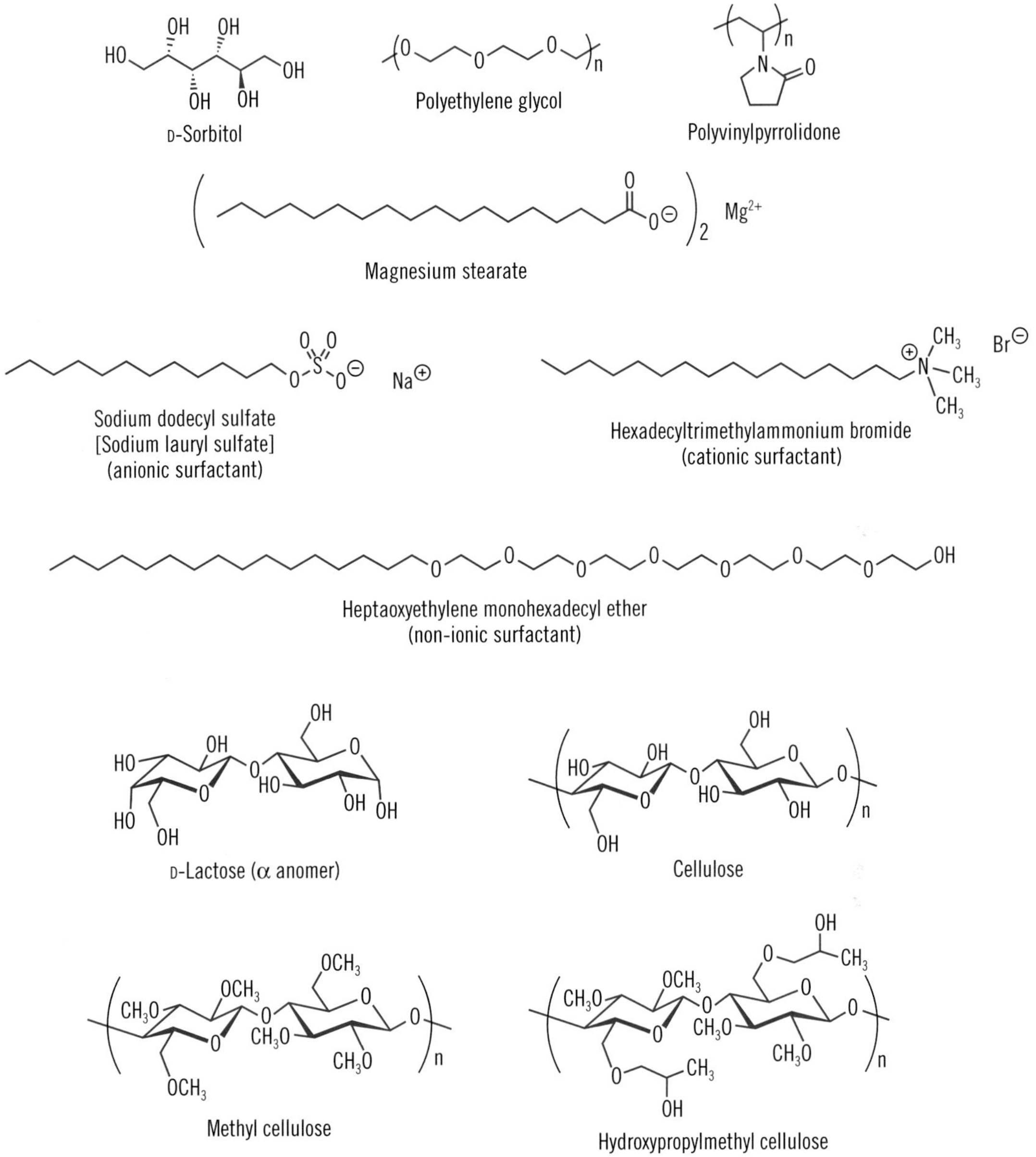

FIGURE 1.3 Molecular structures of selected excipients.

drug administered via intravenous infusion to treat severe pain but this form of treatment may not be tolerated to treat a minor headache, despite its effectiveness.

During release of the drug substance from the dosage form and its delivery to the target receptor, the pharmaceutical compounds (drug substance and excipients) are required to change from one phase to another. For example, a drug substance may change from a solid phase to a solution phase or from an aqueous solution phase to a lipid solution phase. In order to understand drug delivery and the design of effective dosage forms, a basic understanding of these phase changes is essential. An explanation of these phase changes is given in Chapter 3.

1.3.1 Routes of administration

Before an active pharmaceutical ingredient (API) can reach its specific biological target, the patient must first be administered a dosage form containing the API. At the start of this section the ideal properties of a dosage form are described. In Section 1.3.2, the wide variety of dosage forms used to deliver APIs are described. The route by which a dosage form is administered to a patient is referred to as the 'route of administration'. Numerous possible routes of administration can be used and Table 1.2 gives a list of some of the main routes of administration along with examples of drugs delivered by each route. A comprehensive list of routes of administration is given in Appendix A1. General terms are used in relation to routes of administration. The term **parenteral** refers to administration via injection, infusion or implantation. **Peroral** refers to administration through the mouth. The meaning of the term **enteral** varies widely. It can refer to direct administration

TABLE 1.2 A list of some of the main routes of administration with examples of drug substances delivered by each route and their indication.

Term	Description	Example of	Indication
Otic	to the ear	Gentamicin	Infection of the ear
Ophthalmic	to the eye	Cromolyn	Conjunctivitis
Nasal	to the nose	Betamethasone	Inflammation of the nose
Oral	to or by way of the mouth	Paracetamol	General analgesia
Buccal	toward the cheek	Prochlorperazine	Vertigo
Sublingual	beneath the tongue	Glyceryl trinitrate	Angina pectoris
Rectal	to the rectum	Meloxicam	Rheumatoid arthritis
Topical	to the outer surface of the body	Benzoyl peroxide	Acne vulgaris
Subcutaneous	beneath the skin	Insulin	Diabetes mellitus
Transdermal	through the dermal layer to systemic circulation	Fentanyl	Analgesia
Implantation	by implanting	Estradiol	Hormone-replacement therapy
Intravenous	within a vein	Gentamicin	Systemic infections
Intramusuclar	within a muscle	Lorazepam	Acute anxiety states
Vaginal	into the vagina	Clotrimazole	Candidal vaginitis
Intrapulmonary	within the lungs or its bronchi	Salbutamol	Asthma

to the intestines. It is also given the more general meaning of administration to any point of the gastrointestinal tract, from the mouth to the rectum.

The administration of a dosage form via these routes can result in either local or systemic (via the blood stream) delivery or both. The choice of route of administration is primarily influenced by location of the biological target.

Local delivery

If the target receptor is external or easily accessed then local delivery of the medication can be a feasible and effective approach. For example, conjunctivitis, an infection of the conjunctiva, can be treated by the local delivery of eye drops to the surface of the eye. The delivery of bronchodilators to the lungs via inhalers for the treatment of asthma is another example of local delivery. Due to advances in surgical techniques and biomaterials, more sophisticated dosage forms are now available for local delivery. Carmustine, a drug effective at treating cancerous brain tumours, can now be delivered locally to the tumour site. A biodegradable wafer containing carmustine is placed under the skull by surgical implantation. Due to its site-specific delivery characteristics, local delivery is generally the preferred approach. Site-specific delivery minimizes the exposure of non-target biological receptors to the drug substance and thereby reduces the risk of adverse effects.

Systemic delivery

When the target receptor cannot be easily accessed systemic delivery is required. Blood is carried throughout the body through a network of veins, arteries and capillaries. The function of the blood system is to transport materials to and from the tissues throughout the body. Therefore, it is an ideal delivery pathway for a drug substance. The drug substance can enter the systemic circulation via the majority of routes of administration.

Dosage forms can be administered directly into the systemic circulation via veins (**intravenous administration**) or less frequently via arteries (**interarterial administration**). Systemic delivery can also be achieved indirectly by the administration of the dosage form via a less invasive route such as an **oral**, **transdermal** or **pulmonary route**. Indirect systemic drug delivery usually requires the release of a drug substance from the dosage form before it can travel across the biological barrier membranes into the blood system. The process by which it crosses the biological barrier into the blood stream is termed **absorption**. The drug substance is required to change phase during the absorption process, for example change from a solid to a liquid phase. These phase transitions are influenced by the particular physicochemical properties of the drug substance. Greater detail regarding these phase transitions is given in Chapter 3.

Compared to local delivery, a major limitation of systemic delivery is the reduced capability for site-specific delivery and increased potential for associated adverse effects. More sophisticated targeted dosage forms and advanced drug substances are being developed to overcome this limitation.

1.3.2 Pharmaceutical dosage forms

Pharmaceutical dosage forms are designed to facilitate the administration of drug substances. The majority of dosage form types can be classified as solid, semisolid, liquid and

TABLE 1.3 Main types of dosage forms.

Class	Type	Specific types
Solid	Powder	for solution, for suspension, for topical application
	Granule	for solution, for suspension, modified release, effervescent,
	Bead	implantable
	Pellet	modified release, implantable
	Minitablets	modified release, implantable, modified release, multilayered
	Tablet	chewable, modified release, effervescent, for solution, for suspension, multilayered, orally disintegrating
	Capsule	powder filled, liquid filled, semisolid filled, containing pellets, with modified release coating
Semisolid	Ointment	API dissolved, solid API suspended
	Cream	oil in water, water in oil, API dissolved, solid API suspended
	Gel	API dissolved, solid API suspended
Liquid	Solution	API dissolved
	Disperse systems	API solubilized in micelles, API solubilized in liposomes
	Suspensions	solid API suspended
	Emulsions	oil in water, water in oil, API dissolved, solid API suspended
	Foam	API dissolved liquid phase, solid API suspended
Gaseous	Medical gases	API gas at room temperature, API volatile liquid at room temperature
	Aerosols	solid API suspended, API solution droplets suspended

gaseous at room temperature. Semisolid dosage forms possess certain liquid and certain solid properties. **Ointments, creams** and **gels** are examples of semisolid dosage forms. At room temperature, semisolid materials behave like solids and do not flow or flow very slowly when low forces are applied. However, when high forces are applied, semisolid materials flow in a manner similar to liquids. Common pharmaceutical dosage forms are listed according to the classification in Table 1.3.

Dosage forms can be composed of more than one phase; these are termed **disperse systems** and are explained in greater detail in Chapters 5 and 6:

- solid material dispersed in liquid, e.g. pharmaceutical suspension
- solid material dispersed in gas, e.g. dry powder inhalers
- liquid dispersed in gas, e.g. nebulized solution
- one liquid dispersed in an immiscible liquid, e.g. pharmaceutical emulsion.

Controlled-release dosage forms

Having looked at the different types of dosage forms, we now consider the ways by which the release of drug substance from these dosage forms can be controlled after administration. The release or liberation of the drug substance from the dosage form can be controlled in certain types of dosage form. Release can be controlled in two ways, (1) the drug substance is released at a specified rate or (2) the drug substance is released at a specified location. These dosage forms are termed **modified release dosage forms**. In Table 1.3 the dosage form types with modified release capability are indicated.

Extended release and delayed release are the most common type of modified release dosage forms. **Extended release** dosage forms allow the rate of drug-substance release to be controlled in such a manner as to allow a reduction in dosing frequency. The rate of drug-substance release is controlled by the inclusion of excipients that retard drug-substance release. Certain designs of delayed release dosage forms contain functional excipients that release the drug substance in response to a stimulus, such as the presence of a specific enzyme, biological substrate or pH. The remainder of this section describes the different solid, semisolid, liquid and gaseous dosage forms.

Solid dosage forms

The majority of solid dosage forms are composed of **powders** processed to some degree to change their form. These include **granules, beads, pellets** and **tablets**. However, simply blending together powders of one or more drug substance and excipients can constitute a simple solid dosage form, a **pharmaceutical powder**.

Pharmaceutical powders can be administered orally or applied topically. Single doses of oral powders are generally wrapped or filled into sachets for ease of administration by the patient. However, oral powders have been largely replaced by tablet and capsule dosage forms. Medicated dusting powders applied topically are used to treat surface skin conditions. They are required to be sterile when designed for application to open wounds.

Granules are composed of solid particles and to the human eye they can look the same as pharmaceutical powders. However, the difference between a granule and a powder can be observed when viewed under a microscope. The granule particles, on close examination, are composed of a number of individual powder particles stuck together. The powder, on the other hand, consists only of solid particles. If we imagine each powder particle the size of a rice crispie then the granule particles would appear like rice crispie buns. The individual rice crispies, like the individual powder particles, are visible but stuck together in groups by a binding material, in this case chocolate. Granules can be manufactured as an intermediate process step in the production of tablets, capsules, pellets or beads. However, they can constitute a dosage form in their own right. Like powders they can be administered orally or topically. Due to their superior flow properties, uniformity of drug content and size and their reduced dustiness they are often preferred to powders as an oral dosage form.

In addition to direct administration of oral powders and granules, they can be dispersed or dissolved in a liquid prior to administration. **Effervescent** granules contain bicarbonate and citric or tartaric acid. These granules effervesce, releasing CO_2, when in contact with water, (for example, vitamin C tablets often comprise effervescent granules). Effervescence enhances the rate of disintegration and dissolution of the granule in liquid prior to oral administration.

Multiparticulate solid dosage forms include beads, pellets and **minitablets**. These multiparticulate dosage forms can be packaged in an outer capsule shell for oral administration. Beads and pellets can be mixed with other excipients and compressed into a tablet dosage form. Sterile multiparticulate dosage forms can be administered by injection or by implantation. Bead and pellet dosage forms can be used for administration of drug substances to wounds and pathological cavities.

Oral tablets are the most common type of dosage form marketed. Tablets are generally manufactured by compression of powders, granules, beads or pellets with the aid of other excipients. In addition to compression, orally dispersible tablets can be manufactured by moulding and freeze drying. Tablets are moulded by pouring a molten mixture containing the drug substance into pre-formed moulds where they solidify into a tablet shape. Many throat lozenges are prepared in this manner. Freeze-dried tablets are prepared by freezing a solution of the drug substance and excipients in pre-formed tablet moulds and removing water from frozen solution under vacuum. The remaining material after drying is a highly porous tablet-shaped solid. Freeze drying is often used to prepare tablets that are required to disintegrate rapidly in the mouth.

Tablets can be taken into the mouth intact or dispersed in water prior to administration. Tablets that are taken into the mouth intact can be designed to be chewed, sucked, dispersed, dissolved sublingually (under the tongue), buccally (between the cheek and gum) or swallowed. Tablets that are swallowed intact release the drug substance in the gastrointestinal tract where it is absorbed into the blood system.

Capsules are defined as a solid dosage form comprising a shell with filling. The shell can be a single sealed enclosure (soft capsules) or can comprise two halves that fit together (hard capsules). The capsule's shell is usually comprised of gelatine but it can also be composed of a starch- or cellulose-based material. The filling of the capsule can be liquid, semisolid or solid. Solid fillings can be powders, granules, beads, pellets or minitablets.

Capsules are generally administered orally and swallowed intact. Capsule dosage forms are also designed so that the contents are emptied or extruded from the shell prior to administration. For example, Creon® capsules that contain a pancreatin enzyme supplement; can be opened and their contents mixed with soft food prior to oral administration.

All solid dosage forms can be designed with modified release properties. Film coatings with modified release properties can be applied externally to a solid core. The drug substance is normally located in the core but in some circumstances can be located in the coat. Alternatively, modified release excipients can be added to the powder mix during the production of tablets, beads, pellets, minitablets and granules. By mixing with the modified release excipients, the drug-substance particles become entrapped in a matrix of excipients that then control the drug's release. Tablets containing these modified release excipients can be administered less frequently than conventional tablets. The reduction in frequency of dosing helps patients to remember to take tablets when prescribed and improves patient compliance with medication. Figure 1.4 gives some examples of solid dosage forms (tablets and capsules) and solid dosage forms reconstituted with liquid (injection and oral suspension) prior to administration.

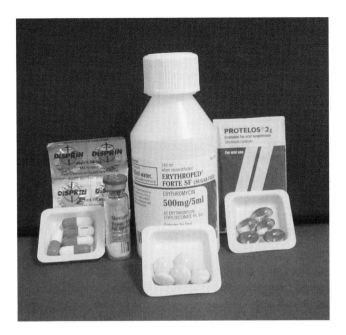

FIGURE 1.4 Solid dosage forms (tablets and capsules) and solid dosage forms reconstituted with liquid (injection and oral suspension) prior to administration.

Semisolid dosage forms

Semisolid dosage forms can exist as single-phase systems (the drug substance in solution in the semisolid material) or as more complex two-phase or multiphase systems. An example of a complex multiphase system is an oil in water emulsion with suspended solid particles. The majority of semisolid dosage forms are applied topically to skin or accessible mucous membranes such as the interior of the nose or rectum. If applied to open wounds or lesions, sterile semisolid dosage forms are required.

Semisolid dosage forms can be divided into a range of different types of preparations. Ointments consist of a waxy, hydrophobic semisolid base to which one or more drug substances can be dispersed or dissolved. Ointment bases are generally composed of hydrocarbons, fixed oil bases or silicones. The term 'paste' is used to describe an ointment in which a large percentage of solid material (20–50%) is dispersed.

Creams are semisolid dosage forms containing two or more phases. They consist of water droplets dispersed in an oil phase or oil droplets dispersed in a water phase. They can be classified as **emulsion** systems, which are discussed in greater detail in Chapters 5 and 6. The drug substance can be dispersed or dissolved in the cream base. Dissolved drug substance can be located in the oil phase, the aqueous phase or both phases of the cream.

Pharmaceutical **gels or jellies** are semisolid dosage forms composed of a liquid phase within a network structure of a solid gelling agent. It is the nature of the network of the gelling agent around the liquid phase that provides the gel with stiffness. Gelling agents can be small inorganic particles, such as aluminium hydroxide, or large organic molecules, polymers. When water is the liquid phase, gels are termed 'hydrogels'. The drug substance can be dispersed or dissolved in the liquid phase in the gel system.

Rigid hydrogels with a low liquid content are referred to as **films**. More detail relating to pharmaceutical gels is given in Chapter 6.

Gel dosage forms can be designed with a particular shape and texture to facilitate administration of drug substances to specific body orifices such as the nose, ear, rectum or vagina. When administered vaginally they are termed **pessaries**; when administered rectally they are termed **suppositories**.

In addition to gel-based systems, other suppository bases can be used. Glyceride-type fatty bases release the drug substance by melting or softening at body temperature. Water-soluble bases release the drug substance by dissolving in the aqueous physiological fluids. The drug substance can be dispersed or dissolved in the suppository base.

Chewing gum can be used as a dosage form to facilitate oral drug delivery. As for confectionary chewing gum, pharmaceutical chewing gum consists of a sweetened and flavoured insoluble plastic material that liberates the drug substance when chewed in the oral cavity.

Modified release semisolid dosage forms can be designed if they contain dispersed solid drug substance with modified release properties. Alternatively, the viscosity and consistency of the semisolid dosage form can be modified to increase or decrease the rate of drug-substance release by modifying the diffusion rate of the drug substance through the semisolid carrier material. The factors that influence this diffusion rate are explained in more detail in Section 3.3.3.

Liquid dosage forms

Compared to semisolid substances, liquids are pourable and conform to the shape of the container they are stored in at room temperature. **Pharmaceutical solutions** are liquid dosage forms that are clear, homogeneous and single-phase systems. They contain one or more drug substance dissolved in one or more solvents. While aqueous solutions are generally preferred for parenteral administration, oil based injections are also used.

Liquid dispersions are two-phase liquid systems. This means that they are composed of one phase dispersed throughout a second phase. The dispersed phase can be solid particles (ranging in size from nanometres to millimetres), **micelles, liposomes**, oil droplets (**oil in water emulsion**) and water droplets (**water in oil emulsion**). **Suspensions** are disperse systems containing the drug substance as solid particles suspended in a suitable liquid vehicle. The structure and characteristics of liquid dispersions are covered in more detail in Chapters 5 and 6. Modified release liquid dispersions can be designed by the inclusion of a solid disperse phase with modified release characteristics. **Foams** can also be classified as two-phase liquid dispersion as they consist of a gas phase dispersed within a liquid. Pharmaceutical foams are useful for the topical, rectal and vaginal delivery of drug substances.

Pharmaceutical solutions and dispersions can be supplied pre-prepared or to be prepared prior to administration by reconstitution with a suitable solvent or vehicle. Liquid dosage forms can be administered by all routes. Liquid dosage forms for parenteral administration are required to be sterile. In addition to sterility, liquid dosage forms for intravenous and intra-arterial administration are required to be free of particles greater than 5 μm. Liquid dosage forms for topical administration must have a viscosity suitable for spreading over the area of administration. The sensory perception (taste and

TABLE 1.4 Various terms used in relation to liquid dosage forms.

Term	Description	Route of administration
Lotion	liquid emulsion, non-greasy	for cutaneous administration
Elixir	liquid solution, clear, flavoured, may contain ethanol	for oral administration
Enema	liquid aqueous solution	for rectal administration
Irrigant	liquid sterile aqueous solution	for topical administration to bathe or flush open wounds or body cavities
Mouthwash	liquid aqueous solution	for oropharyngeal administration
Shampoo	liquid soap or detergent	for administration to the scalp, used as a vehicle for dermatologic agents
Paint	liquid solution or suspension	for cutaneous administration and less frequently to mucous membranes
Collodion	liquid solution of pyroxylin in ether and ethanol	for cutaneous administration, dries on the skin to form a flexible film at the site of administration
Linctus	liquid solution, viscous with high proportion of sucrose, other sugars or polyhydric alcohols	for oral administration in the treatment or relief of cough
Liniment	liquids solution	for cutaneous administration to the unbroken skin with friction

mouth feel) are critical considerations in the formulation of oral liquids. Table 1.4 lists terms commonly used in describing pharmaceutical liquid dosage forms.

Examples of some common liquid dosage forms are shown in Fig. 1.5.

Gaseous dosage forms

Gaseous medical products consist of medical gases and aerosols. Medical gases are ideal for intrapulmonary administration via inhalation generally through a breathing apparatus. Medical gases do not require incorporation of the drug substance into a dosage form to facilitate administration. Aerosols also can be considered as gaseous dosage forms. In aerosols the gaseous phase is the vehicle for administration of the drug substance. Aerosols are stable dispersions of solid particles or liquid droplets in a gaseous medium.

Compared to other dosage forms, aerosols exhibit a high degree of physical instability with time. Therefore, aerosols are generated immediately prior to administration via an aerosol-generating device. The three primary aerosol-generating devices are **metered dose, dry powder inhalers,** and **nebulizers**. Metered dose aerosol devices are used for delivery topically to the skin, into the nose, mouth or lungs. Dry powder inhalers and nebulizers are used primarily for administration to the lungs. More detail in relation to aerosols is given in Chapter 6.

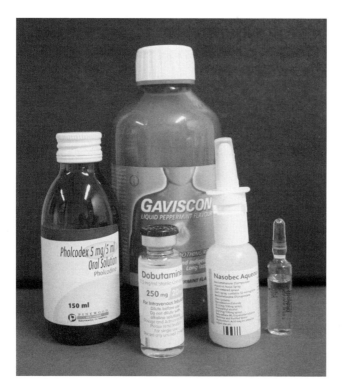

FIGURE 1.5 Some liquid dosage forms.

1.3.3 Factors influencing dosage form choice

The factors that influence the choice of dosage form for a particular drug substance include the patient type, illness type and drug substance's physicochemical properties. Let's consider each of these in turn.

Patient-type factors

Age, cognitive understanding of dosing regimen, consciousness and living conditions must be considered when selecting an appropriate dosage form. Elderly patients and children under the age of 6 years can have difficulties swallowing solid dosage forms. For these patients oral liquids, dispersible tablets and suppositories are preferable. A child's physiology undergoes development up to puberty. The organs and enzyme systems of children can function differently from those of healthy adults. During development, differences in the physiology of routes of administration, such as skin, lungs and gastro-intestinal tract, can result in specific dosage form requirements for children compared to adults. The geriatric population will also exhibit differences in physiology due to aging that may require alterations in dosage forms compared to the younger healthy adults.

   Visual impairment and arthritic conditions are common ailments of the elderly population and can complicate their self-administration of a number of dosage forms. Dosage forms that require a degree of manual dexterity and accurate vision, such as drawing up liquid to a set volume in a syringe, instillation of eye drops and use of metered dose inhalers, may not be best choice of dosage forms for this patient group. Patients with cognitive impairment may have difficulties in complying with dosage regimes that require regular administration. In such cases, **extended release** dosage forms are often favoured. Sustained release intramuscular injections of fluphenazine decanoate, used in the treatment of schizophrenia, have a half-life of seven to ten days and therefore can be administered at two-week intervals. In situations where the patient is unconscious or in a coma, injections, infusions, transdermal patches or suppositories may be required to replace oral dosage forms.

It also should be considered as to whether the medication will be self-administered by the patient or administered by a carer or healthcare professional. Many dosage forms are relatively simple to administer but others require the patient to undergo a level of training or instruction to ensure correct administration. Patients should receive verbal or written instructions before administrating drug substances using dosage forms such as inhalers, nebulizers, eye, nose, ear drops, sublingual tablets, suppositories and pessaries. There are a selection of dosage forms that must be administrated by trained healthcare professionals. These include the majority of parenteral dosage forms, implants and drug-eluting devices.

Illness-type factors

The type of illness that the drug substance is administered to treat or alleviate will also influence the choice of a suitable dosage form. For example, nitroglycerin is used in the treatment of chest pain due to ischaemic heart disease and is available in a range of dosage forms. The intravenous injection, sublingual tablet and spray dosage forms are recommended to treat chest pain and prevent chest pain caused by exertion because of their advantageous rapid onset of action (1–2 min). However, these preparations are unsuitable for long-term maintenance and prophylaxis due to their short duration of action. Nitroglycerin oral extended release tablets and transdermal patches are more suitable for these indications. While they have a slower onset of action, their duration of action is longer.

The severity of the illness or disease for which the drug substance is indicated can also influence the dosage form selected. Paclitaxel, a widely indicated anticancer agent, is formulated as a solution for intravenous injection. Due to its poor water solubility, Cremophor® EL, a non-ionic surfactant is incorporated in the dosage form as a solubility increasing agent. Cremophor® EL is associated with severe allergic reactions when administered intravenously. So severe and distressing is this allergic reaction that it is commonplace that patients are required to be administrated steroids prior to administration of the paclitaxel injection to prevent the occurrence of an allergic reaction. These adverse effects may be tolerated in the treatment of cancer but it is difficult to imagine a patient tolerating them for a more minor ailment. For minor ailments the adverse effects experienced due to the dosage form might outweigh the relief achieved by its administration.

The vast proportion of oral dosage forms on the market compared to parenteral forms indicates patients' preference for oral dosage forms over the more invasive parenteral dosage forms. However, where an alternative effective oral dosage form is not available, patients will tolerate repeated daily subcutaneous injections, as is the case of insulin for the treatment of diabetes.

TABLE 1.5 Physicochemical properties associated with poor absorption outlined in Lipinski's Rule of Five.

Property	Value
Molecular weight	Greater than 500
Partition coefficient Log $P_{octanol/water}$	Greater than 5
Number of H-bond donors	Greater than 5
Number of H-bond acceptors	Greater than 10

Active pharmaceutical ingredient physicochemical properties

The physicochemical properties of the drug substance play a significant role in the choice of dosage form and route of administration. **Lipinski's Rule of Five** (ROF) was developed as a general guide to the physicochemical properties that can cause problems with respect to aqueous solubility in physiological fluids after release from the dosage form and permeability of biological barriers during absorption (Lipinski 2001). The rule states that problems with absorption can be avoided when any two of the four rules listed in Table 1.5 are avoided. The explanation of why these problems occur will be explored in more detail in Chapters 2, 7 and 8.

The Rule of Five is a useful general guide to identifying compounds with API-like properties: aqueous solubility, permeability and chemical and enzymatic stability. It does not address the specific differences in physicochemical requirements for compounds delivered by different routes of delivery to various sites in the body, i.e. across the blood brain barrier or to the lungs. The physicochemical requirements of individual routes are dealt with in greater detail in Chapter 7. The suboptimal physicochemical properties of the active pharmaceutical ingredient can be altered or compensated for by intelligent dosage form design. Simple examples would be in the inclusion of surfactants to increase the solubility of poorly soluble drug substances and more sophisticated approaches would include liposomal formulations detailed in Chapter 6. The remaining chapters of this book will highlight the theory behind these physicochemical properties and how these influence dosage form design, manufacture and performance.

◘ 1.4 SUMMARY

- Pharmaceuticals are the chemical constituents of medicines. Pharmaceuticals can be divided into two categories: active pharmaceutical ingredients (APIs) and excipients. APIs are the actual drug substances that give medicines their therapeutic properties. The main classes of APIs are psychopharmacological agents, pharmacodynamic agents, chemotherapeutic agents, and agents acting on metabolic diseases and endocrine function. The majority of APIs are organic compounds. Many contain one or more cyclic systems in their molecular structures. They also often contain polar functional groups.

- Excipients are the chemical components of medicines other than APIs. They are combined with the API to make dosage forms, or medicines, which can deliver the API to the required site of action in the body. The excipients may act as diluents, surfactants, lubricants, disintegrants, viscosity enhancers, or binders.

- Dosage forms may be administered to the patient by a variety of routes of administration. These include administration by the mouth (oral), by the surface of the body (topical) and by other routes. Delivery may be local or systemic. Types of dosage form include ointments, creams, gels, capsules, granules, beads, pellets and tablets. Two-phase dosage forms include micelles, liposomes, foams and suspensions. The choice of dosage form may depend on patient factors and illness factors.

- The key physical and chemical properties of importance for the activity of pharmaceutical compounds are known as their physicochemical properties.

REFERENCES

FDA (Food and Drug Administration). (2004). Innovation or stagnation: Challenge and opportunity on the critical path to new medical products.

Lipinski, C. A. *et al.* (2001). *Experimental and computational approaches to estimate solubility and permeability in drug discovery and development settings.* Advanced Drug Delivery Review, 46, 3–26.

FURTHER READING

Aulton M. E. (ed.) (2002). *Pharmaceutics, the science of dosage form design* (2nd edn). Edinburgh: Churchill Livingstone.

Florence, A. T. and Attwood, D. (2006). *Physicochemical principles of pharmacy* (4th edn). London: Pharmaceutical Press.

Patrick, G. L. (2005). *An Introduction to Medicinal Chemistry* (3rd edn). Oxford: Oxford University Press.

PHARMACEUTICAL SOLUTIONS

02

Learning objectives

Having studied this chapter, you should be able to:

- appreciate the pharmaceutical importance of solubility and solutions
- use common expressions of solubility
- describe aqueous and lipid media
- understand the factors affecting the solubility of pharmaceutical compounds
- apply quantitative measures of the acidity or basicity of pharmaceutical compounds
- understand the pharmaceutical role of salt formation and select appropriate salt forms
- appreciate the possibility of degradation by hydrolysis.

Solutions, dissolution, and solubility play critical roles in pharmaceutical activity. One important reason for this is that drug molecules usually need to be in solution in the physiological fluid adjacent to their pharmacological targets if they are to produce therapeutic effects. Solutions are also of great importance in, amongst other things, the formulation and the development of new pharmaceutical products. This chapter will examine the main concepts relating to the dissolution and solubility of pharmaceutical compounds.

2.1 DEFINITIONS AND EXPRESSIONS OF SOLUBILITY

A solution is a system of one phase and more than one chemical component. The term 'solution' generally suggests a liquid phase, but solutions may also be solid- or vapour-phase systems. This chapter is concerned exclusively with liquid-phase solutions. A solution consists of one phase, that is, a chemically and physically uniform part of a system, which is separated from all other parts of the system by distinct boundaries. The components of a solution are categorized either as solvent or as solute. The term 'solvent' is used to describe the component or components that are present in greater quantity. The term 'solute' is used to describe the component or components present in lesser quantity. For example, a mixture of water and ethanol could be considered as either a solution of water in ethanol or ethanol in water. Usually, the one present in excess is considered the solvent and that present in lesser quantity is considered the solute. However, it would be less obvious which is the solute or solvent in, say, a 50:50 mixture of water and ethanol. The mixture is really one phase consisting of two chemical components.

Under some circumstances, the assignment of solvent and solute(s) may be somewhat arbitrary. However, for most pharmaceutical solutions, the assignment of solvent and solute(s) is clear, as the solvent is present in much greater quantity than the solute(s). A solution can be prepared by mixing solvent and solute until a homogeneous liquid phase is obtained. Such a mixing is said to involve the **dissolution** of the solute in the solvent. In the resulting solution, the solute is said to be dissolved in the solvent.

The maximum quantity of a particular solute that can be dissolved in a given quantity of a particular solvent under specific conditions is the solubility of that solute in that solvent under those conditions. The **equilibrium solubility** of a solute in a given solvent is achieved when the solution exists in equilibrium with excess undissolved solute. Such a solution is said to be saturated with respect to that solute. For example, at equilibrium, a saturated solution of sodium chloride in water at 0 °C contains no more than 36 g of sodium chloride per 100 mL of water. Under certain circumstances, discussed in Section 4.5.1, solubility in excess of the equilibrium solubility can sometimes be achieved temporarily. Values for solubility are expressed in terms of concentration, that is, in terms of quantity of solute for a given quantity of solvent. These quantities can be masses or volumes of solute or solvent, for example g L^{-1}. Specific expressions of concentration that are in use include mole fraction, molarity, molality, percentage, parts per million, and equivalents. We will explain how the more important of these concentration terms are used, one by one.

Mole fraction

The mole fraction of a solute in a solvent is defined below by eqn 2.1. If we think of a mixture as being made up of some number of moles, the mole fraction of any component of the mixture is the number of moles of that component as a fraction of the total number of moles in the mixture. For a solution consisting of one solvent component and one solute component, the mole fraction of solute, x_{solute}, is given as

$$x_{solute} = \frac{n_{solute}}{n_{solute} + n_{solvent}},$$
(2.1)

in which n_{solute} and $n_{solvent}$ are the number of moles of solute and solvent, respectively. In other words, the mole fraction of the solute is the number of moles of the solute as a fraction of the combined number of moles of solute and solvent. Mole fractions are often used in the construction of phase diagrams, as discussed in Section 3.2.

For example, during the manufacture of an active pharmaceutical ingredient (API), a solution of 75 g of the API in 240 g of ethanol is prepared. The molecular weight of the API is 302 g mol^{-1}, so that the solution contains 75/302 = 0.25 mol of the API. The molecular weight of ethanol is 46 g mol^{-1}, so that the solution contains 240/46 = 5.22 mol of ethanol. The total number of moles in the solution is 0.25 + 5.22 = 5.47 mol. The mole fraction of the API solute is 0.25/5.47 = 0.046; while the mole fraction of the ethanol solvent is 5.22/5.47 = 0.954.

Molarity

The molarity of a solution is the number of moles of solute dissolved in one litre of solution. The units are mol L^{-1} or mol dm^{-3}. A solution of one mole per litre of solvent is known as a **molar solution**. Although expressions of molarity are widely used, one disadvantage is that changes in temperature can result in variations in the densities of solutions, and hence in variations in molarities.

Molality

The molality of a solution is the number of moles of solute dissolved in one kilogram of solvent. The unit of molality is mol kg^{-1}. Note that whereas molarity is determined with respect to the *volume* of the solution, molality is determined with respect to the *mass* of the pure solvent. As the mass of solvent does not vary with temperature, expressions of molality are not temperature dependent. The molality of a solute in a solution is also unaltered if some quantity of a different solute is added, whereas the molarity changes under these circumstances due to an increase in the total number of moles.

Percentages

Percentages express solution concentrations as the mass of solute per 100 mL of solution. The symbol '% w/v' is used to show that we are referring to the number of grams of solute per 100 mL of solution. % w/v is widely used pharmaceutically. For example, a 5% w/v solution of glucose would contain 1 g of glucose dissolved in every 20 mL of solution. As with molarity, variation in solvent density with temperature can result in variation of % w/v with temperature.

The symbol '% w/w' is used to show that we are referring to the number of grams of solute per 100 g of solution.

Parts per million

Parts per million, or 'ppm', express concentration as the number of units of mass of solute per 10^6 units of mass of solution. This is equivalent to

$$\frac{\text{mass solute (g)}}{\text{total mass solution (g)}} \times 10^6 \text{ (ppm)}.$$

Parts per million are used to express the concentrations of very dilute solutions. For solutions of even greater dilution, parts per billion, or 'ppb', can be used. These are equivalent to

$$\frac{\text{mass solute (g)}}{\text{total mass solution (g)}} \times 10^9 \text{ (ppb)}.$$

Other concentration terms used for dilute pharmacological solutions include µg/mL, which refers to the number of micrograms of solute per millilitre of solution.

Equivalents

The concept of equivalents applies to solutions of ionic solutes, and is dependent on the valencies of the ions. An equivalent is equal to one mole of charge. The number of equivalents of an ion can be calculated by multiplying the number of moles of charged particles by the valence.

For monovalent ions, for example Na^+ ions, a solution of one equivalent per litre, or 1 Eq L^{-1}, is the same as a solution of one mole per litre. For multivalent ions, for example SO_4^{2-} ions, the number of equivalents takes into account the valency of the ions. For example, a solution of 2.20 mmol L^{-1} Na_2SO_4 would be equal to 4.40 mEq L^{-1} of SO_4^{2-}. Older usage refers to a solution of 1 Eq L^{-1} as a 'normal' solution. This should not be confused with a 'physiologically normal' sodium chloride solution, which contains 0.9 g NaCl per 100 mL (the concentration of NaCl in the blood), which is not the same concentration as a 1 Eq L^{-1} NaCl solution.

▷ **WORKED EXAMPLE 2.1**

The equilibrium solubility of sucrose in water at 20 °C is 89% w/v. Express the equilibrium solubility of sucrose in water at 20 °C as (i) mole fraction, (ii) molarity, and (iii) molality. The molecular weight of sucrose is 342.3 g mol^{-1}. The density of an 89% w/v solution of sucrose in water at 20 °C is 1.33 g mL^{-1}.

(i) A concentration of 89% w/v sucrose in water implies a mass of 89 g of sucrose for every 100 mL of solution. 100 mL of solution corresponds to a mass of $100 \times 1.33 = 133$ g of solution. Of this 133 g solution, 89 g are of sucrose, hence 44 g is the mass of water in 100 mL of solution. The number of moles of sucrose present is therefore $89/342.3 = 0.26$ mol, while the number of moles of water present is $44/18.0 = 2.44$ mol. The mole fraction of sucrose is therefore

$$\frac{0.26 \text{ mol}}{0.26 \text{ mol} + 2.44 \text{ mol}} = 0.096.$$

(ii) From part (i) above, 100 mL of solution contains 0.26 mol, therefore 1 L of solution contains 2.60 mol, that is the molarity of the solution is 2.60 mol L^{-1}.

(iii) From part (i) above, 100 mL of solution contained 89 g of sucrose and 44 g of water. A quantity of the solution containing 1 kg of water would therefore contain 2.02 kg of sucrose. This mass of sucrose is equivalent to 5.90 mol. Hence, the molality of the solution is 5.90 mol kg^{-1}.

2.1.1 Measurement of solubility and solubility curves

Solubility is a critical pharmaceutical property. If a drug molecule is to be active in a biological tissue, it has to be soluble at least to some extent in the biological fluids that feed that tissue. To measure the solubility of a pharmaceutical solute in a solvent, the solvent and an excess of the solute must be allowed to reach equilibrium. Achieving the equilibrium solubility of a solute in a given solvent requires the solution phase to exist in equilibrium with excess undissolved solute phase. The solution and excess solute phase need to be in contact for sufficient time for equilibrium to be achieved. Agitation of the mixture is usually required. It is essential that the system be maintained at a constant temperature, and that temperature measurements be accurate. This is because solubility varies with temperature, so that solubility measurements are only valid if carried out at some specific temperature.

Once the mixture reaches equilibrium, a period of settling is required to allow the solution phase and the excess solute phase to separate. For example, if the excess solute phase is undissolved solid, this may settle to the bottom of the vessel, leaving the solution phase as a supernatant liquid. The supernatant solution can then be analysed and solute concentration determined. Any suitable analytical method can be used, for example HPLC, titration, electrochemical analysis, spectrophotometry, or other methods. However, any analytical method used must be properly validated; samples should be weighed (rather than measured by volume) and it is also useful to measure the densities of the samples.

There are very many methods of measuring solubility based on these principles. An alternative approach is to use a mixture of solute and solvent of known, fixed proportion. The temperature of such a mixture can be gradually raised until complete solubility is achieved. With repeated experiments, the temperature at which that proportion of solute and solvent form a single solution phase can be determined. An alternative to varying the temperature of the mixture is to gradually add aliquots of solvent at a fixed temperature until solution is achieved.

Plots of the variation of equilibrium solubility with temperature are known as solubility curves. These plots are useful, as they tell us how the solubility of a pharmaceutical might change with changes in temperature (for instance, if the temperature changes from room temperature to body temperature). Figure 2.1 shows the solubility curves for three substances – sucrose, citric acid and sodium chloride – in water.

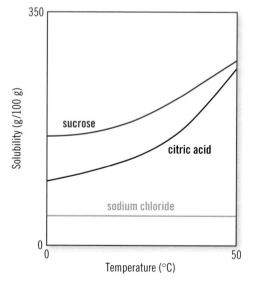

FIGURE 2.1 Solubility curves for sucrose, citric acid and sodium chloride in water [from Roger Davey and John Garside, From Molecules to Crystallizers, An Introduction to Crystallization, 2000, OUP, by permission of Oxford University Press].

The solubility of a solute in a solvent usually increases with temperature. This is illustrated by the curves for sucrose and citric acid in Fig. 2.1. However, for some substances, there is little variation in solubility with temperature. Sodium chloride is an example of such a substance, as indicated by the almost horizontal solubility curve for sodium chloride seen in Fig. 2.1. Variations in pressure generally have little effect on the solubility of liquids and solids. However, variation in pressure can have a significant effect on the solubility of gases. The relationship between pressure and the solubility of gases is described by Henry's law, which is discussed in Section 3.2.2.

2.2 SOLVENT STRUCTURE

Two types of solvent media are significant pharmaceutically. These are water-based or aqueous media, and lipid-based or hydrophobic media. Examples of aqueous media include the gastrointestinal tract fluid, the blood and aqueous suspension or solution formulations. Examples of hydrophobic media include cell membranes and micelles. Most biochemical and pharmacological processes occur in aqueous media. For this reason, the structure of aqueous solvent and solutions are particularly important.

2.2.1 Hydrogen bonding and the structure of water

This section concerns water as a liquid-phase material, and in particular as a solvent. To a large extent, the structure of water as a material derives from the interactive properties of water molecules. These are of the familiar formula H_2O, in which the oxygen atom forms one covalent bond to each of the hydrogen atoms, with an angle of 104.5° between the two bonds as depicted in Fig. 2.2. As there is a large difference in electronegativity between oxygen and hydrogen atoms, the O–H bonds are very polarized, with the hydrogens possessing partial positive charges, symbolized by 'δ^+', and the oxygen atoms possessing partial negative charges, symbolized by 'δ^-'. The O–H bonds therefore possess large bond dipoles. The highly dipolar bonds on neighbouring water molecules can interact with each other. The resulting attractive force is known as a **hydrogen bond**.

Hydrogen bonds can exist between groups X and H–X, where X is oxygen, nitrogen or fluorine. Hence, any molecule that possesses O–H bonds, or similarly polarized bonds such as N–H bonds can form hydrogen bonds. When a hydrogen bond exists between two molecules, the molecule possessing the electronegative atom is known as

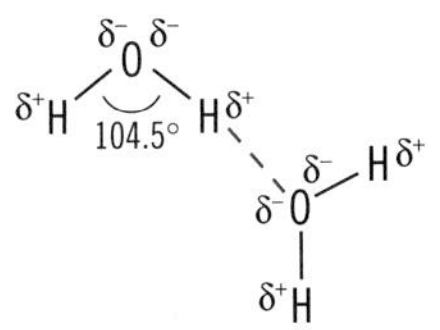

FIGURE 2.2 Molecular structure of water, showing O–H bond dipoles, and a hydrogen bond between two water molecules.

the hydrogen-bond acceptor, and the molecule possessing the polarized hydrogen atom is known as the hydrogen-bond donor. A generalized hydrogen bond can be represented as follows

$$X-H\cdots A$$

H-bond   H-bond
donor   acceptor

in which the atoms X and A are oxygen, nitrogen or fluorine atoms. The range of distances between the H atom and the atom A in hydrogen bonds is 1.2 to 2.2 Å. The angle XHA can range from 120° to 180°.

Any one water molecule can simultaneously donate two hydrogen bonds and accept two hydrogen bonds. Hence, each water molecule in a solid or liquid phase can, in principle, coordinate four other water molecules. Such an arrangement occurs in the structure of ordinary ice, Fig. 2.3. Each oxygen atom shown in Fig. 2.3 is in an approximately tetrahedral environment, in which it is covalently bonded to two hydrogen atoms, and is acting as a hydrogen-bond acceptor for another two. This arrangement provides a well-ordered and stable crystalline structure.

Long-range ordered structure would be expected to exist in the solid ice phase. However, the liquid water phase is fluid and so the molecules in liquid water must have some ability to move relative to each other. Therefore, in liquid water, we may only expect short-range order to be present. Whatever short-range order exists in water might be expected to resemble the ice structure in part. Experimental data suggests that such is the case. However, the structure of liquid water is subject to debate, and a number of

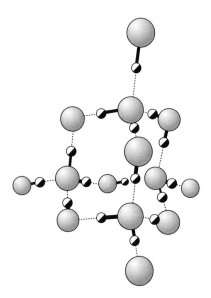

FIGURE 2.3 Structure of ordinary ice. Large spheres represent oxygen atoms, small spheres represent hydrogen atoms. Reproduced from PW Atkins, Physical Chemistry, 2nd edn, 1982, by permission of Oxford University Press.

model structures have been proposed. One model proposes that each H_2O molecule in liquid water continues to be hydrogen bonded to four other H_2O molecules, but that the hydrogen bonds can be distorted to allow for disorder in the liquid phase (Eisenberg and Kauzmann, 1969). An alternative model, known as the flickering cluster model, views liquid water as a mixture of well-ordered, ice-like hydrogen-bonded regions existing in equilibrium with non-hydrogen-bonded H_2O molecules (Frank, 1970).

2.2.2 Lipid-based media

Non-polar organic solvents such as toluene or hexane are examples of hydrophobic solvents. Solvents such as these are not normally part of any biological system. However, for a drug to travel from its site of administration to its site of action, it will generally be necessary for it to travel across one or more cell membranes. The transport of drugs across cell membranes and other biological structures is considered in detail in Chapter 7. However, in the simplest and most common form of transport across membranes – that is, passive diffusion – drug molecules effectively undergo dissolution in the cell membrane. For this reason the structure of these membranes will be considered in outline here.

Cell membranes are composed of lipids, proteins and carbohydrates. Of these, the lipids compose the core of the membranes. The two main types of lipid occurring in cell membranes are phospholipids and cholesterol. Representative structures for these are shown in Fig. 2.4. Notice that the structures shown in Fig. 2.4 are amphipathic, that is, they posses both polar hydrophilic regions, and non-polar hydrophobic regions. The polar hydrophilic regions are the quaternary ammonium and phosphate groups

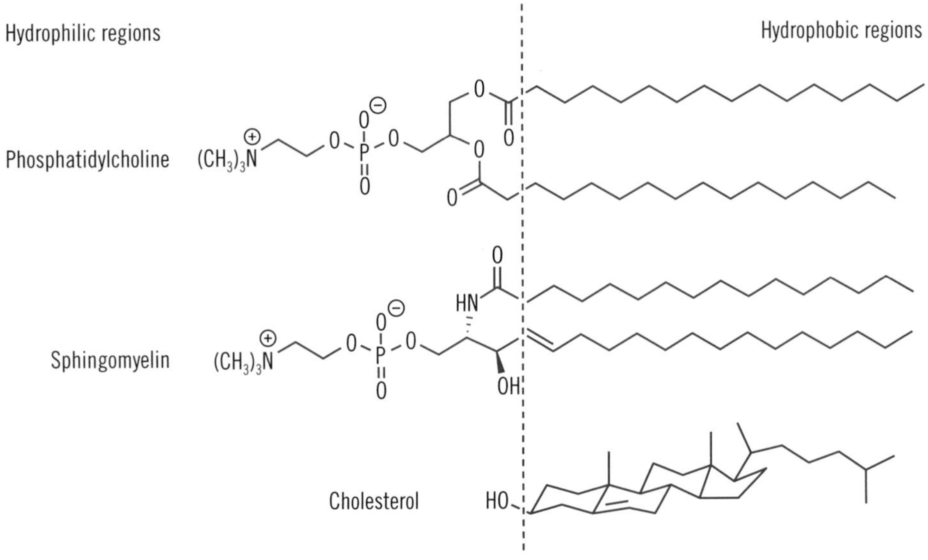

FIGURE 2.4 Molecular structures of selected lipids.

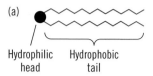

(a)

Hydrophilic   Hydrophobic
head   tail

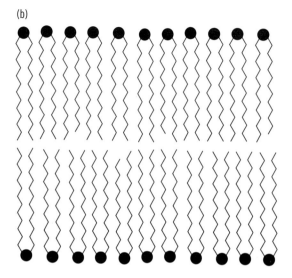

(b)

FIGURE 2.5 (a) Simplified representation of a cell membrane lipid. (b) Schematic representation of a lipid bilayer.

of phosphatidylcholine and sphingomyelin, and the hydroxyl group of cholesterol. The non-polar hydrophobic regions are the long alkyl chains of phosphatidylcholine and sphingomyelin, and the steroidal polycycle and short alkyl chain of cholesterol. For convenience, lipid structures such as those shown in Fig. 2.4 are often represented in a simplified form, such as that shown in Fig. 2.5(a).

In an aqueous environment, the amphipathic lipid molecules form structures in which interaction of the hydrophilic regions with the surrounding aqueous medium is maximized, while interactions of the hydrophobic regions with the aqueous medium are minimized. The hydrophobic parts of the molecules come together, exposing the hydrophilic parts to the aqueous surroundings. This is the principle underlying the formation of micellar systems. Molecules that have both a hydrophilic and a hydrophobic region are known as amphiphiles.

Amphiphiles possessing two alkyl chains, such as the lipids shown in Fig. 2.4, tend to form planar bilayers. Such a bilayer is shown schematically in Fig. 2.5(b). The polar hydrophilic groups are presented to the aqueous exterior, while the non-polar hydrophobic alkyl groups extend into the interior of the bilayer and avoid contact with the aqueous exterior.

Cell membranes consist of lipid bilayers that incorporate protein components. Carbohydrate components are attached to the lipid bilayers on the surface that is exterior to the cell. The structure of cell membranes is discussed in greater detail in Section 7.5. The distance across a lipid bilayer is in the range 2.5 to 3.5 nm. The 'length' of a drug molecule would be of the order of 1 to 2 nm. Hence, while traversing a cell membrane, a typical drug molecule would be completely enclosed within the interior of the cell membrane. In that environment, the drug molecule will be surrounded by hydrophobic alkyl groups.

As mentioned above, a drug molecule is likely to interact with a variety of biological membranes while transporting from the site of administration to the site of action. The membranes are therefore very important hydrophobic, or lipophilic, media. Other lipophilic media of pharmaceutical importance are synthetic micellar structures and hydrocarbon excipients such as paraffin. These are also, essentially, alkyl-rich environments.

2.3 DISSOLUTION AND SOLVATION

Processes of dissolution directly affect the absorption and distribution of drugs. The previous section examined the structure of both water-based and lipid-based media. To reach its site of action in the body, a drug molecule must be soluble at least to some extent in both of these types of media. The majority of drug substances are solids in their pure forms. Many are also administered to patients in solid-state dosage forms, especially as tablets. If the drug substance is to be biologically active, these solid state materials must dissolve to some extent in biological fluids. Dissolution is therefore a critical step in the process by which drug molecules act.

Dissolution of a drug from a solid phase can be considered to occur in the following stages.

- Firstly, drug molecules break away from the solid structure.
- Secondly, a cavity suitable for the accommodation of the drug molecules is generated in the solvent medium.
- Finally, the drug molecule locates within the solvent cavity, that is, it becomes solvated within the solvent, producing a solution.

Figure 2.6 gives a schematic visualization of these processes.

Solvation consists of the generation of an ordered region surrounding the solute drug. Within this region, the solute and solvent molecules interact by weak intermolecular forces. Solutes can be very broadly divided into two categories. These are polar, or hydrophilic, solutes, and non-polar, or hydrophobic, solutes. Drugs are often of intermediate polarity.

Solvation is governed by a general rule: 'Like dissolves like' – that is polar solutes are generally better solvated by polar solvents, and non-polar solutes by non-polar solvents. Water is a polar solvent: it possesses a relatively high dipole moment and dielectric constant, and the ability to donate and accept hydrogen bonds. It is therefore a good solvent for polar solutes, with which it can interact by hydrogen bonding, dipole–dipole interactions, and ion–dipole interactions. Further, the high dielectric constant of water reduces the attraction between oppositely charged ions solvated in water. Regions of

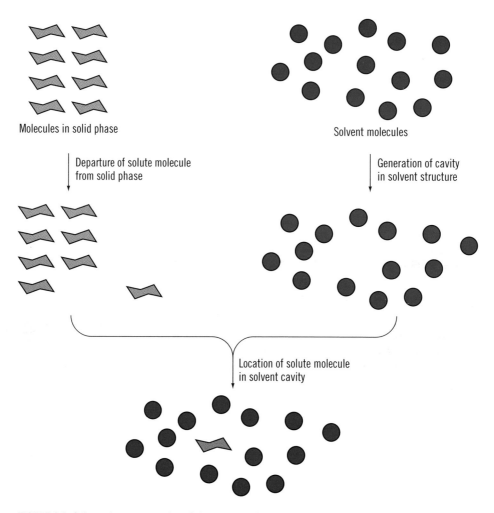

Molecules in solid phase

Solvent molecules

Departure of solute molecule from solid phase

Generation of cavity in solvent structure

Location of solute molecule in solvent cavity

FIGURE 2.6 Schematic representation of the processes involved in dissolution.

solvating water molecules also form around ions, as shown schematically in Fig. 2.7. Within these regions water molecules interact with ions by dipole–ion interactions. The water molecules immediately surrounding the ions have the strongest interaction with them. The water molecules in such a region are highly ordered and are said to form a primary solvation layer. As the ions diffuse through the solvent medium, the primary solvation layer moves with them. The primary solvation layer may be separated from the bulk of the water solvent by a shell of less ordered water molecules known as the secondary solvation layer.

Polar, hydrophilic solutes include ionic compounds such as salts, and compounds with moderate ability to form hydrogen bonds. Hydrocarbon groups, such as alkyl or aromatic groups, tend to reduce polarity and increase hydrophobicity. Often, there is a balance between these effects. An ionic group will impart a strongly hydrophilic character to a molecule. A short hydrocarbon chain may not outweigh this effect. However, if the chain

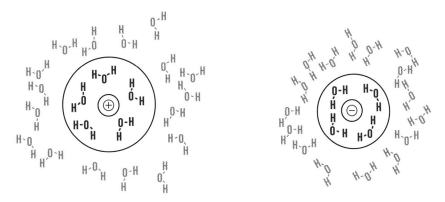

FIGURE 2.7 Regions of solvation about (left) cations and (right) anions in aqueous solution. The regions closest to the ions are the primary solvation layers. These are separated from the bulk aqueous medium by the secondary solvation layer. (Medicinal Chemistry, An Introduction, G. Thomas, 2000, © John Wiley & Sons Ltd., reproduced with permission.)

length increases, a 'tipping point' will eventually be reached at which the hydrophilic character of the polar group is outweighed by the hydrophobic alkyl chain.

Compounds that are extensively hydrocarbon in nature tend to have poor solubility in water, but tend to have good solubility in non-polar solvents and in lipid media. The poor solubility of non-polar groups in water is known as the hydrophobic effect. The hydrophobic effect plays a key role in the structuring of lipid phases. For example, as discussed in Section 2.2.2, lipid molecules form bilayer structures in which interactions of the hydrophobic regions of the lipids with the surrounding aqueous medium are minimized.

Various explanations can be proposed for the hydrophobic effect – for example repulsion between polar water molecules and non-polar solutes, or competing attractive forces between hydrocarbon groups. Attractive forces between hydrocarbons, such as dispersion forces, do exist. However, these are relatively weak and are insufficient to account for the hydrophobic effect. Neither does it appear to be the case that the hydrophobic effect is caused by actual repulsion between water molecules and molecules of non-polar solutes. The main factor in the hydrophobic effect is the hydrogen-bonding interactions between the water molecules themselves, and the structured nature of water, as discussed in Section 2.2.1.

In order to accommodate solute molecules, the structure of water must be distorted to provide a suitable cavity for solvation. This decreases the degree of freedom of the water molecules in the region of the cavity, and hence is unfavourable in terms of entropy. (The concept of entropy is discussed in Section 3.1.2.) In the case of polar solutes, this unfavourable entropy change is overcome in most cases by favourable changes associated with dipole–dipole and ion–dipole interactions between water and solute. However, no such compensating interaction is possible in the case of non-polar, hydrocarbon-rich solutes. For this reason, solvation of hydrophobic solutes in water is energetically unfavourable.

2.4 FACTORS AFFECTING SOLUBILITY

For a drug to travel from its site of administration to its site of action in the body, it must be soluble at least to some extent in both aqueous and lipid environments. Solubility is therefore a vital consideration in drug design. Lipinski's 'Rule of Five', first encountered in Section 1.3.3, summarizes the factors that affect relative solubility in aqueous and lipid media. In this section we discuss how molecular features affect these factors. The molecular features we consider are molecular weight, hydrogen-bonding ability, hydrophilicity/ lipophilicity and ionization.

2.4.1 Molecular weight

In general, increased molecular weight results in decreased solubility, irrespective of solvent. For example, the series of molecules benzene, naphthalene, anthracene, shown in Fig. 2.8, show decreasing solubility in diethyl ether, with benzene being soluble in all proportions, naphthalene very soluble, and anthracene only slightly soluble. The molecular feature that changes throughout this series is the size of the molecules. The increasing size is reflected in increasing molecular weights as we proceed through the series, from 78 g mol^{-1} for benzene, to 128 g mol^{-1} for naphthalene and 178 g mol^{-1} for anthracene. Higher molecular weight molecules often display poor solubility in general. This effect can have a direct impact on the bioavailability of molecules, as higher molecular weight molecules can have poor solubility in both aqueous and lipid media, hindering their absorption from the site of administration and their distribution within the body to their site of action. The general 'rule of thumb' for pharmaceutical molecules is that molecular weights of greater than 500 g mol^{-1} are likely to result in reduced solubility and bioavailability, while molecular weights greater than 1200 g mol^{-1} are to be avoided.

The melting points for benzene, naphthalene and anthracene are 5.5, 82.0, and 218.0 °C, respectively. The increasing melting points reflect the increasing molecular weights of the three compounds. As the size (and molecular weight) of the molecules increase, so too does the extent of weak intermolecular forces between the molecules. These forces are weak electrostatic attractions between the electrons of one molecule and the nuclei of another (London dispersion forces and Van der Waals' forces). The larger the molecules are, the greater is the extent of these weak intermolecular forces, and the greater is the tendency to form condensed phases (liquids and solids), rather than gases. Hence, a general

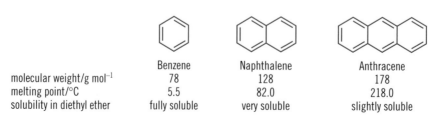

	Benzene	Naphthalene	Anthracene
molecular weight/g mol^{-1}	78	128	178
melting point/°C	5.5	82.0	218.0
solubility in diethyl ether	fully soluble	very soluble	slightly soluble

FIGURE 2.8 Molecular structures of, from left to right, benzene, naphthalene and anthracene. Also given are the molecular weights, melting points and solubility in diethyl ether of the three compounds.

trend is that compounds with higher melting points are likely to be of higher molecular weight and, hence, of lower solubility and bioavailability.

The potency of the drug molecule is also an important consideration. Modern drug design often produces drug structures that are relatively high in molecular weight, for example within the range 500 to 1000 g mol^{-1}, but that are also highly potent, that is, are active in very low concentrations. Lesser quantities of these high-potency drugs are therefore required. This counteracts the poor solubility and bioavailability of these compounds.

2.4.2 Hydrogen bonding

As outlined in Section 2.2.1, hydrogen bonding is the main form of intermolecular bonding in water. Drug molecules that are capable of donating or accepting hydrogen bonds therefore have an ability to interact with water molecules, as water molecules can also donate or accept hydrogen bonds. Hence, the presence on a molecule of hydrogen-bond-donating groups (such as NH or OH groups) or hydrogen-bond-accepting groups (such as N, O or F atoms) allows hydrogen-bonding interactions with water molecules. Therefore, compounds possessing these groups could be expected to have good solubility in water. This is the case to an extent, but there is a limit to the water solubility provided by hydrogen bonding. This is illustrated by the case of the three compounds shown in Fig. 2.9, urea, oxamide and N,N′-dimethyloxamide.

The first and third of these compounds, urea and N,N′-dimethyloxamide, are highly water soluble. However, the second compound, oxamide, is poorly soluble in water. All three compounds are of relatively low molecular weight. The critical factor in this case is the hydrogen-bonding capacity. Urea has four hydrogen-bond-donating N–H groups, and three hydrogen-bond-accepting N or O atoms. N,N′-dimethyloxamide has two hydrogen-bond-accepting N–H groups, and four hydrogen-bond-accepting N or O atoms. The second compound, oxamide, has the highest hydrogen-bonding capacity, having four hydrogen-bond-donating N–H groups, and four hydrogen-bond-accepting N or O atoms. However, this compound has the lowest solubility in water.

This seeming paradox illustrates an important point about hydrogen-bonding capacity: some ability to form hydrogen bonds aids the water solubility of compounds, but too much hydrogen-bonding capacity appears to reduce it. This is because hydrogen bonding is also an important form of intermolecular bonding in the solid state (see Chapter 4). Hence, molecules with extensive hydrogen-bonding capacity have a greater tendency to remain in the solid state, and resist dissolution.

So, a moderate amount of hydrogen-bonding ability is therefore beneficial as it promotes solubility in water (which is essential for pharmaceutical activity), while too much hydrogen bonding is detrimental, as it impairs solubility in any solvent. For the purposes of drug design, a balance is required between these effects. Lipinski's 'Rule of Five'

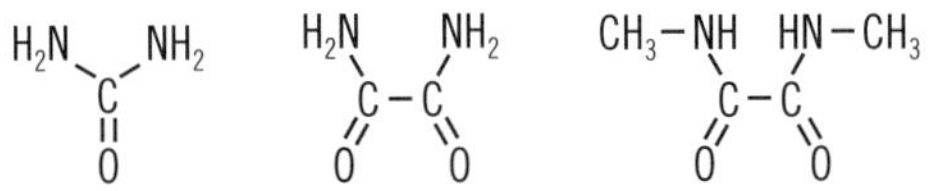

FIGURE 2.9 Molecular structures of, from left to right, urea, oxamide, and N,N′-dimethyloxamide.

(Table 1.5) provides a guideline for optimizing hydrogen-bonding capacity. The guideline is that the number of hydrogen-bond donor groups, that is N–H or O–H groups, should preferably not exceed five, while the number of hydrogen-bond-accepting groups, in particular N or O atoms, should not exceed ten.

2.4.3 Hydrophobic and hydrophilic groups

Most drugs are organic molecules that contain hydrocarbon-like groups that are hydrophobic, and polar functional groups that are hydrophilic. Drug molecules may therefore be divided into various hydrophilic and hydrophobic regions. For example, Fig. 2.10 shows how the antihypertensive drug captopril is composed of hydrophilic and hydrophobic regions.

The solubility of the drug molecule in either aqueous or lipid media may depend on the relative proportions of these regions, and on their positioning on the drug structure. Certain groups are highly hydrophilic, such that their inclusion anywhere on a structure tends to result in increased solubility in water. Other polar groups may be moderately hydrophilic, in which case their effect on aqueous solubility may depend on their number and positioning on the drug structure.

Hydrophobic groups include simple alkyl groups such as methyl groups, ethyl groups, isopropyl groups, etc. These groups are hydrophobic whether they are attached to heteroatoms such as nitrogen, oxygen or sulfur, or whether they are substitutents on larger hydrocarbons units such as aromatic rings. Alkyl groups in the interior of a structure, such as methylene (CH_2) groups, ethylene (CH_2CH_2) groups, etc. also tend to be hydrophobic.

Hydrophilic groups tend to be groups that can interact with water molecules by hydrogen bonding. Groups that can both donate and accept hydrogen bonds tend to be hydrophilic. Examples of such groups include hydroxyl (OH) groups, amino groups (NH_2), primary amide groups ($NH_2C{=}O$), and carboxylic acid groups (CO_2H). Groups that only accept hydrogen bonds tend to be hydrophilic, but to a lesser extent. Examples of these less hydrophilic groups are methoxy (OCH_3) groups, nitro (NO_2) groups, and tertiary amide groups ($NR_2C{=}O$, in which R = alkyl groups).

Water is effective at solvating *ions* by ion–dipole interactions. For this reason, inclusion of ionized groups on a drug structure has a strong tendency to enhance solubility in water. Ionized groups are therefore classed as highly hydrophilic. Examples of such groups include ammonium (NH_3^+) groups, carboxylate (CO_2^-) groups, and sulfonate (SO_3^-) groups.

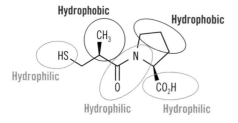

FIGURE 2.10 Hydrophilic and hydrophobic regions of captopril.

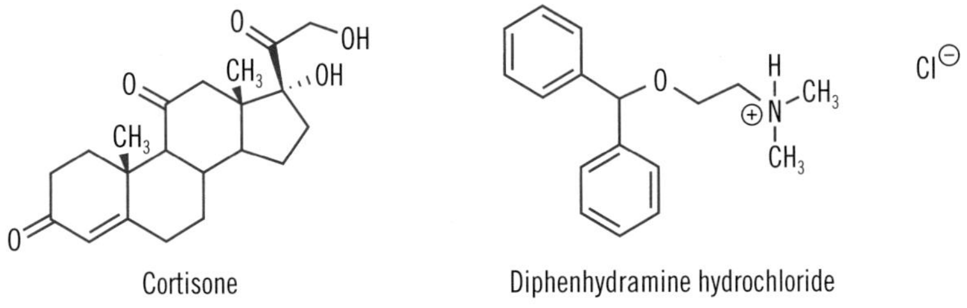

FIGURE 2.11 Molecular structures of cortisone and diphenhydramine hydrochloride.

The examples shown in Fig. 2.11 illustrate some of these effects. Cortisone is a hydrocarbon-rich molecule, with a largely hydrophobic structure. It might therefore be expected to be almost completely insoluble in water. However, it does have some moderately hydrophilic groups, specifically the three ketone oxygens atoms and the two hydroxyl (-OH) groups. Because of the influence of these, cortisone is slightly soluble in water. Diphenhydramine also has a considerably hydrophobic structure. In particular, the two benzene rings are very hydrophobic groups. However, in its hydrochloride salt form, the diphenhydramine molecule has a positively charged ammonium group. This makes the molecule relatively water soluble (1 g in 1 mL of water) even though it is essentially a hydrophobic organic molecule. The effect of salt formation on pharmaceutical solubility will be considered further in Section 2.6.

2.5 ACIDITY AND BASICITY

Many pharmaceuticals are acidic or basic to some extent. This gives them the ability to exist in both neutral and ionized forms. The neutral and ionized forms will have very different solubility that will affect how they behave when absorbed into the body. The extent to which they are ionized will be affected by the pH of the biological medium in which they are dissolved. Hence, ionization critically affects the ability of a drug to arrive at its correct site of action in the body. For this reason, we next need to consider the acid and base properties of pharmaceuticals.

Compounds that possess acidic or basic groups can exist in equilibrium between ionized and un-ionized forms. For example, the following equilibrium exists for compounds possessing carboxylic acid groups

$$R\text{-}CO_2H \rightleftharpoons R\text{-}CO_2^- + H^+. \tag{2.2}$$

A drug molecule possessing a carboxylic acid group therefore exists in equilibrium between a neutral un-ionized form and a charged ionized form. The neutral un-ionized form will be more hydrophobic and will have greater solubility in *lipid* media. By contrast, the charged ionized form will be more hydrophilic, and will have greater solubility in *aqueous* media. Notice how the presence of the acidic group therefore allows the compound to exist in two forms, one of which is relatively hydrophobic and one of which is relatively hydrophilic.

To be pharmaceutically active, a molecule must have at least some solubility in aqueous environments, such as the gastrointestinal tract, and some solubility in lipid environments such as the interior of cell membranes. The presence of acidic groups assists in this, by allowing the molecule to adopt either more hydrophilic or more hydrophobic forms. The same applies to molecules possessing basic groups, as these can exist in the following equilibrium

$$R\text{-}NH_2 + H^+ \rightleftharpoons R\text{-}NH_3^+. \tag{2.3}$$

Again, these molecules can exist in both neutral un-ionized forms, and charged ionized forms.

Hence, the presence of acidic or basic groups on a drug structure assists in achieving sufficient solubility in both aqueous and lipid environments. These groups are therefore beneficial for pharmaceutical activity, and are common features of drug structures. It is also very common that it is the ionized forms of drugs that bind to their pharmacological targets.

The factors that need to be considered when discussing the acidity or basicity of drugs are the position of equilibrium between un-ionized and ionized forms, and the effect on that equilibrium of the local environment. Before tackling these issues, we need to understand how acidity and basicity are quantified. Hence, we need to review the concepts of pK_a and pK_b.

2.5.1 pK_a and pK_b

The quantitative measure of the acidity of a compound is its pK_a. Likewise, the quantitative measure of the basicity of a compound is its pK_b. Both of these are measures of the equilibrium constants for proton (H^+) transfer processes such as those shown in eqns 2.2 and 2.3. The equilibrium position for such processes is very much affected by the medium in which the process is occurring. If the process is occurring in solution, then the solvent needs to be considered. This is because interaction of the solvent with either the neutral or ionized forms can affect the position of equilibrium for proton transfer. Unless otherwise stated, pK_a and pK_b values generally refer to processes occurring in aqueous solution. The equilibrium for a process such as that shown in eqn 2.2 should therefore be rewritten as follows to allow for the role of the water solvent

$$R\text{-}CO_2H + H_2O \rightleftharpoons R\text{-}CO_2^- + H_3O^+. \tag{2.4}$$

In the process illustrated in eqn 2.4, RCO_2H acts as an acid, as it donates a proton. H_2O acts as a base, as it accepts a proton. RCO_2^- acts as a conjugate base, as it accepts a proton in the reverse process. H_3O^+ acts as a conjugate acid, as it donates a proton in the reverse process. The species H_3O^+ is known as the hydronium ion. Research has shown that in water, protons do not exist as free H^+, but rather in hydrated forms for which H_3O^+ is the most straightforward proposed structure (Bell, 1973). An equilibrium constant, K_a', for the process shown in eqn 2.4 would be as follows

$$K_a' = \frac{[RCO_2^-][H_3O^+]}{[RCO_2H][H_2O]}. \tag{2.5}$$

While H_2O acts as a proton acceptor in the above process, it is also the solvent. For that reason, very many more H_2O molecules are present in the system than other species. It is therefore assumed that the concentration of H_2O molecules remains effectively constant relative to the concentrations of the RCO_2H, RCO_2^-, and H_3O^+ species. Consequently, eqn 2.5 can be given in the following form

$$K_a = \frac{[RCO_2^-][H_3O^+]}{[RCO_2H]}. \tag{2.6}$$

K_a is the **acidity constant** and provides a measure of the extent of dissociation of an acid in water, and hence is a measure of acid strength. (Note that the stronger the acid, the greater the K_a value.) For example, penicillin G is a carboxylic acid and so would be relatively acidic. Its K_a is 1.6×10^{-3}. This tells us that the equilibrium

$$R\text{-}CO_2H + H_2O \rightleftharpoons R\text{-}CO_2^- + H_3O^+$$

lies to the left (because $[RCO_2H]$ must be larger than $[RCO_2^-][H_3O^+]$ for K_a to be smaller than 1) but not so far to the left that no dissociation of RCO_2H occurs. By comparison, the antiepileptic drug phenytoin, which is a weakly acid pharmaceutical, has a K_a of 5.0×10^{-9} (a much smaller number). Consequently, the concentration of the un-ionized form at equilibrium for phenytoin must be higher than for penicillin G – that is, the equilibrium lies further to the left, and so phenytoin is less ready than penicillin G to give up a proton (to act as an acid).

The range of K_a values for drug-like compounds is vast, and actual values for specific compounds are often very large or very small. For convenience, therefore, a modified form of the K_a is used for routine comparison of acid strengths. This modified form is the pK_a. By analogy with the definition of pH, pK_a is defined as follows

$$pK_a = -\log_{10} K_a. \tag{2.7}$$

(A brief review of the use of logarithms is given in Appendix A2.) The pK_a converts the K_a values to a logarithmic scale, which encompasses a narrower range of values, making a comparison between values more straightforward. Note that the stronger the acid, the greater the K_a value and the lower the pK_a value. Table 2.1 compares the K_a and pK_a values for compounds of varying acidity.

For bases, an equivalent process to eqn 2.4 would be

$$RNH_2 + H_2O \rightleftharpoons RNH_3^+ + HO^-. \tag{2.8}$$

From eqn 2.8, the following equilibrium constant can be obtained

$$K_b = \frac{[RNH_3^+][HO^-]}{[RNH_2]}. \tag{2.9}$$

Note that, as with the acidity constant, K_a, K_b factors in the term for the concentration of water, which is taken to be effectively constant. K_b values provide a measure of the relative strengths of different bases in water. As with K_a values, K_b values are usually used in a more convenient form, that is, as the pK_b, defined as follows

$$pK_b = -\log_{10} K_b. \tag{2.10}$$

TABLE 2.1 Comparison of the K_a and pK_a values for compounds of varying acidity.

Compound	Acidity	K_a	pK_a
Sulfuric acid	Strong acid	1.0×10^5	−5.0
Methanesulfonic acid	Strong acid	1.0×10^2	−2.0
Aspirin	Moderate acid	3.2×10^{-4}	3.5
Sulfathiazole	Weak acid	7.9×10^{-8}	7.1
para-Chlorophenol	Weak acid	6.3×10^{-10}	9.2
Caffeine	Very weak acid	1.0×10^{-14}	14.0
Acetone	Very weak acid	1.0×10^{-20}	20.0

A commonly used alternative to pK_b values is to use the pK_a of the conjugate acid. In this way, the relative strengths of both acids and bases can be compared on a single pK_a scale. Equation 2.8 can be rewritten as the acid dissociation of the conjugate acid of RNH_2, as follows

$$RNH_3^+ + H_2O \rightleftharpoons RNH_2 + H_3O^+. \tag{2.11}$$

The K_a for the equilibrium shown in eqn 2.11 can be given as follows

$$K_a = \frac{[RNH_2][H_3O^+]}{[RNH_3^+]}. \tag{2.12}$$

The *stronger* the base RNH_2, the *weaker* the conjugate acid RNH_3^+, and hence the lower the value of the K_a and the greater the value of the pK_a.

The pK_b of a base and the pK_a of its conjugate acid are related to each other as follows

$$pK_b + pK_a = pK_w = 14.0 \text{ at } 25\,°C. \tag{2.13}$$

Equation 2.13 allows us to relate the pK_b of a base to the pK_a of its conjugate acid, and vice versa, so that if one is known, the other can be worked out. (The justification for eqn 2.13 is given in Box 2.1.) Figure 2.12 shows the molecular structures of some pharmaceutical acids and bases, as well as their pK_a values.

▷ **WORKED EXAMPLE 2.2**

In Fig. 2.12, the pK_a of pethidine is given as 8.7. What is the equilibrium that defines this pK_a value? What is the corresponding pK_b value at 25 °C?

Pethidine is an amine and is therefore basic. The pK_a value quoted must therefore be the pK_a of the corresponding conjugate acid. The equilibrium defining the pK_a is therefore that shown in Fig. 2.13.

BOX 2.1 Relationship between the pK_b of a base and the pK_a of its conjugate acid.

The pK_b of a base and the pK_a of its conjugate acid are related to each other. If one is known, the other can be worked out. Equation 2.9 above gives the equilibrium constant for a base accepting a proton. Equation 2.12 gives the equilibrium constant for the conjugate base donating a proton. Multiplication of eqn 2.9 by eqn 2.12 gives

$$K_b.K_a = \frac{[RNH_3^+][HO^-]}{[RNH_2]} \cdot \frac{[RNH_2][H_3O^+]}{[RNH_3^+]} = [HO^-][H_3O^+].$$

The term $[HO^-][H_3O^+]$ is just the ion-product constant for water, K_w. Hence

$$K_b.K_a = K_w.$$

At 25 °C, K_w is equal to 10^{-14}. This equation can be rewritten as

$$pK_b + pK_a = pK_w = 14.0 \text{ at } 25\ °C. \qquad (2.13)$$

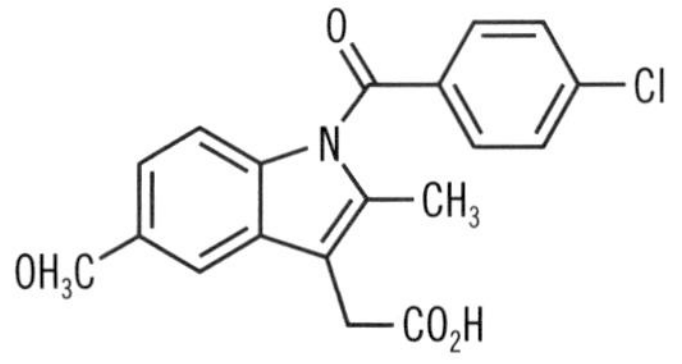

Indomethacin (4.5)

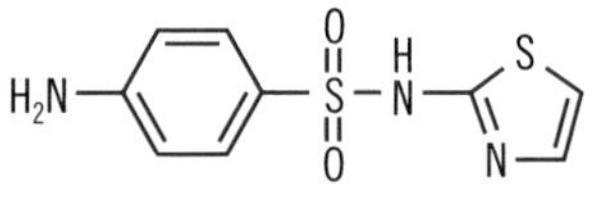

Sulfathiazole (7.2)

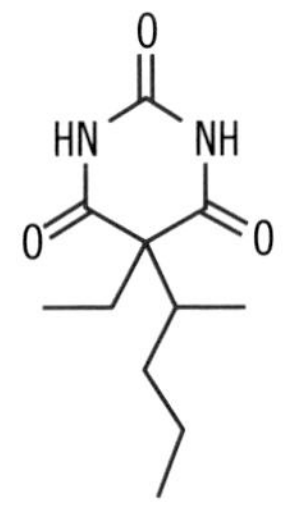

Pentobarbital (8.1)

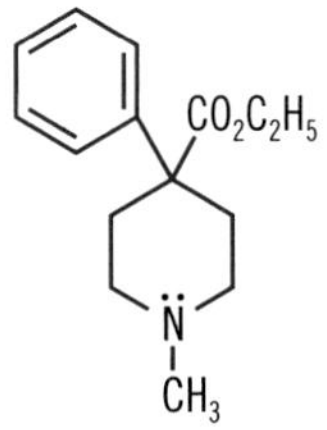

Pethidine (8.7)

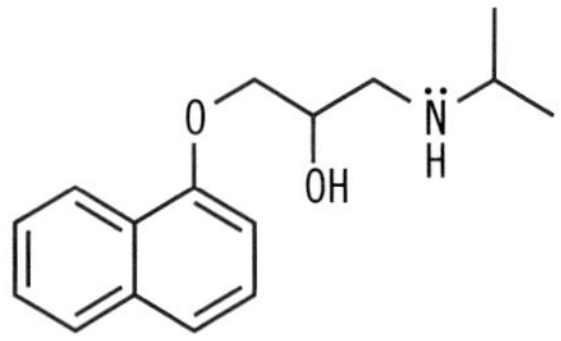

Propranolol (9.5)

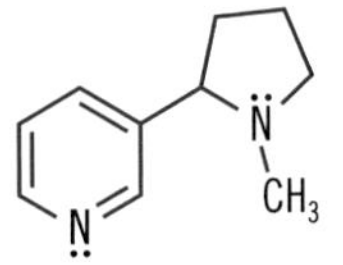

Nicotine (7.9, 3.1)

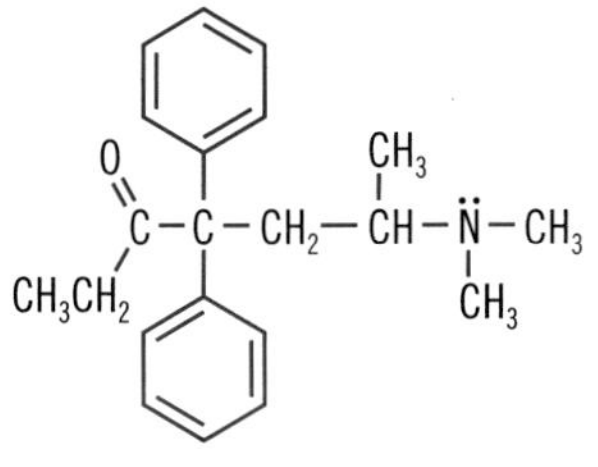

Methadone (8.3)

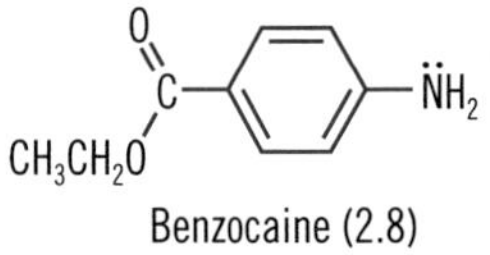

Benzocaine (2.8)

FIGURE 2.12 Molecular structures and pK_a values for some pharmaceutical compounds.

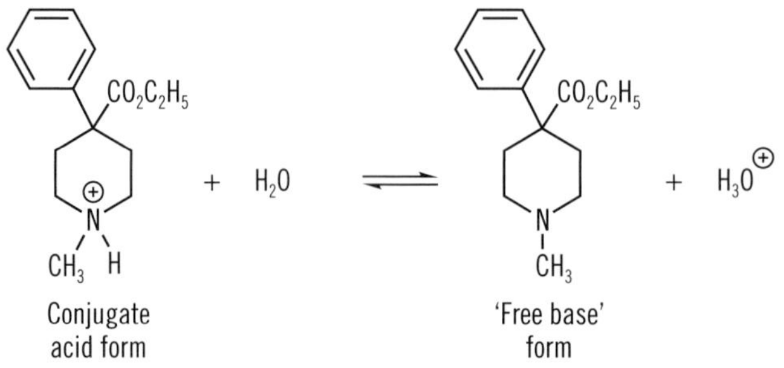

FIGURE 2.13 Acid dissociation of the conjugate acid form of pethidine.

The pK_a value of 8.7 is therefore the pK_a of the conjugate acid form of pethidine. To quantify the base strength of the 'free base' form at 25 °C, use eqn 2.13 as follows

$pK_b = pK_w - pK_a$; hence $pK_b = 14.0 - 8.7 = 5.3$ at 25 °C.

2.5.2 Acidity and environment

The position of equilibrium for the processes shown in eqns 2.2 and 2.3 will be affected by the introduction of protons from other sources into the system. This is very significant pharmaceutically, as many physiological environments, such as the gastrointestinal tract, have complex mechanisms for the maintenance of local pH. A compound may be exposed to conditions of different pH as it travels through the body. Consequently, it may be ionized to different extents depending on its location. The impact of local pH on the acidity or basicity of pharmaceuticals therefore has to be considered. This can be done using a Henderson–Hasselbalch equation.

The Henderson–Hasselbalch equations are equations that relate pH with acidity or basicity, in the form of pK_a or pK_b values. There are different precise Henderson–Hasselbalch equations depending on whether acids, bases or mixtures of these in buffer solutions are being considered. Henderson–Hasselbalch equations can be derived from the equilibrium constant for the relevant process. Although this involves a small amount of mathematical manipulation, it is worth going through, as it helps to explain why there are different forms of the Henderson–Hasselbalch equation, and identifies which form is the right one for any particular process.

For example, an equilibrium for an acid dissociation process can be given generally as follows

$$HX \rightleftharpoons X^- + H^+. \tag{2.14}$$

An acidity constant for this process can be given as

$$K_a = \frac{[X^-][H^+]}{[HX]}, \text{ or } [H^+] = \frac{K_a[HX]}{[X^-]}. \tag{2.15}$$

Converting each term in this equation into the negative of the logarithm gives

$$-\log_{10}[H^+] = -\log_{10}K_a - \log_{10}\frac{[HX]}{[X^-]}. \tag{2.16}$$

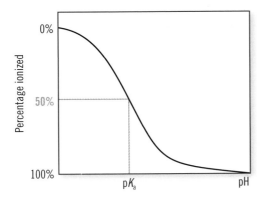

FIGURE 2.14 Variation of extent of ionization with pH for an acidic pharmaceutical.

Rearrangement by adding $\log_{10}\dfrac{[\mathrm{HX}]}{[\mathrm{X}^-]}$ to both sides gives

$$-\log_{10} K_a = -\log_{10}[\mathrm{H}^+] + \log_{10}\frac{[\mathrm{HX}]}{[\mathrm{X}^-]}.$$

Remembering that $\mathrm{pH} = -\log_{10}[\mathrm{H}^+]$ and $pK_a = -\log_{10} K_a$ we can rewrite this to find that

$$pK_a = \mathrm{pH} + \log_{10}\frac{[\mathrm{HX}]}{[\mathrm{X}^-]}. \tag{2.17}$$

Equation 2.17 is the form of the Henderson–Hasselbalch equation to be used for an acid.

Equation 2.17 can be presented graphically. Figure 2.14 shows the variation of the percentage of a dose of an acidic API that is ionized as a function of pH. The pH value at which the acid is 50% un-ionized and 50% ionized is equal to the pK_a of the acid.

▶ WORKED EXAMPLE 2.3

Figure 2.12 shows that the pK_a of the non-steroidal anti-inflammatory drug indomethacin is 4.5. If the pH of the stomach after fasting is 2.0, and after a full meal is 5.0, use eqn 2.17 to calculate the percentage of a dose of indomethacin that would remain un-ionized in the stomach in the fasting state and after a meal.

If '% Un-ionized' stands for the percentage un-ionized, then the percentage ionized must be equal to [100 – % Un-ionized] (as the percentage ionized plus the percentage un-ionized equals 100% of the dose.). For the fasting state, that is, at pH 2.0, eqn 2.17 gives the following values.

$$4.5 = 2.0 + \log_{10}\frac{[\%\ \mathrm{Un\text{-}ionized}]}{[100 - \%\mathrm{Un\text{-}ionized}]}, \text{ so that } 2.5 = \log_{10}\frac{[\%\ \mathrm{Un\text{-}ionized}]}{[100 - \%\mathrm{Un\text{-}ionized}]}.$$

This means that

$$316.2 = \frac{[\%\ \mathrm{Un\text{-}ionized}]}{[100 - \%\mathrm{Un\text{-}ionized}]};$$

or

$$31\,620 - 316.2[\% \text{ Un-ionized}] = [\% \text{ Un-ionized}], \text{ or } 31\,620 = 317.2[\% \text{ Un-ionized}].$$

So,

$$[\% \text{ Un-ionized}] = \frac{31\,620}{317.2} = 99.7\% \text{ at pH 2.0.}$$

In the fed state, at pH 5.0, use of eqn 2.17 gives

$$4.5 = 5.0 + \log_{10}\frac{[\% \text{ Un-ionized}]}{[100 - \%\text{Un-ionized}]}, \text{ so that } -0.5 = \log_{10}\frac{[\% \text{ Un-ionized}]}{[100 - \%\text{Un-ionized}]}.$$

This means that

$$0.316 = \frac{[\% \text{ Un-ionized}]}{[100 - \%\text{Un-ionized}]};$$

or

$$31.6 - 0.316[\% \text{ Un-ionized}] = [\% \text{ Un-ionized}], \text{ or } 31.6 = 1.316[\% \text{ Un-ionized}],$$

which implies that the [% Un-ionized] at pH 5.0 = 24.0%

These answers make chemical sense. Indomethacin has a pK_a of 4.5, which implies that it is a moderate acid. A medium at pH 2.0 is quite strongly acidic with a lot of H^+ present. The high H^+ concentration pushes the acid dissociation equilibrium to the left, that is, towards un-ionized carboxylic acid. Hence, the majority of a dose of indomethacin would be expected to be un-ionized at that pH. Our calculation supports this intuition. A pH of 5.0 is much closer to the pK_a of indomethacin, so that a finer balance between the percentages ionized and un-ionized would be expected, with the majority of the dose being ionized. Our calculated value of 24% un-ionized (which means 76% ionized) is in line with this expectation. The un-ionized form would be expected to be better absorbed across the gut wall into the bloodstream, so greater absorption of the dose from the stomach would be expected at pH 2.0 than at pH 5.0.

If the pK_a value is the pK_a of the conjugate acid of a base, then the following form of eqn 2.17 would be applicable.

$$pK_a = pH + \log_{10}\frac{[BH^+]}{[B]}, \tag{2.18}$$

in which 'B' represents the free base form, and 'BH$^+$' the conjugate acid form of the compound.

Modified versions of these equations can be derived that allow for limited solubility. For example, in the process shown in eqn 2.14, it can be assumed that the ionized form, X^-, is fully soluble in water. The un-ionized form, HX, is assumed to have limited solubility in water. The limit of solubility of the un-ionized form is represented by S_0. Equation 2.15 can be rearranged to give

$$\frac{K_a}{[H^+]} = \frac{[X^-]}{[HX]}. \tag{2.19}$$

If it is assumed that the concentration of the un-ionized form, [HX], is equal to its limiting solubility, S_0, eqn 2.19 can be given as

$$\frac{K_a}{[H^+]} = \frac{[X^-]}{S_0}.$$

(2.20)

Both the un-ionized form HX and the ionized form X^- of an acidic drug are pharmaceutically relevant. The total concentration of pharmaceutically relevant species present in solution, S, is therefore equal to $[HX] + [X^-]$. It has already been assumed that [HX] equals S_0, therefore $S = [X^-] + S_0$, and so $[X^-] = S - S_0$. Substituting $S - S_0$ for X^- in eqn 2.20 gives

$$\frac{K_a}{[H^+]} = \frac{S - S_0}{S_0}.$$

(2.21)

Written in terms of pH and pK_a, eqn 2.21 becomes

$$pH - pK_a = \log_{10}\left[\frac{S - S_0}{S_0}\right].$$

(2.22)

The importance of eqn 2.22 is that it relates the pK_a and solubility of a drug substance to the local pH. It shows us how close to the pK_a value of the drug can the local pH value approach before the concentration of the poorly soluble un-ionized form approaches its limiting solubility. This allows us to determine the conditions under which a compound might *fail* to be fully soluble, as so to precipitate as solid.

An equivalent equation to 2.22 for basic drugs would be

$$pH - pK_a = \log_{10}\left[\frac{S_0}{S - S_0}\right].$$

(2.23)

▶ **WORKED EXAMPLE 2.4**

From Fig. 2.12, the pK_a of pentobarbital is given as 8.1. This worked example uses the equation described above to solve three separate questions, (i), (ii) and (iii), about the ionization and solubility of pentobarbital.

(i) What is the equilibrium from which this pK_a value is defined? Pentobarbital is an acidic compound. The acid dissociation process is illustrated in Fig. 2.15.

(ii) Determine the percentage of pentobarbital that would remain un-ionized at a pH of 7.4.

Equation 2.17 relates the pK_a and degree of dissociation of an acidic compound with the pH of its environment. In this case,

$$8.1 = 7.4 + \log_{10}\frac{[HX]}{[X^-]}; \text{ hence } \log_{10}\frac{[HX]}{[X^-]} = 0.7.$$

The un-ionized form, HX, is the un-ionized form of pentobarbital, while the ionized form, X^-, corresponds to the ionized conjugate base, as shown in Fig. 2.15. Hence

$$0.7 = \log_{10}\frac{[\text{Un-ionized form}]}{[\text{Ionized form}]}.$$

(2.24)

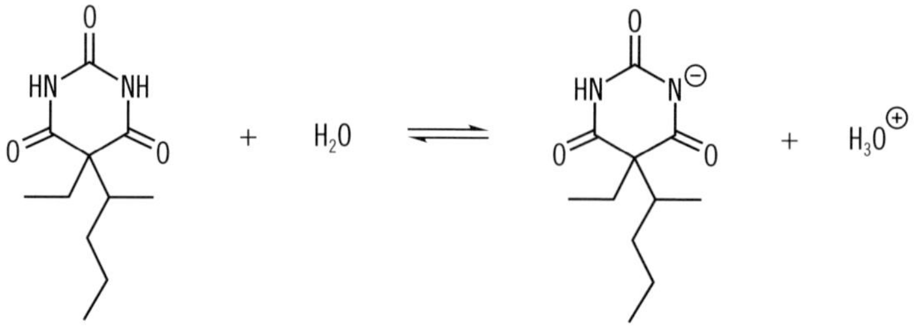

FIGURE 2.15 Acid dissociation of pentobarbital.

The question asks for the percentage of the dose of pentobarbital remaining un-ionized to be calculated. If the percentage un-ionized is given by '[% Un-ionized]', the percentage of the dose that is ionized is equal to $100 - [\%\ \text{Un-ionized}]$. Substituting into eqn 2.24 above gives

$$0.7 = \log_{10}\left[\frac{(\%\text{Un-ionized})}{100 - (\%\text{Un-ionized})}\right].$$

Therefore,

$$\frac{(\%\text{Un-ionized})}{100 - (\%\text{Un-ionized})} = 5.01;\ \text{hence [\% Un-ionized]} = 83.36\%.$$

(iii) The solubility of pentobarbital in water is $0.003\ \text{mol dm}^{-3}$. What is the maximum total solubility of pentobarbital that can be achieved at a pH of 7.4?

Equation 2.22 relates pH, pK_a and solubility for an acidic compound. In this case

$$7.4 - 8.1 = \log_{10}\left[\frac{S - 0.003(\text{mol dm}^{-3})}{0.003(\text{mol dm}^{-3})}\right].$$

Hence,

$$-0.7 = \log_{10}\left[\frac{S - 0.003(\text{mol dm}^{-3})}{0.003(\text{mol dm}^{-3})}\right];\ 0.1995 = \left[\frac{S - 0.003(\text{mol dm}^{-3})}{0.003(\text{mol dm}^{-3})}\right].$$

Therefore, $S - 0.003\ (\text{mol dm}^{-3}) = 0.000599\ (\text{mol dm}^{-3})$; $S = 0.003599\ (\text{mol dm}^{-3})$.

This implies that a $0.0036\ \text{mol dm}^{-3}$ solution of, for example, pentobarbital sodium (the sodium salt of the ionized form) would begin to form a precipitate once the pH of the solution was lowered below 7.4.

2.5.3 Buffer solutions

Solutions of weak acids, or bases, and their conjugate salts play important roles in biochemistry and dosage formulation. These solutions have the ability to absorb quantities of added acid (H_3O^+) or base (HO^-) without large changes in the pH of the solution. This effect is known as *buffering* and these solutions are known as *buffer solutions*. Buffer solutions allow us to control the pH of a pharmaceutical solution or suspension so as to, for example, minimize degradation of the active pharmaceutical ingredient. In physiological fluids such as blood, buffering maintains pH within ranges acceptable for proper functioning of enzymes and other biochemical processes.

A buffer solution will contain either a weak acid or base, and the conjugate salt of the acid or base. For example, a buffer solution could contain equal concentrations of the weak acid acetic acid (CH_3CO_2H) and a conjugate base of acetic acid such as sodium acetate (CH_3CO_2Na). If base (say NaOH) is added to the solution, it will react with the weak acid as follows

$$CH_3CO_2H + NaOH \rightleftharpoons CH_3CO_2Na + H_2O. \tag{2.25}$$

If acid (say HCl) is added to the solution, it will react with the conjugate base as follows

$$CH_3CO_2Na + HCl \rightleftharpoons CH_3CO_2H + NaCl. \tag{2.26}$$

The weak acid/conjugate base mixture therefore has the ability to absorb quantities of added acid or base so that the overall change in the pH of the solution is small.

The relationship between the weak acid (CH_3CO_2H) and the conjugate base ($CH_3CO_2^-$) is described by the following equilibrium

$$CH_3CO_2H + H_2O \rightleftharpoons H_3O^+ + CH_3CO_2^-. \tag{2.27}$$

The acidity constant, K_a, relating to eqn 2.27 can be written in a similar manner to eqn 2.6, giving

$$K_a = \frac{[H_3O^+][CH_3CO_2^-]}{[CH_3CO_2H]}. \tag{2.28}$$

The pH of the solution is determined by the concentration of H_3O^+. To see how this is affected by the concentrations of weak acid and conjugate base, eqn 2.28 can be rearranged as follows

$$[H_3O^+] = \frac{K_a[CH_3CO_2H]}{[CH_3CO_2^-]}. \tag{2.29}$$

Because CH_3CO_2H is a weak acid, K_a is a very small number (1.8×10^{-5}, in fact). Provided the ratio $[CH_3CO_2H]/[CH_3CO_2^-]$ maintains a value close to one (in practice, between 0.10 and 10.0 will suffice), any change in the overall value for $[H_3O^+]$ will be relatively small, and the resulting change in pH (which is $-\log_{10}[H_3O^+]$) will be small. Hence, the buffer solution can absorb quantities of added acid or base with only minor changes to pH, provided the ratio $[CH_3CO_2H]/[CH_3CO_2^-]$ stays within the value range 0.10 to 10.0.

Equation 2.29 can be generalized for any weak acid/conjugate base buffer mixture, as follows

$$[H_3O^+] = \frac{K_a[\text{weak acid}]}{[\text{conjugate base}]}. \tag{2.30}$$

This can be converted into a Henderson–Hasselbalch type equation by taking the negative logarithms for the terms, giving

$$pH = pK_a + \log_{10}\frac{[\text{conjugate base}]}{[\text{weak acid}]}. \tag{2.31}$$

Equation 2.31 is a Henderson–Hasselbalch equation for relating the pH of a buffer solution consisting of a weak acid and a conjugate base of the acid to the concentrations of the acid and base, and the pK_a of the weak acid.

The acetic acid/sodium acetate buffer described above will maintain a solution within the pH range 3.7 to 5.6 at 25 °C. Another example is KH_2PO_3/K_2HPO_3, which maintains a pH range of 5.3 to 8.0 at 25 °C. Blood plasma is maintained at a pH of 7.4 by buffer systems such as HCO_3^-/H_2CO_3.

2.6 SALT SELECTION AND FORMATION

One of the first chemical reactions we tend to be introduced to is 'acid and base gives salt and water'. By this description, a salt is a neutral compound composed of positively and negatively charged ions, neither of which is H^+ or HO^-. Acid and base can exist in equilibrium between ionized and un-ionized form. However, salts, provided they are stable, exist permanently in ionized forms at neutral pH. This property can have many advantages in pharmaceutical manufacturing and dosage form design. Salt forms of pharmaceutical chemical entities are often used to improve aqueous solubility and to optimize physicochemical parameters.

As ionic materials, salts generally have greater aqueous solubility than the neutral free acid or base forms. For example, the 'free acid' or un-ionized form of a carboxylic acid, RCO_2H, consists of a single neutral molecular species. By contrast, a salt form, say $RCO_2^- Na^+$, consists of two ionized or charged species, a cation and an anion. Likewise, a 'free base' consists of a neutral molecular species, say RNH_2. By contrast, a salt form, say a hydrochloride $RNH_3^+ Cl^-$, consists of two ionized cationic and anionic species. In addition, a salt form may have a melting point that is 50–100 °C higher than that of the free acid or base form. Salt formation may therefore be used specifically to convert low melting or oily material into crystalline solid. As higher melting crystalline solids, salt forms may have more desirable physical properties. For example, they may be more suitable for particle-size reduction by milling.

The formula weight of a salt form is always greater than that of the free acid or base. This is because the mass of the counteranion or cation adds to the overall mass of the compound. Hence, a given mass of a salt form contains fewer moles of API than the same mass of the un-ionized form. The proportion of pharmaceutically active content is therefore *decreased*. Consequently, relatively high doses may become necessary to achieve clinical effects. Salt forms also have a greater propensity to form hydrates, solvates and other multiple crystal forms. As discussed in Chapter 4, formation of multiple crystal forms is a serious problem in pharmaceutical manufacturing.

Formation of a stable salt requires a sufficiently large pK_a difference between the proton donor and the conjugate acid formed by proton transfer. This is to ensure that the reaction 'acid plus base gives salt plus water', converts completely into products, and does not give an equilibrium mixture of products and starting materials. As a 'rule of thumb', a difference of at least three pK_a units is required.

For example, to form a stable salt of an acid of pK_a 3.5, the acid should be reacted with a base that forms a conjugate acid with a pK_a of at least 6.5. To form a stable salt of a

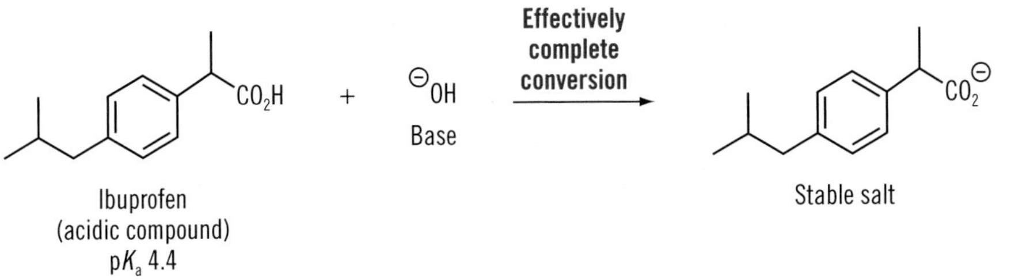

FIGURE 2.16 Stable salt formation of an acidic compound.

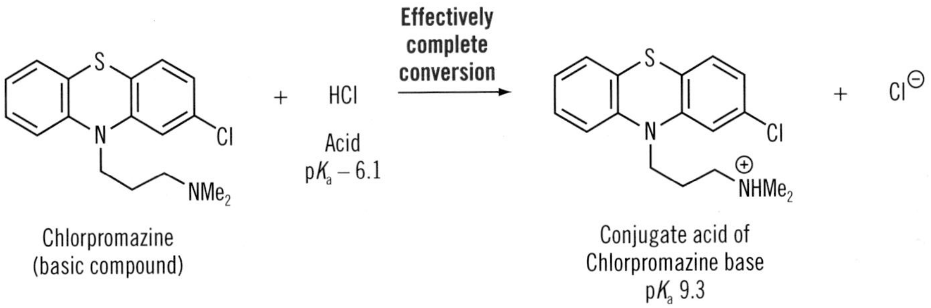

FIGURE 2.17 Stable salt formation of a basic compound.

base that has a conjugate acid of pK_a 10.0, reaction with an acid of pK_a 7 or less is required. For instance, an hydroxide ion forms a stable salt upon reaction with ibuprofen. The conjugate acid of an hydroxide ion (that is, water) has a pK_a of 15.7, which is greater by 11.3 pK_a units than the pK_a of ibuprofen (Fig. 2.16). In the case of the basic compound chlorpromazine (Fig. 2.17), a stable salt is formed by reaction with hydrochloric acid, as the pK_a of hydrochloric acid is 15.4 pK_a units less than the pK_a of the conjugate acid of chlorpromazine.

Salts formed from strong acids or bases will generally have good aqueous solubility. However, they are also more likely to give rise to hygroscopic material (that is, material that has a tendency to absorb water). This can result in instability in certain dosage formulations due to absorption of moisture. Salt selection will be influenced by factors such as particle size and shape, powder properties and crystal-form stability. The counterion also needs to be free of any pharmaceutically unacceptable toxic effects. The following list shows some of the commonly used pharmaceutical salt forms.

Pharmaceutically acceptable salt forms for basic compounds

The acids listed in Table 2.2 are suitable for preparing salts of basic pharmaceuticals. The types of salts formed from each are also listed.

The mineral and sulfonic acids are strong acids, typically with pK_a values < 0. The carboxylic acids are moderate acids, with pK_a values in the range 2 to 5.

Hydrochloride salts

Hydrochlorides are the most widely used salt forms for basic pharmaceutical compounds. As hydrochloric acid is a strong acid, stable hydrochloride salts can be formed from most

TABLE 2.2 Pharmaceutically acceptable salt formers for basic compounds.

Class	Acids	Salt forms produced
Strong mineral acids	Hydrochloric acid	Hydrochlorides
	Hydrobromic acid	Hydrobromides
	Sulfuric acid	Sulfates
	Nitric acid	Nitrates
	Phosphoric acid	Phosphates
Sulfonic acids	Methanesulfonic acid (CH_3SO_3H)	Methanesulfonates (mesylates)
	Ethanesulfonic acid ($CH_3CH_2SO_3H$)	Ethanesulfonates
	2-Hydroxyethanesulfonic acid	Isethionates
	4-Toluenesulfonic acid	4-Toluenesulfonates (Tosylates)
	Benzenesulfonic acid	Benzenesulfonates (Besylates)
Carboxylic acids	Acetic acid (CH_3CO_2H)	Acetates
	Propionic acid ($CH_3CH_2CO_2H$)	Propionates
	Benzoic acid ($C_6H_5CO_2H$)	Benzoates
	Hexanoic acid ($CH_3(CH_2)_4CO_2H$)	Hexanoates
	Octanoic acid ($CH_3(CH_2)_6CO_2H$)	Octanoates
	Decanoic acid ($CH_3(CH_2)_8CO_2H$)	Decanoates
	Stearic acid ($CH_3(CH_2)_{16}CO_2H$)	Stearates
	Succinic acid ($HO_2CCH_2CH_2CO_2H$)	Succinates
	Glycolic acid ($HOCH_2CO_2H$)	Glycolates
	Lactic acid ($CH_3CH(OH)CO_2H$)	Lactates
	Tartaric acid ($[CH(OH)CO_2H]_2$)	Tartrates
	Glutamic acid	Glutamates
	Aspartic acid	Aspartates
	Maleic acid	Maleates
	Fumaric acid	Fumarates
	Oleic acid	Oleates
	Citric acid	Citrates
	Salicylic acid	Salicylates

basic compounds, including weakly basic compounds. However, hydrochloride salts can suffer from a number of drawbacks. For example, hydrochlorides can produce solution formulations of high acidity. These may react unfavourably with other components in dosage forms. They may also cause harm to the stomach when used in oral dosage forms.

Hydrochloride salts often have less than optimal solubility in gastric juice due the abundance of chloride ions in the gastric medium. These excess chloride ions push the equilibrium for dissociation of the hydrochloride salts, i.e.

$$RNH_2 + HCl \rightleftharpoons RNH_3^{\oplus} \ Cl^{\ominus}$$

to the left-hand side, toward the less soluble un-ionized form. This is an example of a common-ion effect, that is, when the dissociation of a weak electrolyte is decreased by the presence of a strong electrolyte that has an ion in common with the weak electrolyte.

Pharmaceutically acceptable salt forms for acidic compounds

Salts of acidic compounds can be formed by reaction with the following.

- *Metals (metal hydroxides/metal carbonates)*: sodium, potassium, calcium, magnesium, zinc.
- *Amines*: triethylamine [$(C_2H_5)_3N$], ethanolamine [$HOCH_2CH_2NH_2$], triethanolamine [$(HOCH_2CH_2)_3N$], ethylenediamine [$H_2NCH_2CH_2NH_2$], lysine [$H_2NCH(CO_2H)CH_2CH_2CH_2CH_2NH_2$], histidine, arginine, *N*-methyl-D-glucamine (meglumine).
- *Quaternary ammonium hydroxides*: choline hydroxide.

The metal hydroxides, choline hydroxide and arginine are strong bases. The metal carbonates and other amines are moderate bases.

2.7 HYDROLYTIC DEGRADATION

In addition to acting as a solvent, water can also act as a nucleophilic reagent. If a solute molecule contains electrophilic functional groups, attack by water molecules can result in degradation of the solute molecule. The result is cleavage, or 'lysis', of the solute molecules by the action of water – that is, **hydrolysis**. If drug molecules contain electrophilic groups, hydrolysis when in aqueous solution is always a possibility. This means that the drug molecule could degrade in solution, resulting in loss of activity. Examples of electrophilic groups include esters, amides, epoxides and alkyl halides. Esters and amides are common functional groups in pharmaceutical molecules. Of these, esters are especially labile to degradation by hydrolysis.

The hydrolysis of a drug substance is a chemical reaction – degradation by water. A critical point to consider is the **rate** of the reaction. If hydrolysis is very slow at physiological temperature, it may not be something we have to be concerned about. However, some species, especially acids or bases, may speed up the rate of reaction. These species are then acting as **catalysts**. In order to say whether the rate of hydrolysis or the presence of catalytic species is significant, we need to able to think about these factors in quantitative terms. This requires use of reaction **kinetics** and **rate constants**.

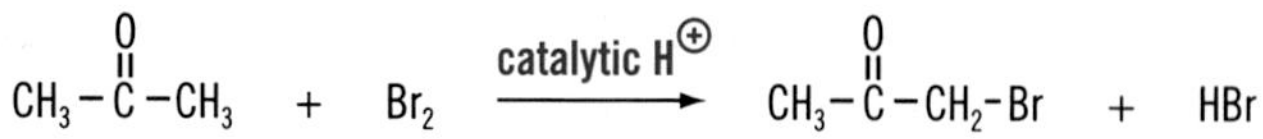

$$CH_3-\overset{\overset{\displaystyle O}{\|}}{C}-CH_3 \ + \ Br_2 \ \xrightarrow{\text{catalytic } H^{\oplus}} \ CH_3-\overset{\overset{\displaystyle O}{\|}}{C}-CH_2-Br \ + \ HBr$$

FIGURE 2.18 Reaction of acetone and bromine in the presence of acid catalyst.

Kinetics and rate constants

The rate of a reaction is the change in concentration of the reactants (or products) over time. Say we have the following reaction

A + B → products.

The rate of the reaction would be given by the following equation

$$Reaction\ rate = k[A]^x[B]^y. \tag{2.32}$$

In this equation, k is known as the **rate constant**. A key point about eqn 2.32 is that the values of the rate constant k and of the powers x and y have to be determined experimentally at specific temperatures. They cannot be predicted from the stoichiometry of the reaction. The reaction of acetone and bromine in the presence of acid catalyst shown in Fig. 2.18 provides an example.

The rate equation for the reaction shown in Fig. 2.18 can be stated as follows

$$Reaction\ rate = k[CH_3COCH_3]^x[Br_2]^y[H^+]^z. \tag{2.33}$$

Equation 2.33 includes terms for all the species involved in the reaction shown in Fig. 2.18, including the acid catalyst. What needs to be found out by experiment (and can only be found out that way) is the values of the powers x, y and z. If we carry out kinetic experiments on the reaction, we find that the values of x, y and z are, respectively, 1, 0 and 1. In other words, the experimentally established version of eqn 2.33 is

$$Reaction\ rate = k[CH_3COCH_3][H^+]. \tag{2.34}$$

Note that as $[Br_2]^0$ equals 1, we omit the bromine concentration term from the equation. What eqn 2.34 tells us is that variation in the concentration of bromine molecules has no impact on the rate of the reaction, but that the concentrations of acetone molecules and of catalytic protons do have an effect. This is not something we could have predicted in advance. It could only have been discovered by experiment.

Another way of expressing the information given in eqn 2.34 is to say that the reaction rate is **zero order** with respect to bromine ($y = 0$), and is **first order** with respect to acetone and acid (x, $z = 1$). The reaction is said to be **second order** overall, as the sum of the powers x, y and z is 2.

Acid and base catalysis

The rate of degradation by hydrolysis can often be enhanced by acid or base catalysis. In these cases, the experimentally measured rate constant for the hydrolytic degradation, k_{obs}, may be described as follows

$$k_{obs} = k_0 + k_{H+}[H^+] + k_{HO-}[HO^-]. \tag{2.35}$$

What eqn 2.35 means is that the measured rate constant, k_{obs}, is made up of three contributions. $k_{H+}[H^+]$ is the contribution made by acid-catalysed hydrolysis. $k_{HO-}[HO^-]$ is the contribution made by base-catalysed hydrolysis. In both of these contributions, there is an acid or base concentration term ([H^+] or [HO^-]) and a rate constant (k_{H+} or k_{HO-}). k_0 is the rate constant for the reaction in the absence of both acid and base catalysis. If we can assign values to the rate constants k_0, k_{H+} and k_{HO-}, it allows us to say how susceptible to hydrolysis the drug will be under various conditions.

In the absence of acid or base (that is, both [H^+] and [HO^- equal zero), $k_{obs} = k_0$. In that case, hydrolysis is occurring without the aid of catalysis. The value k_{H+} is known as the rate constant for specific acid catalysis, while the term k_{HO-} is known as the rate constant for specific base catalysis. Specific acid catalysis is said to occur when the rate-determining step involves a protonated species. The actual proton transfer occurs in a rapid step that proceeds the rate-determining step. The rate of the reaction is therefore not affected by the source of protons. Likewise, in cases of specific base catalysis, the rate-determining step involves a deprotonated species. The actual deprotonation occurs in a rapid step that proceeds the rate-determining step.

Reactions in which actual proton transfer is rate determining are said to involve general acid or base catalysis. The rate of such reactions may be affected by the presence of various proton donors or acceptors, such as buffer solution components. Equation 2.35 can be extended to allow for general acid or base catalysis

$$k_{obs} = k_0 + k_{H+}[H^+] + k_{HO-}[HO^-] + k_{HX}[HX] + k_{X-}[X^-]. \tag{2.36}$$

The additional term 'k_{HX}' symbolizes the general acid-catalysed rate constant, while the term 'k_{X-}' symbolizes the general base-catalysed rate constant. Careful kinetic studies on pharmaceutical compounds would allow us to assign values to the rate constants k_0, k_{H+}, k_{HO-}, k_{HX} and k_{X-} for individual compounds. This allows us to say whether a compound has a greater tendency to hydrolyse under acidic or basic conditions, or even whether it tends to hydrolyse under neutral conditions (significant k_0). Compounds that tend to undergo general acid or base catalysis might be degraded by buffer solution components. This information would be used to guide dosage form design, so as to avoid conditions that might cause hydrolysis of the active ingredient.

Hydrolysis of aspirin

An example of a compound that undergoes hydrolysis in aqueous solution is acetylsalicylic acid, also known as aspirin [Garrett, 1957]. Acetylsalicylic acid is an ester. Hydrolysis of the ester group gives salicylic acid and acetic acid as products (Fig. 2.19). Figure 2.19 shows the observed rate constant for the hydrolysis of acetylsalicylic acid as a function of pH. The hydrolysis reaction is slowest at pH 2.5. At this pH, the acetylsalicylic acid molecules are all un-ionized. They are not protonated and the hydroxide ion is minimal so neither acid nor base catalysis is occurring. Instead, *uncatalysed* hydrolysis is occurring, also known as spontaneous hydrolysis. Although this pH provides the conditions under which acetylsalicylic acid is least susceptible to hydrolysis, even under these conditions some hydrolysis is still occurring. For this reason, degradation by hydrolysis is a problem for solution formulations of aspirin.

At pH values less than 2.0, the rate of acetylsalicylic acid hydrolysis is proportional to the concentration of acid, so that an approximately linear increase in log of the rate

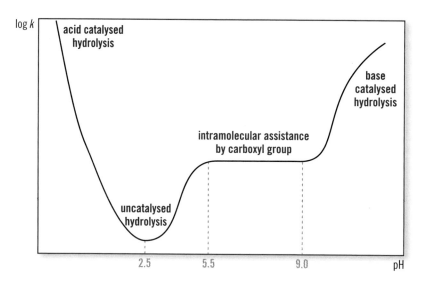

FIGURE 2.19 Hydrolysis of acetylsalicyclic acid and plot of the logarithm of the observed rate constant, k, for the hydrolysis as a function of pH. Reprinted with permission Garrett, E. R. *J. Am. Chem. Soc.*, 1957, **79**, 3401, Copyright (1957) American Chemical Society.

constant is observed with increasing acidity (decreasing pH). This corresponds to hydrolysis of a protonated form of acetylsalicylic acid. At pH values greater than 9.0, the rate of hydrolysis is proportional to the concentration of hydroxide ion. This corresponds to hydrolysis by hydroxide ion. In the region of the plot between pH values 5.5 and 9.0, the rate of hydrolysis remains constant (that is, is independent of pH). Hydrolysis within this pH range is thought to proceed by a mechanism that involves the assistance of the carboxylic acid group.

◙ 2.8 SUMMARY

Solubility and solutions are of key importance pharmaceutically as drug molecules need to be in solution in the biological fluids surrounding the site of action to achieve a therapeutic effect. We usually define solutions by describing the components present in greater quantity as the solvents, and those present in lesser quantity as the solutes. Common expressions of solubility for pharmaceutical uses include mole fractions, molarity, molality, percentages, parts per million and equivalents. Solubility can vary with temperature. Plots of solubility as a function of temperature are known as solubility curves.

Two types of solvent medium exist in the body: water-based and lipid-based. Liquid water consists of a mixture of well-ordered, extensively hydrogen-bonded regions existing in equilibrium with non-hydrogen-bonded regions. Lipid-based media consist of amphipathic phospholipids assembled into lipid bilayer structures. During the dissolution process, solute molecules leave their previous, usually solid, state, and become solvated within suitable cavities in the solvent medium. The solubility of compounds in particular compounds is influenced by various factors including molecular weight, hydrogen-bonding capacity, and the presence of hydrophilic or hydrophobic groups.

Many pharmaceuticals are acids or bases, and so can exist in equilibrium between ionized and un-ionized forms. The extent of ionization is dependent on the pH of the medium, and can be calculated from the pK_a of the compound using Henderson–Hasselbalch equations. Acidic or basic pharmaceuticals can be completely converted into salts by choice of suitable counterions.

While in aqueous solution, drug compounds may be susceptible to degradation by hydrolysis.

☰ REFERENCES

Bell, R. P. (1973) *The Proton in Chemistry* (2nd edn) London: Chapman and Hall.

Eisenberg, D., and Kauzmann, W. (1969). *The Structure and Properties of Water*. Oxford University Press.

Frank, H. S. (1970). *Science*, **169**, 635.

Garrett, E. R. (1957). *J. Am. Chem. Soc.*, **79**, 3401.

❓ EXERCISES

2.1   At 75 °C, an equilibrium aqueous solution of the artificial sweetener saccharin contains 1.3 g of saccharin per 100 g of water. Given that the molecular weight of saccharin is 183.1 g mol^{-1}, calculate the molality and the mole fraction of saccharin in the solution at this temperature.

2.2   Figure 2.12 gives the pK_a of propranolol as 9.5. Given that propranolol is an amine, write the equilibrium that defines this pK_a value and calculate the corresponding pK_b value at 25 °C.

2.3   The pH of the stomach can vary between 2.0 and 5.0, approximately, depending on how much has been eaten and how recently. If the pK_a of the non-steroidal anti-inflammatory drug ibuprofen is 4.4, calculate the percentage of a dose of ibuprofen that would be ionized in the stomach at pH 3.5.

2.4   The solubility of sulfapyridine (pK_a 8.4) in water at room temperature is 1.163×10^{-3} mol dm^{-3}. Calculate the pH at which sulfapyridine will begin to precipitate from a 5.0×10^{-3} mol dm^{-3} solution of the sodium salt of sulfapyridine.

PHARMACEUTICAL EQUILIBRIA

03

Learning objectives

Having studied this chapter, you should be able to:

- appreciate the pharmaceutical importance of equilibria and thermodynamics
- understand thermodynamic concepts such as internal energy, enthalpy, entropy, free energy and chemical potential
- understand the relationship between chemical potential, free energy and equilibrium
- describe multiphase pharmaceutical systems and interpret phase diagrams
- understand the pharmaceutical importance of mass-transfer processes
- apply quantitative approaches to diffusion and partition.

The solubility of a drug substance in various solvent systems influences the drug substance's development into a medicine and its activity after administration to a patient. During delivery to a target receptor, a drug molecule may be required to be soluble in a range of solvent systems such as the gastrointestinal fluids, the intestinal barrier, plasma, the cell membrane and the fluid surrounding the target receptor. In Chapter 2, the theory of solubility and the factors that influence the solubility of a drug in solvent systems are explained.

Another issue that was discussed in Chapter 2 is the different ionic species of a drug that can exist in solution. It is also pointed out in that chapter that a drug in solution and undissolved drug exist in equilibrium. A drug can exist in a number of equilibrium states. Drugs may exist in many different physical states (liquid/solution, solid, gaseous) and environments (liquid and solvent media of varying composition) during the various stages of their manufacture and after administration to a patient. These states are most often at equilibrium. For example, a medicine in a suspension formulation will consist of a solid active ingredient suspended in a liquid medium. The liquid medium may be composed of many substances, such as solvents and emulsifiers. The medicine therefore consists of several different chemical components and different physical states (solid and liquid states in this case) existing in **equilibrium**. At certain points, the pharmaceutical system may be disturbed from an equilibrium state, such as when the suspension is taken by a patient. Eventually, the drug arrives at another state of equilibrium, for instance when it binds to a receptor to cause a physiological response.

Hence, pharmaceutical systems are often either at equilibrium or arriving at equilibrium. It is the disturbance of equilibrium and process of arriving at a new equilibrium state that results in the movement of drug molecules from the dosage form to the physiological fluids and through a series of different physiological fluids to the target receptor. The processes of diffusion, dissolution and partition are driven by disturbances in equilibrium and the system's movement towards a new equilibrium.

The law of mass action and equilibrium constants

Any equilibrium process is subject to the **law of mass action**. Say an equilibrium exists that can be written as follows

$$a\,A + b\,B \rightleftharpoons c\,C + d\,D.$$

In this equation, a, b, c and d are the reaction stoichiometries of reactants A and B, and products C and D, respectively. By the law of mass action, we can state that the following relationship holds

$$K_{eq} = \frac{[C]^c [D]^d}{[A]^a [B]^b}, \tag{3.1}$$

in which K_{eq} is known as the **equilibrium constant** for the reaction or process. For a specific process, the value of K_{eq} should remain constant under specific conditions (especially temperature).

For example, the reaction of hydrogen and oxygen to produce water proceeds in the presence of a catalyst according to the following balanced equation

$$H_2 + \tfrac{1}{2} O_2 \rightleftharpoons H_2O.$$

The equilibrium constant relationship for this process would be as follows

$$K_{eq} = \frac{[H_2O]}{[H_2][O_2]^{1/2}}.$$

From experiments, we know that at 25 °C, K_{eq} for this process is equal to 10^{40}.

▷ **WORKED EXAMPLE 3.1**

Mixtures of acetic acid (ethanoic acid) and ethanol were heated to 100 °C to form ethyl acetate and water. The mixtures were analysed once the process had reached equilibrium. The data in Table 3.1 gives some typical results of these analyses. Write the expression for the equilibrium constant for the reaction, and use the data in Table 3.1 to calculate the equilibrium constant.

TABLE 3.1 Concentrations of acetic (ethanoic) acid, ethanol, ethyl acetate and water in a mixture that has reached equilibrium at 100 °C.

Acetic acid/mol L^{-1}	Ethanol/mol L^{-1}	Ethyl acetate/mol L^{-1}	water/mol L^{-1}
0.142	1.142	0.858	0.858

The reaction occurring is

$$CH_3CO_2H + CH_3CH_2OH \rightleftharpoons CH_3CO_2CH_2CH_3 + H_2O.$$

The equilibrium constant for this process would be

$$K_{eq} = \frac{[CH_3CO_2CH_2CH_3][H_2O]}{[CH_3CO_2H][CH_3CH_2OH]}.$$

Using the data from Table 3.1 gives the following value for the equilibrium constant at 100 °C.

$$K_{eq} = \frac{[0.858 \text{ mol L}^{-1}][0.858 \text{ mol L}^{-1}]}{[0.142 \text{ mol L}^{-1}][1.142 \text{ mol L}^{-1}]} = 4.5.$$

For any system, the position of equilibrium is determined by the energetics of the system. An understanding of the energetics of pharmaceutical processes is therefore critical. Thermodynamics is the science concerning the different kinds of energy associated with chemical systems, including pharmaceutical systems. For this reason, before proceeding further with examining pharmaceutical equilibria, discussion of the essential concepts of thermodynamics is required.

To apply the concepts of thermodynamics to real-world issues such as pharmaceuticals, we need to be able use them in a quantitative way. In other words, we need to be able to carry out thermodynamic calculations, and for that, we need to be prepared to grasp thermodynamic concepts in their mathematical forms. The first section of this chapter will cover the essential concepts of thermodynamics, this will be followed by the theory of phase equilibria and the final section will show how they apply to the processes of diffusion, dissolution and partitioning.

3.1 ESSENTIAL CONCEPTS IN THERMODYNAMICS

Energy changes accompany all chemical and physical processes. These include pharmaceutical process such as drug dissolution or transfer of drugs across membranes. Thermodynamics concerns the different kinds of energy involved in processes of change. So, for pharmaceutical systems, thermodynamics provides an insight into what processes are likely to occur, what physical states pharmaceutical substances may exist in, and what the chemical composition of pharmaceutical materials may be in different environments.

Energy is the capacity to do work. In general terms, whenever a quantity of energy of one form is consumed, an exactly equivalent quantity of some other form of energy is produced. This concept is what we call the first law of thermodynamics.

Work is done when energy is used to move an object against a force. For example, rolling an object to the top of a hill requires energy (in the form of work) to move the object against the force of gravity. However, once at the top of the hill, the object has potential energy (the potential to do work): the potential energy is equal to the work done to get the object to the top of the hill in the first place. That is, the energy used to get the object to the top of the hill has been *converted* into potential energy. The energy required to move an object can be presented mathematically by the following equation in which the work, *w*, done in moving an object equals the opposing force, *F*, multiplied by the distance, *d*, that the object is moved

$$w = F\,d. \tag{3.2}$$

If the object is then allowed to roll back down the hill, its potential energy becomes zero when it reaches the bottom and rolls to a halt. While it is rolling down the hill, the object is said to possess a form of energy associated with moving bodies – that is, **kinetic energy**. The total kinetic energy of the object during its fall is equal to its potential energy while at the top of the hill, minus any energy dissipated as heat due to friction with the atmosphere. Potential energy and kinetic energy are both forms of mechanical energy. Light is another form of energy, as is electrical energy.

▶ **WORKED EXAMPLE 3.2**

The force exerted by gravity on an object of 1 kg mass is 9.81 N (Newtons). How much work is done in lifting an object of 1 kg mass by 0.5 m?

From equation 3.2, $w = F\,d$,

so $w = 9.81\ \text{N} \times 0.5\ \text{m} = 4.91\ \text{N m} = 4.91\ \text{J}$ (Joules, as $1\ \text{J} = 1\ \text{N m}$)

3.1.1 Internal energy, enthalpy and the first law of thermodynamics

Energy changes accompany all physical or chemical changes. For example, when a tablet is taken by a patient, a number of physical and chemical processes occur. The tablet disintegrates in the stomach, and the drug begins to dissolve in the gastric fluid. All the processes occurring, such as the disintegration of the tablet and the dissolution of the drug compound, are critically affected by the accompanying changes in energy.

If we imagine an assembly of molecules, the total energy contained in the assembly is called the internal energy and is given the symbol *U*. The internal energy of an assembly of molecules consists of the sum of all the nuclear, electronic, vibrational, rotational and translational energies of the molecules, plus the energies of the intermolecular interaction between the molecules.

Now, imagine that the assembly undergoes a process of change, from some initial state A to some new state B. An example might be when a sample of a drug substance is mixed with a solvent to produce a solution. Remembering that energy changes accompany all physical or chemical changes, the assembly of molecules in state A possesses different internal energy to the assembly in state B. This change in energy manifests itself as either the absorption of energy from the surroundings, or transmission of energy *to* the surroundings. This transfer of energy to or from the surroundings is what we know as 'heat'.

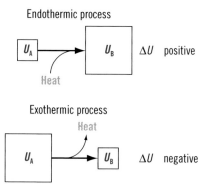

FIGURE 3.1 Endothermic and exothermic processes.

We can say that the internal energy of the assembly in state A is U_A, and that the internal energy of the assembly in state B is U_B. If the process of change is carried out with no change in the volume of the assembly, the change in internal energy, ΔU, is given by

$$\Delta U = U_B - U_A.$$

The symbol 'Δ' indicates the change in the value of U – that is, the value of the quantity in the final state minus the value of the quantity in the initial state.

When energy is absorbed by the assembly, U_B will be greater than U_A, and ΔU will have a positive value. When heat is evolved by the assembly, then U_B will be less than U_A, and ΔU will have a negative value. Processes that absorb heat from the surroundings are known as endothermic processes, while processes that release heat into the surroundings are known as exothermic processes. The concepts of endothermic and exothermic processes are illustrated schematically in Fig. 3.1.

When a system undergoes a physical or chemical change then apart from absorbing or evolving heat, it may also perform work. If a gas undergoes a change in volume ΔV against a constant pressure P, the work done, w, is given by

$$w = -P\Delta V. \tag{3.3}$$

We see from this equation that the greater the pressure and/or the greater the increase in volume, the more work that has been done.

The minus sign appears in eqn 3.3 for the following reason. If the gas has expanded, the volume has increased so that ΔV will be positive. This means that the gas will have done work on its surroundings, and so its internal energy will have decreased. The work done is therefore energy that has been *lost* by the system, and so has a negative value. If the gas has contracted, the volume will have decreased so that ΔV will be negative, indicating that work has been done on the gas by its surroundings. The internal energy of the system has been increased, so that the work done is positive.

So overall, if the gas expands (positive ΔV), it has done work and lost internal energy (negative w). If the gas contracts (negative ΔV), work has been done on it and it has gained interval energy (positive w).

The internal energy, U, of a system can therefore be *increased* by processes that transfer heat to the system from its surroundings, or by having work done on the system by its surroundings. The internal energy, U, of a system can be *decreased* by processes that transfer heat from the system to its surroundings, or by the system doing work on its surroundings. In other words, if a sample is heated or physically manipulated, say by stirring, energy is inputted into the system and its internal energy increases. If a sample gives off heat, say as a consequence of a chemical reaction, or expands to force back the surrounding atmosphere, it loses energy to its surroundings and its internal energy decreases.

If the heat transferred to the system from its surroundings is given the symbol q, the above statements can be summarized mathematically as

$$\Delta U = q + w. \tag{3.4}$$

Equation 3.4 can be regarded as a mathematical statement of the first law of thermodynamics. The equation tells us that the internal energy of a system can be changed by transferring energy either as heat, or as work, or both.

▶ **WORKED EXAMPLE 3.3**

A propellant gas expands, doing 126 J of work. At the same time it absorbs 137 J of heat. What is the change in the internal energy of the gas?

Work has been done by the propellant gas on its surroundings. The work therefore has a negative value, as it constitutes energy lost by the system. Hence, $w = -126$ J. Heat has been absorbed by the gas, so this is energy gained by the system. Hence, the heat absorbed, q, has a positive value ($q = 137$ J).

We use eqn 3.4 to work out the change in internal energy of the system. By eqn 3.4

$$\Delta U = q + w = (137 \text{ J}) + (-126 \text{ J}) = 11 \text{ J}.$$

If a process takes place at constant volume, then ΔV will be equal to zero. Considering the relationship $w = -P\Delta V$, this means that w equals zero, so that eqn 3.4 will reduce to

$$\Delta U = q_v, \tag{3.5}$$

in which q_v is the heat absorbed or evolved at constant volume. We see that, under conditions of constant volume, the change in internal energy is equal to the transfer of energy (as heat) to or from the system.

Most pharmacological changes occur under conditions of constant pressure. If a process takes place under constant pressure, contraction or expansion may occur. (Expansion means that ΔV will be positive; contraction means that ΔV will be negative.) Remembering that $w = -P\Delta V$, the relationship $\Delta U = q + w$ becomes $\Delta U = q + (-P\Delta V)$, or

$$\Delta U = q_p - P\Delta V, \tag{3.6}$$

in which q_p is the heat absorbed or evolved at constant pressure, and $-P\Delta V$ is equal to the work w done by the system.

If U_i and V_i are the *initial* internal energy and volume of the system, and U_f and V_f are the *final* internal energy and volume of the system, eqn 3.6 can be rewritten as

$$\underbrace{(U_f - U_i)}_{\Delta U} = q_p - P\underbrace{(V_f - V_i)}_{\Delta V}. \tag{3.7}$$

This can be rearranged to give

$$q_p = (U_f + PV_f) - (U_i + PV_i). \tag{3.8}$$

Equation 3.8 shows that the heat q_p absorbed or evolved by a process occurring under constant pressure is equal to the difference in the terms $U + PV$ for the initial and final states. The term '$U + PV$' is known as the enthalpy, H – that is

$$H = U + PV. \tag{3.9}$$

When systems change their structure or their form (for instance when drug molecules dissolve in the gastrointestinal tract) the change in enthalpy for that process will reflect the relative strengths of the bonds within and between molecules before and after the change.

So, how does the enthalpy of a system relate to q, the heat absorbed or evolved? The heat q_p absorbed or evolved by a process occurring under constant pressure can now be given as

$$q_p = H_f - H_i = \Delta H. \tag{3.10}$$

This tells us that the transfer of energy (as heat) equals the difference in enthalpy at the start and end of the process. As most pharmaceutical processes occur under constant-pressure conditions, and as heat transfer under these conditions is equal to the enthalpy change, enthalpy is a more widely used (and maybe more familiar) concept than internal energy.

Internal energy U and enthalpy H are both said to be thermodynamic state functions. The thermodynamic state of the system is defined in terms of four so-called variables of state: pressure P, volume V, temperature T, and composition n. Thermodynamic state functions are determined by the values of P, V, T, and n for a system at any particular instant. They are independent of events prior to that instant. A property of a thermodynamic state function is that any change in its value depends only on the initial and final states of the system, and independent of how the change was achieved (that is, it is independent of the path taken).

By contrast, work is not a thermodynamic state property, as the amount of work done during a change in the state of the system depends on how the process of change occurred, and not just on the values of P, V, T, and n in the initial and final states of the system. Work is done only during a change in the state of the system and its surroundings. It is only energy, and not work, which is determined by the initial and final states of the system.

3.1.2 Entropy and the second and third laws of thermodynamics

The second law of thermodynamics states that isolated systems become more disordered during processes of *spontaneous* change. Everyday experience tells us that some changes occur spontaneously while others do not. In other words, some things 'just happen' and other thing do not. Changes that result in more disorder seem more likely to happen 'by themselves' than changes that result in some structuring or ordering; these usually require our input.

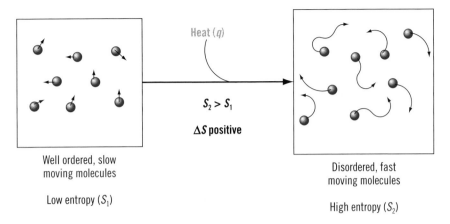

Heat (q)

$S_2 > S_1$

ΔS positive

Well ordered, slow
moving molecules

Low entropy (S_1)

Disordered, fast
moving molecules

High entropy (S_2)

FIGURE 3.2 Increase in the entropy, S, of a system upon absorption of heat.

At the conclusion of a spontaneous process, the system is in equilibrium with its surroundings. At the commencement of the process, the system and its surrounding are not at equilibrium. There is therefore a difference in potential energy between the initial and final states of the system and its surroundings. For example, an object posed at the top of a hill possesses potential energy equal to the work done in raising it to that point against the force of gravity. When the object rolls to the bottom of the hill and stops, that potential energy is dissipated through heat and friction, and a state of equilibrium is achieved between the object and all the forces acting on it.

The example of an object at the top of a hill involves mechanical potential energy. When a process of spontaneous change involves transfer of energy as heat, the energy flows from a body at a higher temperature to one at a lower temperature.

The factor that describes the capacity for the transfer of heat is a thermodynamic state function known as the entropy, S. Entropy is a measure of the *disorder* of a system. As a system absorbs energy, it becomes more disordered. For example, at a low temperature, the molecules in a sample will move slowly. If the sample is heated, the molecules absorb energy and begin to move faster. There are more ways in which the kinetic energies of the faster moving molecules can be distributed, and so the sample is more disordered. An ordered system has low entropy, and so has the capacity to absorb heat energy. A disordered system has high entropy, and has less capacity to absorb heat energy. The concept of entropy is illustrated in Fig. 3.2.

Mathematically, entropy, S, can be defined as follows

$$\Delta S = \frac{q_{rev}}{T},\tag{3.11}$$

in which q_{rev} is the heat transferred reversibly by a system to the surroundings at some temperature T. Note that when using eqn 3.11, temperature T has to be given in degrees absolute (K).

A body gaining heat has undergone a *positive* entropy change, while a body losing heat has undergone a *negative* entropy change. The heat transferred reversibly is the maximum heat that can be transferred to or from the system at that temperature. During a reversible heat transfer, the system and its surroundings are allowed to reach equilibrium after each instant of the transfer. Heat transferred irreversibly can be less than or equal to, but not greater than, q_{rev}.

Imagine that heat is being reversibly transferred to a sample of a pure liquid from its surroundings at constant temperature T and pressure P. As the heat transfer occurs under conditions of constant pressure P, q_{rev} is equal to the change in enthalpy, because $q_p = \Delta H$ (eqn 3.10). The effect of transfer of heat is to cause some of the liquid to vaporize. So, in this case, the enthalpy in question is the enthalpy of vaporization (or ΔH_{vap}). We can write

$$\Delta S = \frac{\Delta H_{vap}}{T}.$$
(3.12)

Similarly, when a crystalline pharmaceutical exists in equilibrium with its liquid phase at a temperature T (and pressure P), the entropy change for the solid to liquid phase transition can be given as

$$\Delta S = \frac{\Delta H_{fus}}{T},$$
(3.13)

in which ΔH_{fus} is the enthalpy of fusion for the solid. (The enthalpy of fusion is the enthalpy change accompanying melting of a solid.) Enthalpies of vaporization and fusion for some pharmaceutical materials are given in Table 3.2.

Enthalpy values such as these provide important information, because they allow us to say how much heat is required for physical changes such as vaporization or melting to happen at specific temperatures. This information may determine, for example, the storage conditions for a pharmaceutical material.

TABLE 3.2 Enthalpies of vaporization and fusion for some pharmaceutical materials.

Compound	ΔH_{vap}/kJ mol^{-1} $(T/$ K$)^a$	ΔH_{fus}/kJ mol^{-1} $(T/$ K$)^a$
H_2O	40.7 (373)	6.0 (273)
Ethanol	38.8 (351)	–
Ethanol	42.4 (298)	–
Glycerol	91.8 (298)	–
Paracetamol (form I)	–	28.1 (441)
Auranofin (form A)	–	37.8 (–)

[a] Temperature T at which the ΔH is measured. For ΔH_{vap} this is often the boiling point of the liquid, while ΔH_{fus} is often measured at the melting point of the crystalline solid. All the above values were recorded at 1 atm pressure.

Reproduced with permission from: Chickos J. S. and Acree W. E., Enthalpies of Vaporization of Organic and Organometallic Compounds 1880–2002, *J. Phys. Chem. Ref. Data*, 2003, **32**, 519–879, with permission of the American Institute of Physics; Burger, A., Zur Interpretation von Polymorphie-Untersuchungen, *Acta Pharm. Tech.*, 1982, **28**, 1–20, with permission of Wissenschaltliche Verlagsgesellschaft mbH; Lindenbaum, S., Rattie, E. S., Zuber, G. E., Miller, M. E., and Ravin, L. J., *Int. J. Pharm.*, 1985, **26**, 123–132, Copyright (1985) Elsevier.

▶ **WORKED EXAMPLE 3.4**

According to Table 3.2, the enthalpy of vaporization of water at 100 °C is 40.7 kJ mol⁻¹. Calculate the change in entropy when one mole of water is converted into steam at 100 °C.

Apply eqn 3.12. First remember that when using eqn 3.12, temperature T has to be given in degrees absolute (that is, in Kelvin, K). 100 °C is equal to 373 K.

Table 3.2 gives the enthalpy of vaporization, ΔH_{vap}, of water at 100 °C (or 373 K) as 40.7 kJ mol⁻¹. Values of entropy are usually given in units of J K⁻¹ mol⁻¹, so the enthalpy value in kJ mol⁻¹ should be converted to a value in J mol⁻¹, that is 40 700 J. The change in entropy, ΔS, is then given by

$$\Delta S = \frac{\Delta H_{vap}}{T} = (40\ 700\ \text{J mol}^{-1})\ /\ (373\ \text{K}) = 109\ \text{J K}^{-1}\ \text{mol}^{-1}.$$

Entropy is often explained in terms of changes in the level of order of a system. A system gaining heat undergoes a positive change in entropy and becomes more disordered. The increase in disorder can be understood in terms of increased molecular motion with increasing temperature. When a process of spontaneous change involves transfer of heat energy, heat flows from a body at a higher temperature to one at a lower temperature. This demonstrates the following statement of the second law of thermodynamics: that *the entropy of an isolated system increases in a spontaneous change.*

Third law of thermodynamics

The third law of thermodynamics places a lower limit on values of entropy. The entropy of a system is a measure of how disordered the system is. If a system becomes more disordered, for example through being heated, its entropy increases. On the other hand, if a system becomes more ordered, its entropy decreases. But how ordered can any system actually become? Could we imagine a system that is so perfectly arranged that it is not possible to make it any more 'ordered'? If this is the case, it might be that there is a minimum possible value for the entropy of a system. This minimum entropy value is reached when the system is so perfectly ordered that no further improvements in its 'ordering' are possible.

A perfectly arranged crystalline solid in which every individual atom is completely motionless would be such a system. In fact, this is the type of system that is used to define the third law. To ensure that the atoms in the crystal are completely motionless, the crystal would have to be at the lowest possible temperature: absolute zero (0 K). A statement of the third law would be as follows: *every substance has a finite positive entropy, but at absolute zero the entropy may become zero and does so in the case of a perfectly crystalline substance.* The third law is inferred from experimental measurements of entropy changes for processes at reduced temperatures. While absolute zero itself is not attainable, very low temperatures within four degrees above absolute zero can be attained experimentally.

3.1.3 Free energy, chemical potential and equilibrium

Neither ΔH nor ΔS taken alone provide a reliable guide to the possibility of a process taking place. Systems undergo change so as to either minimize enthalpy, maximize entropy, or both. A thermodynamic property is required that takes into account both ΔH and ΔS for a system undergoing change. This property is the free energy, also known as the Gibbs free energy, and is given the symbol G. The following equation defines free energy G

$$G = H - TS. \tag{3.14}$$

For a process occurring at constant temperature and pressure, the change in free energy, ΔG, is given by

$$\Delta G = \Delta H - T\Delta S. \tag{3.15}$$

Equation 3.15 tells us in which direction a process will move spontaneously. A highly exothermic process (large negative ΔH) will release energy from the system to its surroundings. The surroundings will absorb this energy and so the *entropy* of the surroundings will increase. This is a favourable outcome that will tend to occur spontaneously. Alternatively, if the entropy of the *system* increases significantly, the $(-T\Delta S)$ term will also be large and negative. A large increase in the entropy of the system would also be favourable and would tend to occur spontaneously.

Large negative values of either ΔH or $-T\Delta S$ indicate favourable processes that tend to occur spontaneously. Positive values for either of these terms will tend to disfavour a process. Equation 3.15 shows that what is required overall for a spontaneous process is a negative ΔG value, as the ΔG value sums the contributions of both the ΔH and the $-T\Delta S$ terms. *The spontaneous direction of a process, therefore, is that which is associated with the negative value for ΔG.*

Chemical potential

Free energy is a thermodynamic state property. If n moles of a *pure* substance are present in a sample, the total free energy of the sample would be n times G. This is only the case for a pure substance – that is, a substance consisting of only one component – in which case the free energy is the *molar* free energy of the substance.

However, if a substance consists of several components (for example compounds X, Y and Z) it is the case that the total free energy of the substance equals $G_X + G_Y + G_Z$ only if the individual components X, Y and Z exhibit ideal behaviour. Ideal behaviour exists when the thermodynamic properties of any particular component are not affected by the presence of the other components. However, in real systems this is not the case, and the thermodynamic properties of any one component are affected by the presence of the other components. This is a reflection of the fact that components generally interact with each other, which affects the physical properties of the components, including thermodynamic properties. This is most certainly the case with mixtures of pharmaceutical components. For this reason, if we are to be able to apply the concept of free energy to complex pharmaceutical systems, we need a method of allowing for the interaction between components.

To do this, we define a property known as the chemical potential, μ. The chemical potential of a component is the contribution that component makes to the overall free energy of the system. The concept of chemical potential recognizes that a component of a complex mixture, for example a drug in a biological tissue, does not exist in isolation from all the other components, but interacts with them. Each component of the system contributes to the overall free energy of the system, but that contribution is affected by the many interactions between the components.

The mathematical definition of chemical potential is usually presented as follows

$$\mu_i = \left(\frac{\partial G}{\partial n_i} \right)_{T,P,n_1,n_2,...,n_{i-1}} . \tag{3.16}$$

Equation 3.16 may appear complex. What it means is that the chemical potential (μ_i) of any one component (component i) of a mixture is determined by how the free energy (G) of the system varies with the amount of it present (n_i), while the temperature, pressure and amounts of all other components stay the same. The component could be, for example, a drug substance dissolved in the interior of a cell. The other components would be water and all the biochemical species present in the cell. If more of the drug substance travels into the cell, its chemical potential changes.

The derivation of eqn 3.16 is not as difficult as one might think. It involves using mathematical quantities known as partial derivatives. Partial derivatives and the derivation of eqn 3.16 are presented separately in Box 3.1. While it is not essential to know the material in Box 3.1, it would give a better understanding of the concept of chemical potential.

BOX 3.1 Partial derivatives and chemical potential.

In a real system consisting of a number, i, of components, numbered 1, 2, 3, etc. up to i, the total free energy of the system is a function of the temperature, pressure and composition of the system. The composition of the system can be described in terms of the respective number of moles n of each component 1, 2, 3, . . . , i, that is by n_1, n_2, n_3, . . . , n_i.

As the system is not ideal, a change in any one of these variables will affect the values of the others. This can be treated mathematically by using partial derivatives. Before doing this, a brief explanation of partial derivatives will be given.

If z is a function of x and y, changes in x by dx and changes in y by dy result in changes in z by dz as follows

$$dz = \left(\frac{\partial z}{\partial x} \right)_y dx + \left(\frac{\partial z}{\partial y} \right)_x dy,$$

in which, for example, the term

$$\left(\frac{\partial z}{\partial x} \right)_y$$

stands for the derivative of z with respect to x at constant y. This approach can be generalized in the following manner. If z is a function of m independent variables x_1, x_2,..., x_m, the differential of z is composed of contributions from each independent variable as follows

$$dz = \left(\frac{\partial z}{\partial x_1} \right)_{x_2,...,x_m} dx_1 + \left(\frac{\partial z}{\partial x_2} \right)_{x_1,...,x_m} dx_2 + ... + \left(\frac{\partial z}{\partial x_m} \right)_{x_1,x_2,...,x_{m-1}} dx_m.$$

In this treatment, each term such as, for example,

$$\left(\frac{\partial z}{\partial x_1}\right)_{x_2,\dots,n_m}$$

gives the variation in z as x_1 changes, while the remaining variables x_2 up to x_m stay constant.

Using this approach, a small change in free energy G can be given as

$$dG = \left(\frac{\partial G}{\partial T}\right)_{P,n_1,n_2,\dots,n_i} dT + \left(\frac{\partial G}{\partial P}\right)_{T,n_1,n_2,\dots,n_i} dP + \left(\frac{\partial G}{\partial n_1}\right)_{T,P,n_2,\dots,n_i} dn_1 + \left(\frac{\partial G}{\partial n_2}\right)_{T,P,n_1,\dots,n_i} dn_2$$

$$+ \dots + \left(\frac{\partial G}{\partial n_i}\right)_{T,P,n_1,n_2,\dots,n_{i-1}} dn_i,$$

in which T is temperature, P is pressure and the number of moles of each component 1, 2, 3, etc. is given by n_1, n_2, n_3, etc.

For a process at constant temperature and pressure, dT and dP are both equal to zero, giving

$$dG_{T,P} = \left(\frac{\partial G}{\partial n_1}\right)_{T,P,n_2,\dots,n_i} dn_1 + \left(\frac{\partial G}{\partial n_2}\right)_{T,P,n_1,\dots,n_i} dn_2 + \dots + \left(\frac{\partial G}{\partial n_i}\right)_{T,P,n_1,n_2,\dots,n_{i-1}} dn_i,$$

in which the subscript "T,P" in "$dG_{T,P}$" indicates a process occurring at constant temperature and pressure. The term

$$\left(\frac{\partial G}{\partial n_i}\right)_{T,P,n_1,n_2,\dots,n_{i-1}}$$

gives the change in the total free energy as the amount of component i changes, while the temperature, pressure and the amounts of all the other components remain constant. This quantity is known as the chemical potential of component i, and is given the symbol μ_i, that is

$$\mu_i = \left(\frac{\partial G}{\partial n_i}\right)_{T,P,n_1,n_2,\dots,n_{i-1}}.$$

The overall free energy of a multicomponent system, such as a pharmaceutical formulation, is made up of the contributions made by the chemical potentials of all of the components. The chemical potential of each component varies if the quantity of that component varies. This can be presented mathematically by the following equation

$$dG_{T,P} = \mu_1 dn_1 + \mu_2 dn_2 + \dots + \mu_i dn_i. \tag{3.17}$$

As has been mentioned before, many important pharmaceutical systems, such as the binding of a drug molecule to a biochemical receptor, are equilibrium systems. Systems at equilibrium have no tendency to change, either in the forward or reverse direction.

Energetically, this means that for systems at equilibrium, $dG_{T,P}$ is equal to zero. In that case, eqn 3.17 can be written as

$$\mu_1 dn_1 + \mu_2 dn_2 + \ldots + \mu_i dn_i = 0. \tag{3.18}$$

An alternative shorthand way of writing eqn 3.18 is as follows

$$\sum_{i=1}^{i=i} \mu_i dn_i = 0. \tag{3.19}$$

Equation 3.19 describes the thermodynamic conditions for a system to exist in equilibrium. It is therefore very important in pharmaceutical terms, as the processes involving the absorption, distribution and action of drugs are equilibrium processes. Equation 3.19 tells us that for a pharmaceutical system to be at equilibrium, infinitesimally small changes in the quantities of the components of the system result in no change in the overall free energy of the system. It does not mean that an equilibrium system is 'frozen'; an equilibrium system can still be dynamic, provided that the condition imposed by eqn 3.19 is still met.

Chemical potential and activity

To make use of the concept of chemical potential, we need to be able to relate chemical potential to the quantities of active components present in a system. To do this, G. N. Lewis proposed the following equation to hold for real substances existing as components in gaseous, liquid or solid systems:

$$\mu = \mu_i^{\circ} + RT \ln a_i. \tag{3.20}$$

μ_i° is the chemical potential of component i in the standard state, that is, at 1 bar pressure. The quantity a_i is known as the activity of component i of the system. At the standard state, a_i equals 1, so that $\mu = \mu_i^{\circ}$ in the standard state. The activity allows us to quantify how the chemical potential of a pure substance changes *when it becomes a component in a mixture*. For a component i of a liquid or solid system,

$$a_i = \gamma_i m_i, \tag{3.21}$$

in which m_i equals the concentration of component i of the system, and γ_i is known as the activity coefficient of the same component. The concept of activity is required because it has been found that the amount of a substance in a system that is active (that is which is causing an effect of some sort) is not always identical to the concentration. This is often the case for pharmaceutical systems, especially for pharmaceutical salts. Equation 3.21 allows us to link the somewhat abstract quantity of activity, a, to a very concrete quantity, concentration (m). The activity coefficient, γ, relates the two.

These equations can now be applied to cases in which the components of a system undergo chemical reaction at constant temperature and pressure. A great many pharmaceutical processes fall under this heading, including processes as fundamental as binding of a substrate to an enzyme. A general chemical reaction can be given as follows

$$a\text{A} + b\text{B} + \ldots \rightleftharpoons z\text{Z} + y\text{Y} \ldots, \tag{3.22}$$

in which a moles of reactant A, b moles of reactant B, etc. react to give z moles of product Z, y moles of product Y, etc. The free energy change for the reaction, ΔG, is given by

$$\Delta G = G_{products} - G_{reactants}. \tag{3.23}$$

Equation 3.23 shows how to calculate the change in free energy, ΔG, when a system undergoes a reaction. Equation 3.17 above gave an equation for the overall free energy, G, of a complex system in terms of chemical potentials of the components. Equation 3.20 allows us to present chemical potentials in terms of the activities, a, of components. We can combine all of these equations to give an equation that relates the change in free energy, ΔG, to the activities of the components undergoing reaction. The actual derivation of this equation is a little cumbersome, and so is given separately in Box 3.2. It is not essential to know the derivation given in Box 3.2, but a better understanding will be obtained if it is studied. The equation that is obtained by combining eqns 3.17, 3.20 and 3.23 is the following

$$\Delta G = \Delta G^\circ + RT \ln \left(\frac{a_Z^z \cdot a_Y^y \cdots}{a_A^a \cdot a_B^b \cdots} \right). \tag{3.24}$$

(Note that R stands for the gas constant: 8.314 J K^{-1} mol^{-1}.) The term inside the brackets in eqn 3.24 has the same form as the equilibrium constant, K, for the reaction given in eqn 3.22. It is very similar to the equilibrium constant expression given in eqn 3.1, except with activities instead of concentrations. When the activities of the reactants and products reach equilibrium, the term inside the brackets becomes equal to the equilibrium constant. At equilibrium, the free energy of the system, ΔG, is equal to zero, so that eqn 3.24 becomes

$$0 = \Delta G^\circ + RT \ln K,$$

or

$$\Delta G^\circ = -RT \ln K. \tag{3.25}$$

If ΔG° is positive, $\ln K$ will be negative and, hence, K will be less than one, indicating that the position of equilibrium lies more to the reactants than to the products of the reaction. Conversely, if ΔG° is negative, $\ln K$ will be positive and, hence, K will be greater than one, indicting that the position of equilibrium lies closer to the products than to the reactants. These relationships are illustrated schematically in Fig. 3.3.

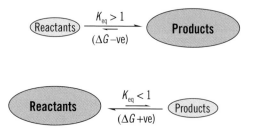

FIGURE 3.3 Schematic illustration of the relationships between ΔG (positive or negative), K_{eq} (less than one or greater than one), and whether the position of equilibrium lies toward the reactants or the products.

BOX 3.2 Free energy and equilibrium

A general chemical reaction can be given as follows

$$aA + bB + \ldots \rightleftharpoons zZ + yY \ldots$$

The general reaction process shown constitutes a real multiple-component system. Hence, the free energies of the component reactants and products must be expressed as chemical potentials, as follows

$$G_{\text{reactants}} = (a\mu_A + b\mu_B + \ldots), \text{ and}$$

$$G_{\text{products}} = (z\mu_Z + y\mu_Y + \ldots),$$

in which, for example,

$$a\mu_A = a\mu_A^\circ + RT \ln a_A^a.$$

Similar equations can be written for $b\mu_B$, $z\mu_Z$, etc. The change in free energy for the overall process is given by

$$\Delta G = G_{\text{products}} - G_{\text{reactants}}.$$

Substitution into this equation gives the following

$$\Delta G = (z\mu_Z^\circ + RT \ln a_Z^z + y\mu_Y^\circ + RT \ln a_Y^y + \ldots) - (a\mu_A^\circ + RT \ln a_A^a + b\mu_B^\circ + RT \ln a_B^b + \ldots).$$

This equation can be rearranged to give

$$\Delta G = [(z\mu_Z^\circ + y\mu_Y^\circ + \ldots) - (a\mu_A^\circ + b\mu_B^\circ + \ldots)] + RT [(\ln a_Z^z + \ln a_Y^y + \ldots) - (\ln a_A^a + \ln a_B^b + \ldots)].$$

The standard free energy for the reaction, ΔG°, is given by

$$\Delta G^\circ = (z\mu_Z^\circ + y\mu_Y^\circ + \ldots) - (a\mu_A^\circ + b\mu_B^\circ + \ldots).$$

Hence,

$$\Delta G = \Delta G^\circ + RT [(\ln a_Z^z + \ln a_Y^y + \ldots) - (\ln a_A^a + \ln a_B^b + \ldots)].$$

This equation can be written to give

$$\Delta G = \Delta G^\circ + RT \ln \left(\frac{a_Z^z \cdot a_Y^y \ldots}{a_A^a \cdot a_B^b \ldots} \right).$$

The term inside the bracket, that is

$$\left(\frac{a_Z^z \cdot a_Y^y \ldots}{a_A^a \cdot a_B^b \ldots} \right)$$

is equivalent to the equilibrium constant, K_{eq}, for the reaction.

WORKED EXAMPLE 3.5

Sodium cholate is a salt of one of the so-called bile acids. These are steroid-like acids that play an important role in the functioning of the gastrointestinal tract. In aqueous solution, sodium cholate exists in equilibrium with a dimeric form, as follows.

2 sodium cholate $\rightleftharpoons$ (sodium cholate)$_2$.

(i) What is the correct expression for the equilibrium constant, K_{eq}, for the dimerization of cholic acid, and (ii) if the equilibrium constant for this process is 2.2 at 25 °C, what is the free energy change, ΔG, for the process at equilibrium?

The law of mass is used to construct the expression for the equilibrium constant (for example, see eqn 3.1). For the dimerization of sodium cholate, the equilibrium constant is given by

$$K_{eq} = \frac{[(\text{sodium cholate})_2]}{[\text{sodium cholate}]^2}.$$

To determine the free energy change, ΔG, for the process at equilibrium, eqn 3.25 is required. To apply eqn 3.25, we need to know the gas constant, R (which is 8.314 J K^{-1} mol^{-1}), and the temperature, T, in degrees Kelvin at which the process is occurring (25 °C equal 298 K). Equation 3.25 then gives

$$\Delta G° = -RT \ln K_{eq} = -(8.314 \text{ J K}^{-1} \text{ mol}^{-1}) \, (298 \text{ K}) \, (\ln 2.2)$$
$$= -(8.314 \text{ J K}^{-1} \text{ mol}^{-1}) \, (298 \text{ K}) \, (0.788) = -1952.3 \text{ J mol}^{-1} \, [\text{or } -1.952 \text{ kJ mol}^{-1}].$$

The van't Hoff reaction isochore

The equilibrium constant, K_{eq}, is temperature dependent. If a reaction is endothermic (heat absorbed by the system, ΔH positive), an increase in temperature should favour the conversion of reactants into products. If the reaction is exothermic (heat released by the system, ΔH negative), increased temperature should disfavour conversion of reactants into products. Equation 3.25 may be used to obtain an expression of the temperature dependence of the equilibrium constant. This expression is known as the van't Hoff reaction isochore. (The derivation of the van't Hoff reaction isochore is given in Box 3.3.) The variation of equilibrium constant with temperature is important, as temperature variations may occur at various stages in pharmaceutical processes, for example, when a medicine stored at room temperature is ingested by a patient, whose body is at physiological temperature. The following equation is one form of the van't Hoff reaction isochore

$$\ln K_2 - \ln K_1 = -\frac{\Delta H°}{R} \left[\frac{1}{T_2} - \frac{1}{T_1} \right], \tag{3.26}$$

in which K_1 is the equilibrium constant at temperature T_1, and K_2 that at temperature T_2. An alternative form is

$$\ln \frac{K_2}{K_1} = \frac{\Delta H°}{R} \left[\frac{T_2 - T_1}{T_2 T_1} \right].$$

BOX 3.3 The van't Hoff reaction isochore.

For a process at equilibrium

$$\ln K = -\frac{\Delta G^\circ}{RT}.$$

At some temperature T_1

$$\ln K_1 = -\frac{\Delta G^\circ}{RT_1},$$

while at another temperature T_2

$$\ln K_2 = -\frac{\Delta G^\circ}{RT_2}.$$

Given that $\Delta G^\circ = \Delta H^\circ - T\Delta S^\circ$, and assuming that ΔH° and ΔS° are independent of temperature, $\ln K_2 - \ln K_1$ equals

$$-\frac{\Delta G^\circ}{RT_2} + \frac{\Delta G^\circ}{RT_1} = -\frac{\Delta H^\circ}{RT_2} + \frac{\Delta S^\circ}{R} + \frac{\Delta H^\circ}{RT_1} - \frac{\Delta S^\circ}{R} = -\frac{\Delta H^\circ}{R}\left[\frac{1}{T_2} - \frac{1}{T_1}\right].$$

This equation is the van't Hoff reaction isochore.

A more general form of these equations is

$$\ln K = -\frac{\Delta H^\circ}{RT} + C, \tag{3.27}$$

in which C is a constant. Plots of $\ln K$ versus $1/T$ should give straight lines with slopes equal to $-\Delta H^\circ/RT$, from which values of ΔH° may be determined. (This is because eqn 3.27 fits the general formula for a straight line, $y = mx + C$; with $y = \ln K$, $x = 1/T$ and $m = -\Delta H^\circ/RT$.)

3.2 PHASE EQUILIBRIA

Phase and equilibrium between phases are central to the pharmaceutical sciences. Issues affected by phase equilibria include whether an interface exists between two liquids, such as between aqueous and lipid phases in a cell structure or in a dosage formulation. The distribution of a drug substance between different media is another type of phase equilibrium. The existence of different phases and of equilibria between them affects a great range of pharmaceutical activities, from dosage-form design to absorption and distribution within the body. An ability to understand and interpret information on phase equilibria is therefore important.

3.2.1 One-component systems and the phase rule

A phase is a chemically and physically uniform part of a system that is separated from all other parts of the system by distinct boundaries. In principle, a phase of a system can be mechanically separated from the rest of the system. For example, a mixture of oil and water with a distinct interface between the two liquids is a system of two phases. The oil and water phases could be mechanically separated using, for example, a separating funnel.

If matter or energy can be exchanged between phases in a system, or if the volumes of different phases can be changed while the overall volume of the system remains the same, then the energies of the different phases of the system are *not* independent. We will see below that for substances to be transferred between phases, the critical parameter is the chemical potential, μ. Specifically, we will see that substances will transfer from a phase in which the substance has higher chemical potential to one in which it has lower chemical potential.

Pharmaceutical systems, such as oil-in-water emulsion formulations, are equilibrium systems. If two pharmaceutical phases are to exist at equilibrium, the temperatures of the two phases must be equal and the pressures of the two phases must be equal. If there is a solute that is dissolved in both phases, the distribution of the solute between the two phases is subject to constraints at equilibrium.

Section 3.1.3 discussed the case of a system of definite composition existing at equilibrium in terms of the chemical potential, μ_i, and the number of moles, n_i, of each component i. If the composition of a component (n_i) changes, the contribution of that component to the total free energy of the system changes. For the system to stay at equilibrium, the overall free energy of the system has to remain unchanged. This condition is summarized by eqn 3.19, which described the condition for such a system to exist at equilibrium at constant temperature and pressure, as follows (eqn 3.19)

$$\sum_{i=1}^{i=i} \mu_i dn_i = 0.$$

Equation 3.19 tells us that when a system exists in a state of dynamic equilibrium, the sum of all the energy changes due to changes of individual components is zero. Some component i could be a solute that is dissolved in both phases of a two-phase system. If the two phases are labelled A and B, reversible transfer of an infinitesimal quantity of solute i from phase A to phase B would decrease the amount of solute i in phase A by dn_i and would increase the amount of solute i in phase B by dn_i. If the chemical potential of solute i in phase A is given by μ_{iA}, and that in phase B by μ_{iB}, eqn 3.19 requires that, for a system at equilibrium,

$$-\mu_{iA}\, dn_i + \mu_{iB}\, dn_i = 0.$$

In other words, what has been taken from A in terms of energy has been added to B, so the net change across A and B combined is zero. Hence, for a system of two phases A and B to be at equilibrium, the chemical potential of any component i of the system must be equal in both phases, that is

$$\mu_{iA} = \mu_{iB}. \tag{3.28}$$

Equation 3.28 applies to all components of the system, and can be expanded to any number of phases.

If $\mu_{iA} > \mu_{iB}$, then

$$-\mu_{iA}\,dn_i + \mu_{iB}\,dn_i < 0,$$

in which case quantities of component i must be transferred from phase A to phase B to approach equilibrium. *This implies that a substance will transfer spontaneously from a phase in which it has higher chemical potential to a phase in which it has lower chemical potential.*

Phase diagrams

In the above discussion, terms such as 'component', 'solute' and 'species' are used interchangeably. However, more explanation is required of the term 'component' as used in discussion of pharmaceutical phase equilibria. The number of components of a system is defined as the minimum number of chemical species required to specify the composition of each phase of the system. For example, in a system consisting only of pure water, the number of components is just one, that is, water. However, a substance such as the sedative-hypnotic chloral hydrate is formed in aqueous solution according to the process shown in Fig. 3.4.

At least three species could therefore be present in solutions of chloral hydrate: water molecules, chloral molecules and chloral hydrate molecules. Because the concentrations of these three species are related by the equilibrium illustrated in Fig. 3.4, knowledge of the concentration of two of them is sufficient to describe the composition of the system. The number of components of that system is therefore two.

Information on the conditions necessary for the existence of different phases of a system of one component can be conveniently presented in plots of pressure versus temperature. Such plots are examples of phase diagrams. Figure 3.5(a) shows a phase diagram for water.

The diagram is partitioned into three regions, each of which describes the ranges of temperatures and pressures at which water exists in one of three physical forms, these being solid ice, liquid water, and gaseous water vapour. At any point on any of the partitioning lines, the system is an equilibrium mixture of the two phases separated by the line. The curves partitioning the three phases meet at the triple point, which occurs at 0.01 °C and 6.1 mbar. At this temperature and pressure, water exists as an equilibrium mixture of ice, liquid and vapour phases.

Water is an unusual substance, in that it is less dense in the solid state than in the liquid state. It therefore melts at slightly lower temperatures with increased pressure. This is indicated by the slightly negative slope of the line dividing the solid and liquid phases in Fig. 3.5(a).

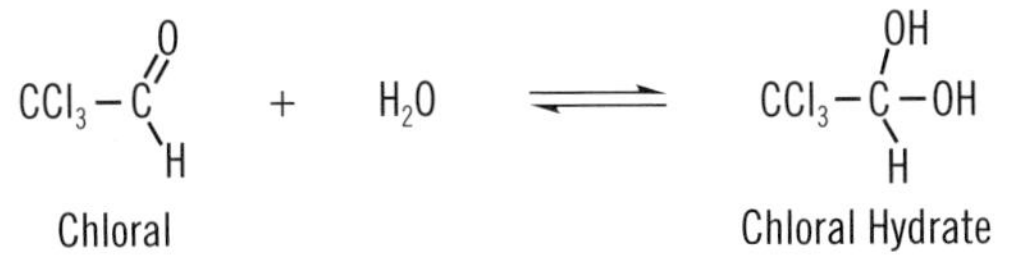

FIGURE 3.4 Reaction of chloral and water to form chloral hydrate.

(a)

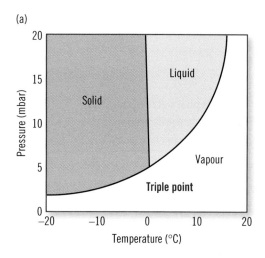

(b)

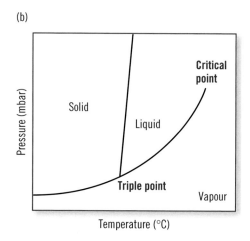

FIGURE 3.5 Phase diagrams for (a) water, and (b) carbon dioxide. Note the slightly negative slope of the line dividing the solid and liquid phase in Fig. 3.5(a).

Figure 3.5(b) shows a phase diagram for carbon dioxide. The diagram is largely partitioned into three regions, again describing the ranges of temperatures and pressures at which carbon dioxide exists as one of the solid, liquid or vapour phases. The triple point for carbon dioxide, at which all three phases co-exist at equilibrium, is at −57 °C and 5.1 bar. Below this pressure, which is approximately 5 atm, solid carbon dioxide transforms directly into carbon dioxide gas with increasing temperature. This is observed in everyday life as the sublimation of 'dry ice'.

Figure 3.5(b) also shows a point, occurring at 31.1 °C and 73.8 bar, referred to as the 'critical point'. Above this temperature and pressure, carbon dioxide exists in a phase that cannot be designated as either vapour or liquid. This phase is known as 'supercritical carbon dioxide'. Supercritical fluids can exhibit useful properties, such as low viscosities, low surface tensions and high diffusivities. Supercritical water occurs at 374.2 °C and 217.6 bar. Hence, the supercritical point for water does not occur within the temperature and pressure ranges shown in Fig. 3.5(a).

For pharmaceutical purposes, the most important supercritical fluid is supercritical carbon dioxide. The values for the critical temperature and pressure for carbon dioxide are relatively low, making supercritical carbon dioxide relatively easy to generate. It is also non-toxic, non-flammable, non-corrosive and inexpensive. The solvating power of supercritical carbon dioxide is relatively low (it is comparable to that of hydrocarbons). Crystallization processes involving supercritical carbon dioxide generally exploit its low solvating power, i.e. it is usually used as an antisolvent (see Section 4.5.1). Use of supercritical carbon dioxide as an antisolvent in pharmaceutical crystallizations allows fine control over crystal size and size distribution.

The phase rule

In both of the phase diagrams shown in Fig. 3.5, one component is present – that is, these are phase diagrams of pure substances. The diagrams are plots of two variables, temperature and pressure. Temperature and pressure are examples of intensive properties, since they are independent of the size of the system. By comparison, properties such as volume and mass are extensive properties, since they depend on the size of the system.

If one particular phase of one of the systems shown in Fig. 3.5 is selected (say the vapour phase) then a range of temperatures and pressures could be selected at which that system would exist in that phase. For example, from Fig. 3.5(a) water exists in the liquid phase at 10 bar and 4 °C, or 17 bar and 11 °C, or at many other combinations of temperature and pressure. However, if we wished to identify conditions at which two phases of the systems in Fig. 3.5 exist in equilibrium, then values of temperature and pressure must be selected that lie along the curves that divide the phases. Hence, if a value of one of the variables is selected, say temperature, the necessary pressure is that which co-ordinates with that temperature on one of the dividing curves.

If we wished to identify conditions at which all three phases exist in equilibrium, the temperatures and pressures selected can only be those that correspond to the triple points. Hence, for a system of one component, there is a relationship between the number of variables that can be adjusted, and the number of phases that are present. If one phase is present, two variables may be adjusted, or it is said that there are two 'degrees of freedom'. If two phases are present, then there is one degree of freedom. If three phases are present, there are zero degrees of freedom. The number of degrees of freedom can be defined as the minimum number of intensive variables that must be specified to completely describe the state of the system.

This finding can be extended to systems of many components to give the phase rule, first formulated by Gibbs in 1876. The phase rule tells us the number of degrees of freedom (F) we have for any particular number of components (C) and phases (P) in any system is as follows

$$F = C - P + 2. \tag{3.29}$$

In the water and carbon dioxide systems illustrated in Fig. 3.5, the number of components C equals one. If one phase is present, P equals 1, giving F equal to two. This means that in a system consisting of one phase of a pure substance, two intensive variables, normally temperature and pressure, can be altered without generating a new phase. If two phases are present, P equals two, giving F equal to one. Hence, if two phases are present in such a system, only one intensive variable can be adjusted without generating a new phase. If three phases are present, F equals zero, which occurs at the triple points. At triple points, we have no degrees of freedom; the temperature, pressure and composition all have to be 'just so' for the three phases to co-exist in equilibrium.

▶ **WORKED EXAMPLE 3.6**

Calcium carbonate is used as the active pharmaceutical ingredient in many antacids, and also as a base for creams and pastes. The following equilibrium exists for calcium carbonate

$$CaCO_3 \text{ (s)} \rightleftharpoons CaO \text{ (s)} + CO_2 \text{ (g)}.$$

Determine (i) how many components and phases are present in this system, and (ii) how many degrees of freedom there are when all phases are present at equilibrium?

(i) Three different chemical species exist: $CaCO_3$, CaO, and CO_2. However, knowledge of the equilibrium constant allows us to describe the composition of any phase in terms of only CaO and CO_2 (because any $CaCO_3$ can be considered to be formed from CaO and CO_2).

A phase is a chemically and physically uniform part of a system, which is, in principle, mechanically separable from the rest of the system. A heterogeneous mixture of two solids constitutes two phases. Overall, therefore, there are three phases in the system: two solid phases and a vapour phase.

(ii) From the above, $C = 2$ and $P = 3$. Hence, from the phase rule, $C - P + 2 = 1$, that is there is one degree of freedom. Therefore, one of either temperature or pressure can be varied independently. If both are varied independently, new phases will form.

3.2.2 Two-component systems

From a physicochemical viewpoint, important two-phase pharmaceutical systems to consider are liquid–vapour systems. These occur, for example, in the use of inhalation anaesthetics and some inhalation devices. To deal with the physical properties of these systems, we must understand how composition and pressure determine whether the system exists in the vapour or liquid phase. Two laws relating these factors will be discussed: Raoult's law and Henry's law.

Before looking at Raoult's law and Henry's law, it is worth remembering that when dealing with gases, values of *pressure* are used instead of values of concentration. The 'concentration' of a gas that is part of a mixture with other gases is known as its *partial pressure*. It might also be worthwhile looking again at the explanation of *mole fractions* given in Section 2.1. In the following sections, mole fractions will be used to express concentrations.

Raoult's law

In an ideal gas mixture, the partial pressure, P_i, of any component i, is defined as the mole fraction, x_i, of that component by the total pressure P of the gas, that is

$$P_i = x_i P. \tag{3.30}$$

For certain systems consisting of solution and vapour phases in equilibrium, Raoult found that the partial pressure of each component was equal to the mole fraction of that component *in solution* multiplied by the vapour pressure of the pure component, P_i^*. Raoult's law can be described mathematically as follows,

$$P_i = y_i P_i^*, \tag{3.31}$$

in which y_i is the mole fraction of component i in the *solution* phase (as against x_i, which is the mole fraction of component i in the gas phase). Raoult's law is often an approximate relationship. The advantage of using Raoult's law is that the mole fraction of a component in solution is relatively easy to measure, compared to its mole fraction in the gas phase. Systems consisting of similar components, for example mixtures of benzene and toluene, display the best compliance with Raoult's law. Raoult's law can also be expressed by combination of eqns 3.30 and 3.31 to give

$$x_i P = y_i P_i^*. \tag{3.32}$$

Raoult's law allows prediction of the phase diagram for an ideal liquid vapour mixture of two components. The total pressure P of the vapour is given by

$$P = y_1 P_1^* + y_2 P_2^*.$$

However, from the definition of mole fraction, $y_1 + y_2$ equals one. This allows us to write the above equation in terms of the amount of one component only, which is much more convenient experimentally, as follows

$$P = y_1 P_1^* + (1 - y_1) P_2^*, \text{ or}$$

$$P = P_2^* + (P_1^* - P_2^*) y_1. \tag{3.33}$$

Equation 3.33 is plotted in Fig. 3.6(a). At y_1 equal to zero, only component 2 is present, and so the total pressure of the vapour phase is equal to the vapour pressure of pure component 2. At y_1 equal to one, only component 1 is present, and so the total pressure of the vapour phase is equal to the vapour pressure of pure component 1.

The line shown in Fig. 3.6(a) is known as the bubble point line. At pressures above P_2^*, mixtures of any proportions of the two components will exist in the liquid phase. If the pressure of a liquid-phase mixture of the two components is reduced, when the pressure reaches the bubble point line, the pressure of the system will then equal the total vapour pressure for the system at that composition. Below that pressure, bubbles of vapour begin to form at the surface of the liquid.

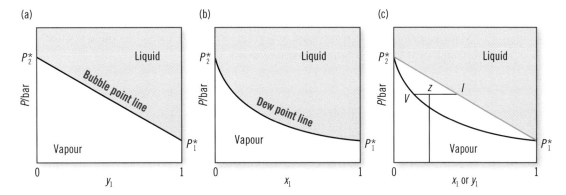

FIGURE 3.6 (a) Plot of total pressure, P, versus solution-phase mole fraction, y_1, of component 1 of an ideal mixture of two components. (b) Plot of total pressure, P, versus vapour-phase mole fraction, x_1, of component 1 of an ideal mixture of two components. (c) Plot of total pressure, P, versus mole fraction showing both the bubble point line and the dew point line. Reprinted from Alberty, R. A. and Silbey R. J., Physical Chemistry, 1992, with permission of John Wiley & Sons, Inc.

We can also derive an expression relating the total pressure P to the mole fraction of component 1 in the vapour phase (x_1). The derivation is given in Box 3.4. The resulting expression is

$$P = \frac{P_1^* P_2^*}{P_1^* + (P_2^* - P_1^*)x_1},$$
(3.34)

in which x_1 is the mole fraction of component 1 in the vapour phase. The value of eqn 3.34 is that it tells us at what composition and pressure a vapour turns into liquid. This can be important pharmaceutically in, for example, inhalation devices.

A plot of eqn 3.34 is shown in Fig. 3.6(b). The curve produced is known as the dew point line. Within the region below the dew point line, the system exists in the vapour phase. If the pressure of a vapour-phase mixture of two components at some specific composition is raised, at pressures just above the dew point line droplets of liquid condensate will begin to form.

BOX 3.4 Relationship between total pressure P and the mole fraction of component 1 in the vapour phase in a two-component vapour–liquid system.

Raoult's law can be expressed by

$$x_i P = y_i P_i^*.$$

For a two-component mixture, this equation can be given as

$$x_1 = \frac{y_1 P_1^*}{P}.$$

Using eqn 3.33 to substitute for P gives

$$x_1 = \frac{y_1 P_1^*}{P_2^* + (P_1^* - P_2^*)y_1}.$$

This equation can be rearranged to give an expression for y_1, the mole fraction of component 1 in the solution phase, as follows

$$y_1 = \frac{x_1 P_2^*}{P_1^* + (P_2^* - P_1^*)x_1}.$$

Raoult's law can also be given as

$$P = \frac{y_1 P_1^*}{x_1}.$$

Substitution of the previous expression for y_1 gives

$$P = \frac{P_1^* P_2^*}{P_1^* + (P_2^* - P_1^*)x_1},$$

in which x_1 is the mole fraction of component 1 in the vapour phase.

Figure 3.6(c) presents both the bubble point line and the dew point line on the same plot. Within the region between the two curves, the liquid and vapour phases exist in equilibrium. Points that occur on both the bubble point line and on the dew point line at the same pressure are connected by horizontal lines known as tie lines. One such tie line, connecting points v and l, is shown in Fig. 3.6(c). These points represent compositions of vapour and liquid phases that are in equilibrium.

From Fig. 3.6(c), if the mole fraction of component 1 is z, the relative compositions of the vapour and liquid phases are given by the level arm rule as follows.

$$\frac{\text{no. of moles liquid}}{\text{no. of moles vapour}} = \frac{z-v}{l-z}.$$

The plots shown in Fig. 3.6 are phase diagrams. These phase diagrams are constructed at some constant temperature. A 'reduced' phase rule therefore applies to these diagrams, that is

$$F = C - P + 1. \tag{3.35}$$

The systems illustrated in Fig. 3.6 have two components. If both liquid and vapour phases are to exist in equilibrium, eqn 3.35 implies that there is only one degree of freedom. In other words, if both phases are to co-exist, either the pressure or the composition can be varied independently, but not both.

▷ **WORKED EXAMPLE 3.7**

At 20 °C, the vapour pressures of pure samples of the inhalation anaesthetics halothane ($CF_3CHClBr$) and methoxyflurane ($CHCl_2CF_2OCH_3$) are 0.325 bar and 0.027 bar, respectively. (i) Determine the partial pressures of halothane and methoxyflurane for a solution with 0.55 mole fraction methoxyflurane. (ii) Give the equations of the bubble point line and the dew point line of a mixture of halothane and methoxyflurane. (iii) Calculate the mole fraction of methoxyflurane in both the solution phase and the vapour phase when the two phases are at equilibrium at a total pressure of 0.20 bar. (iv) Determine the ratio of moles in the liquid phase to moles in the vapour phase at a total pressure of 0.20 bar if the overall mole fraction of methoxyflurane in the system is 0.15.

(i) Using Raoult's law in the form of eqn 3.31, and taking methoxyflurane to be component 1, the partial pressure of methoxyflurane is given by

$$P_1 = (0.55)(0.027 \text{ bar}) = 0.0149 \text{ bar},$$

and that of halothane by

$$P_2 = (0.45)(0.325 \text{ bar}) = 0.146 \text{ bar}.$$

(ii) Equation 3.33 is the general equation for a bubble point line. Taking $P_2^* = 0.325$ bar, and $P_1^* = 0.027$ bar, the bubble line equation is

$$P = 0.325 \text{ bar} + (0.027 \text{ bar} - 0.325 \text{ bar})y_1 = 0.325 \text{ bar} - (0.298 \text{ bar})y_1.$$

Equation 3.34 gives the general equation for a dew point line. Inserting $P_2^* = 0.325$ bar, and $P_1^* = 0.027$ bar into eqn 3.34 gives

$$P = \frac{(0.027 \text{ bar})(0.325 \text{ bar})}{0.027 \text{ bar} + (0.325 \text{ bar} - 0.027 \text{ bar})x_1} = \frac{0.00878 \text{ bar}^2}{0.027 \text{ bar} + (0.298 \text{ bar})x_1}.$$

(iii) Use the equation for the bubble point line to determine the mole fraction of methoxyflurane in the solution phase at a total pressure, P, of 0.20 bar.

0.200 bar = 0.325 bar − (0.298 bar)y_1, hence $y_1 = 0.419$.

Use the dew line equation to determine the mole fraction of methoxyflurane in the vapour phase at a total pressure of 0.20 bar

$$0.200 \text{ bar} = \frac{0.00878 \text{ bar}^2}{0.027 \text{ bar} + (0.298 \text{ bar})x_1}, \text{ hence } x_1 = 0.057.$$

(iv) Use the lever arm rule as follows

$$\frac{\text{no. of moles liquid}}{\text{no. of moles vapour}} = \frac{z - v}{l - z} = \frac{0.150 - 0.057}{0.419 - 0.150} = \frac{0.093}{0.269}.$$

Henry's law

Mixtures that do not behave ideally deviate from Raoult's law. These deviations occur when vapour-phase interactions between molecules of different components are significantly stronger or weaker than interactions between molecules of the same component. When interactions between components are stronger, the vapour pressures of the components of the mixture are less than would be the case with ideal behaviour. This is observed as negative deviation from Raoult's law, Fig. 3.7. When interactions between components are significantly weaker, positive deviations from Raoult's law are observed, Fig. 3.7.

In cases of non-ideal behaviour, agreement with Raoult's law is observed for a component as its mole fraction approaches one. As the mole fraction of a component approaches zero, its partial pressure tends to agree with

$$P_i = k_i y_i, \tag{3.36}$$

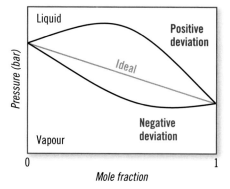

FIGURE 3.7 Positive and negative deviations from Raoult's law.

in which k_i is a constant. This is known as Henry's law. Henry's law tell us that at low concentration, the partial pressure of a component is directly proportional to its mole fraction in solution. This is a very convenient relationship, as it allows direct relation of solution concentration to gas-phase concentration. However, it only 'holds good' at low concentrations.

Values for the Henry's law constant are obtained by plotting P_i/y_i versus y_i and extrapolating to y_i equal to zero. The basis for Henry's law is that at concentrations below some dilute concentration, the environment of molecules of the dilute component remains effectively constant, and hence its partial pressure is proportional to its mole fraction. Henry's law is often used to express the solubility of gases in liquids. At reasonable temperatures and pressures, for example at 1 bar and 25 °C, acceptable compliance with Henry's law is observed for mole fractions less than 0.05.

▷ **WORKED EXAMPLE 3.8**

The Henry's law constant for oxygen, $k(O_2)$, in water at 25 °C is 4.9×10^4 atm (atmospheres). Given that the partial pressure of oxygen in air is 0.22 atm, calculate the mole fraction of oxygen (O_2) in water at 25 °C at an atmospheric pressure of 1 atm.

Use eqn 3.36 ($P_i = k_i y_i$). In this case, we want to calculate the mole fraction of oxygen molecules $y(O_2)$, so we will use the equation in the following form

$$P(O_2) / k(O_2) = y(O_2).$$

This gives $(0.22 \text{ atm}) / (4.9 \times 10^4 \text{ atm}) = 4.4 \times 10^{-6}$.

This shows us that the amount of dissolved oxygen present in an equilibrium water/air system at 25 °C is very low. A mole fraction of 4.4×10^{-6} corresponds to an oxygen concentration of 2.4×10^{-4} mol L^{-1}.

For an ideal gas, pressure P, volume V, temperature T, and number of moles n, are related by $PV = nRT$, or

$$P = \frac{nRT}{V}.$$

For real gases, corrections are required to this equation, in particular for the finite volumes occupied by real gases, and for the attractive forces between real gas molecules. Van der Waals introduced two constants, a and b, to allow for these factors. As attractive forces between molecules affect pairs of molecules, the effect of attractive forces increases with the square of the number of molecules per unit volume, that is with $(n/V)^2$. The correction for reduction in pressure due to molecular attraction is therefore $-n^2a/V^2$. The correction for finite volume occupied by real gas molecules is $-nb$. Van der Waals applied these corrections as follows

$$P = \frac{nRT}{V - nb} - \frac{n^2a}{V^2},$$

in which the volume correction is applied in the first term and the correction for attractive forces in the second term. This equation is often presented in the following form

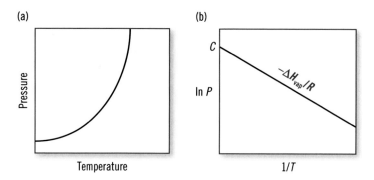

FIGURE 3.8 (a) General form of plots of vapour pressure versus temperature. (b) Plots of ln P versus $1/T$ of slope $-\Delta H_{vap}/R$ and intercept C.

$$\left(P + \frac{n^2 a}{V^2}\right)(V - nb) = nRT. \tag{3.37}$$

Examples of van der Waals constants are, for water, $a = 5.536$ L^2 bar mol^{-2} and $b = 0.0305$ L mol^{-1}, and for nitrous oxide, $a = 3.831$ L^2 bar mol^{-2} and $b = 0.0442$ L mol^{-1}. [A. T. Florence and D. Attwood, *Physicochemical Principles of Pharmacy*, 3rd edn, 1998, Palgrave, reproduced with permission of Palgrave Macmillan.]

Plots of vapour pressure of gases as a function of temperature have the general appearance shown in Fig. 3.8(a). These plots consist of curves showing rapid increases in vapour pressure with increasing temperature. The relationship between vapour pressure and temperature is given by the Clausius–Clapeyron equation, which is

$$\ln P = \frac{-\Delta H_{vap}}{RT} + C, \tag{3.38}$$

in which ΔH_{vap} is the molar enthalpy of vaporization, and C is a constant. Plots of ln P versus $1/T$ should give straight line graphs of slope $-\Delta H_{vap}/R$ and intercept C, as in Fig. 3.8(b).

3.2.3 Three-component systems

For three-component systems, the phase rule (eqn 3.29) gives

$$F = 3 - P + 2.$$

This places limitations on our ability to construct two-dimensional phase diagrams. This is unfortunate, as phase diagrams are a very useful way of summarizing a great deal of thermodynamic information. If only one phase is present, then $P = 1$ and $F = 4$. So, a four-dimensional representation would be required for a full phase diagram of such a system. However, if either one or both of temperature or pressure remain constant, the number of degrees of freedom is reduced. If both are constant, then

$$F = 3 - P,$$

and two-dimensional representations may be possible. A system of three components has two independent composition variables, for example, the mole fractions of any two

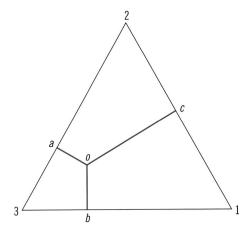

FIGURE 3.9 Equilateral triangular three-component phase diagram.

components. These could be plotted in the conventional manner. However, it is more useful to use triangular diagrams, because these allow us to see immediately how variation of any one of the components impacts on the other two.

A common method uses equilateral triangles. Each apex represents a pure component, that is, a mole fraction equal to one for either component 1, 2 or 3. A point on any side of the triangle represents a mixture of two components, with no contribution from the component represented by the excluded apex. Any point in the interior of the triangle represents a mixture of all three components. The relative quantities of the three components in a given mixture are equal to the relative perpendicular distances of the point representing the mixture from the sides of the triangle. For example, point o in Fig. 3.9 represents a mixture of components 1, 2 and 3 in which the relative proportions of 1:2:3 is given by the relative distances $oa:ob:oc$.

Figure 3.10 shows a three-component system consisting of three liquids. Two pairs of components, water/alcohol and alcohol/oil, are fully miscible in all proportions. The third pairing, oil/water, is only partially miscible. Note that the phase diagram is constructed at some constant temperature and pressure, for example at 25 °C and atmospheric pressure. When water is mixed with oil, the composition of the mixture lies along the line bc in Fig. 3.10. When small proportions of oil, for example, are added to much larger proportions of water, the composition lies along line bc very close to apex c. The oil dissolves in the water to give a single liquid phase. However, as more oil is added, the solution becomes saturated at point q. Two liquid phases then form, oil and water. The water phase is saturated in oil and has composition q. The oil phase is saturated in water and has composition o. As more oil is added, the relative amounts of the two phases alter, but the compositions of the phases remain the same. When sufficient oil is added to reach composition point o, one phase again forms, this time a solution of water in oil.

The alcohol is completely miscible with both oil and water in all proportions. If alcohol is also present in the mixture, the number of liquid phases present depends on whether or not the total composition is represented by a point within the arc opq, or not.

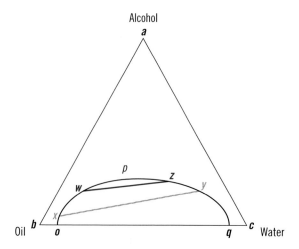

FIGURE 3.10 Triangular three-component phase diagram of an oil/water/alcohol mixture at constant temperature *T* and pressure *P*. Adapted from Alberty, R. A. and Silbey R. J., *Physical Chemistry*, 1992, with permission of John Wiley & Sons, Inc.

Say that oil, water and alcohol are present in proportions so as to give a composition at some point along the line *xy* in Fig. 3.10. Two liquid phases at equilibrium will then be present, one with composition *x* and one with composition *y*. Similarly, if the total composition point lies on line *wz*, two liquid phases at equilibrium will be present, one with composition *w* and one with composition *z*. Lines *xy* and *wz* are tie lines. In triangular three-component phase diagrams, the tie lines are not, in general, parallel to any axis or to each other. Their existence must be determined empirically. As the proportion of alcohol increases, the tie lines become shorter and eventually approach point *p*, known as the critical point. Further addition of alcohol beyond point *p* will produce a single homogeneous three-component liquid phase.

3.3 DRUG DELIVERY: PHASE TRANSITIONS

In this section we will explain in greater detail the relevance of the concepts of phase equilibrium and thermodynamics to drug delivery. Post-administration, a drug must undergo a number of phase transitions before it reaches the target receptor. As stated previously, the capacity of a drug to undergo these phase transitions is dependent upon the drug's physicochemical properties and the path it must take after administration. The phase transitions are also driven by the disturbance of a system's equilibrium due to the addition of new components or changes in components' concentrations.

The first stage of this drug-delivery pathway involves release of the drug from the dosage form into solution in the physiological fluids at the site of administration. Table 3.3 lists the phase transitions involved in the release of drug into solution in physiological fluids when administered in different phases by different routes.

The following section gives examples of the common phase transitions observed during drug delivery.

TABLE 3.3 Phase transitions after administration of API by various routes.

Route of administration	Drug phase	Phase transitions involved in dissolving in physiological fluids
Intravenous	Liquid (aqueous solution)	None or precipitation and dissolution
	Solid (suspension)	Dissolution
Intramuscular	Liquid (aqueous solution)	None or precipitation and dissolution
Subcutaneous	Liquid (non-aqueous solution)	Partitioning or precipitation and dissolution
	Solid (suspension)	Dissolution
Oral, Sublingual	Liquid (aqueous solution)	None or precipitation and dissolution
Buccal, Rectal, Vaginal	Liquid (non-aqueous solution)	Partitioning or precipitation and dissolution
	Solid (suspension, tablet, etc.)	Dissolution
Topical	Liquid (aqueous solution)	Partitioning
	Solid (suspension)	Dissolution and partitioning
Intrapulmonary	Gaseous	Gas absorption
	Liquid (aqueous solution)	None or precipitation and dissolution
	Solid (dispersion)	Dissolution

3.3.1 Examples of phase transitions during drug delivery

Liquid–solid phase transitions

Certain factors can result in the precipitation (a liquid–solid phase transition) of a drug from solution after administration. These include pH, interactions between the drug and constituents of the physiological fluid, common-ion effects and antisolvent effects. For example, a drug may precipitate from an oral solution due to the acidic pH of the stomach. Drugs such as ciprofloxacin can chelate with divalent ions present in ingested dairy products, antacids and dietary supplements and precipitate. After precipitation of the drug from solution, dissolution (a solid–liquid phase transition) of the solid drug back into the solution state is required prior to absorption.

Solid–liquid phase transitions

Drugs that are administered in a solid phase, for example a tablet, must dissolve in physiological fluids prior to absorption. Drug solubility, the composition and pH of the physiological fluid, the dose of drug and the available volume of dissolution medium all influence the rate of dissolution. Drug concentration in physiological fluids after administration is constantly changing due to drug input via dissolution and drug output, due to absorption in biological membranes at the site of administration (Fig. 3.11).

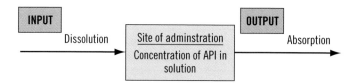

FIGURE 3.11 Factors that affect drug concentration in physiological fluids after administration in the solid phase.

In situations where the volume of dissolution media is small, as in the case of rectal and intramuscular administration, drug dissolution may occur over a prolonged period of time. However, even with large dissolution volumes it may be difficult to achieve rapid dissolution of the required drug dose when a drug exhibits poor aqueous solubility.

The composition of the physiological dissolution medium can also impact on the drug dissolution. This is particularly evident in the gastrointestinal tract, the components and pH of the stomach and small intestine contents fluctuate greatly between the fed or fasted states. In the fed state the small intestine contains higher amounts of bile salts compared to the fasted state. Lipophilic molecules exhibit higher dissolution rates in the fed state due to the presence of these components. Figure 3.12 summarizes the phase transitions involved in the dissolution of drug into solution in aqueous physiological medium.

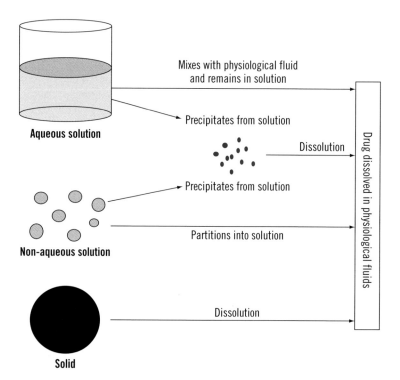

FIGURE 3.12 Phase transitions experienced during the dissolution of a drug into solution in aqueous physiological fluids.

Liquid–liquid phase transitions

Drugs such as norethisterone enantate are administered as a non-aqueous solution, which is administered intramuscularly by injection. The drug is released from these dosage forms by partitioning – a liquid–liquid phase transition. Another example, drug release from an ointment into the upper layer of skin, the stratum cornea, requires the partitioning of the drug from solution in the cream into the stratum cornea. Intrapulmonary drug delivery requires the drug to be soluble in the lung fluid prior to transport through the pulmonary epithelium by partitioning.

Gas–liquid phase transitions

Gaseous drug molecules are rapidly absorbed across the pulmonary epithelium into the blood. Consequently, gaseous drugs with high gas absorption coefficients have faster onsets of a pharmacological effect than inhaled solutions and solid drugs.

Further phase transitions occur during the absorption, distribution, metabolism and elimination after release of drug from the dosage form, these are explained in Chapter 8. More examples of drug phase transitions during transport across biological membranes are given in Chapter 7.

3.3.2 Using phase transitions to understand drug delivery

In Chapter 1, it was pointed out that in order to exert a therapeutic effect, a drug must reach the specific target receptor and interact with it. Target receptors can be present dispersed or dissolved in physiological fluids, located on cellular membranes or within a cell. When target receptors are located remote from the site of administration, systemic delivery is required. The drug must be in solution in the physiological fluids adjacent to the target receptor for the interaction between the drug and target receptor to occur. A drug in solution is considered to be a single liquid phase as it is homogeneous down to the molecular level. Therefore, a drug must be in a liquid phase adjacent to the target receptor before it can exert a pharmacological effect.

The overall process by which a drug reaches the target receptor in the physiological system can appear very complex. That is why it is easier to break it down into a number of smaller processes rather than look at it as a single process. Consider that the overall process of drug delivery can be divided into two separate processes (1) the preparation of dosage form containing the drug and (2) the transport of the drug in the physiological system from the dosage form to the target receptor. These processes are still quite large and complex so it is useful to further break them down into smaller processes.

For example, the preparation of a pharmaceutical oral solution can be divided into the processes of (1) dissolving the drug in a suitable solvent to form a solution and (2) adding excipients such as flavourings, colourings and preservatives to the drug solution. Now that the process is simplified into smaller blocks they are easier to deal with and can be more easily understood. Dissolving a solid drug in a suitable solvent requires the drug to change from a solid phase to a liquid phase. The liquid phase formed (the solution) is a two-component system (drug and solvent molecules). Adding other excipients may also involve solid–liquid phase transitions.

The liquid system is now changed from a two-component system to a multicomponent system. These two processes can appear simple but to explain why they occur and to predict how the multicomponent system formed will behave in response to the application of external energies such as agitation, light and heat is not so simple. To achieve a fundamental understanding of these processes and of the subsequent dosage form, an understanding of the concepts of phase equilibrium and thermodynamics explained earlier is required, in addition to an understanding of the chemistry of the multicomponent systems involved.

Similarly, the processes involved in the transport of the drug molecules from an oral solution after administration to a target receptor in the physiological system can be broken down into a number of component processes. For this example a bisphosphonate drug used to treat osteoporosis by binding to the hydroxyapatite crystals located in the bone (target receptors) will be considered. This process can be broken down into the processes of (1) dissolving of the drug in the gastrointestinal fluids, (2) transport of the drug across the gastrointestinal barrier into the blood system, (3) transport of the drug across the wall of the blood vessel, and (4) attachment of the drug molecule to the hydroxyapatite crystals in the bone. Again, each process involves phase changes and multicomponent systems. To understand these processes and the extent to which they occur again requires an understanding of the chemistry of the multicomponent physiological systems and the concepts of phase equilibrium and thermodynamics explained earlier.

Pharmaceutical phase transitions

Drugs can be administered in liquid, solid and gaseous phases. Semisolid material is also referred to: these materials can be considered to have phase properties intermediate between solid and liquid phases. Drugs undergo a number of phase changes between the time of administration and their arrival in solution at the target receptor. The most common types of phase changes are **dissolution** (solid–liquid), **precipitation** (liquid–solid), **partitioning** (liquid–liquid) and **gas absorption** (gas–liquid). The capacity of a drug to undergo these phase changes is dependent upon (1) the drug's physicochemical properties, (2) the physicochemical properties of the dosage form components, (3) the path it must take after administration to the target receptor and (4) the physiochemical properties of the physiological systems encountered en route to the target receptor.

As the term suggests, phase changes such as dissolution, precipitation, partitioning and gas absorption involve the movement of a drug from one phase to another phase. For example, dissolution involves the movement of molecules from a solid phase into a liquid phase. In the following section, the phase transitions involved in drug delivery are explained in greater detail. Before explaining the pharmaceutically important phase transitions, it is worthwhile re-emphasizing some of the relevant concepts covered in the earlier sections.

Phase transitions are a result of a disturbance of a system's equilibrium. In pharmaceutical drug delivery this is usually caused by the addition of a new component to the system or a change in the concentration of a component of the system, for example the drug. The disturbance of equilibrium is due to a difference of free energy (ΔG) within a phase or between two phases in contact. The free energy of a system (G) is composed of the enthalpy (H) and the entropy (S) of the system (eqn 3.15)

$$\Delta G = \Delta H - T\Delta S.$$

A free-energy difference within a system results in the system not being at equilibrium ($\Delta G \neq 0$). Systems that are not at equilibrium are thermodynamically unstable systems. Negative free-energy differences can arise in pharmaceutical systems during dosage-form preparation and drug delivery, giving rise to spontaneous changes. The rearrangement of assemblies of molecules within the system is a common example of the type of spontaneous change that occurs during drug delivery.

A change or process that results in the spontaneous rearrangement of assemblies of molecules in a system is called a mass-transfer process. A mass-transfer process is one in which material moves from one location to another. The driving force for the movement of material within a system is the negative free-energy change associated with the movement. Material movement will occur until the driving force is eliminated and the system reaches equilibrium ($\Delta G = 0$). Phase transitions are classified as mass-transfer processes as they involve the movement of materials between phases. The movement of a material within a phase is also classified as a mass-transfer process.

3.3.3 Diffusion (molecular movement within a phase)

Before movement of material from one phase to another can be understood, the movement of a substance within a phase must be explained. Diffusion is the spontaneous movement of matter from a region of high concentration to a region of low concentration. A system's mass transfer involved in diffusion is quantified as the flux. Flux is the amount of material that flows through a unit area of the system per unit time. An example of commonly used units to describe flux are mol m^{-2} s^{-1}.

Diffusion and chemical potential

The driving force for diffusion is the difference in chemical potential within the system. As explained in Section 3.1.3, the chemical potential of a component (μ_i) is the contribution that component makes to the overall free energy of the system (G). In other words, the presence of a molecule in solution contributes to the overall free energy of the solution. In Section 3.1.3, it was also explained that the free energy of a system varies as the concentration of each component in the system varies. This was described mathematically by

$$dG_{T,P} = \mu_1 dn_1 + \mu_2 dn_2 + \ldots + \mu_i dn_i. \tag{3.17}$$

It can be seen from this equation that as the relative amount or concentration of a component changes there will be a change in the overall free energy of the system.

Now consider a solution where there are regions of high and low concentrations of a component solute molecule. These differences in concentration result in regions of higher and lower chemical potential and therefore higher and lower free energy. The free-energy difference of the system (ΔG) results in a thermodynamically unstable system. Due to the unstable nature of the system there is a spontaneous rearrangement of molecules within the system to eliminate the free-energy difference ($\Delta G = 0$). The spontaneous rearrangement of a system's molecules involves the movement of solute molecules from the regions of higher concentration to lower concentrations until the system has a homogeneous molecular distribution as shown in Fig. 3.13. This process of spontaneous movement of molecules within a phase is called **diffusion**.

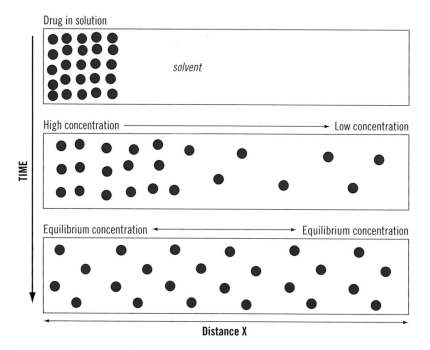

Drug in solution

solvent

High concentration ⟶ Low concentration

TIME

Equilibrium concentration ⟷ Equilibrium concentration

Distance X

FIGURE 3.13 Schematic diagram of the movement of drug molecules in a solvent by diffusion across a concentration gradient.

Therefore, it is easier to think about the driving force for diffusion in terms of concentration differences in the solution (dc) rather than chemical potential. The term concentration gradient is often referred to in relation to diffusion. The term 'gradient' describes the variation of concentration as a function of distance in the x-direction in Fig. 3.13. The gradient itself is the rate of change of the concentration with distance (dc/dx).

Diffusion coefficient

The concentration gradient (dc/dx) is the driving force for diffusion and the flux of molecules is proportional to the concentration gradient. The flux of molecules due to diffusion is also proportional to a second factor called the diffusion coefficient (or diffusivity, D). The diffusion coefficient of a specific molecule is dependent on the properties of the molecule, in particular the size of the diffusing molecule. Higher molecular weight and bulky molecules diffuse slower than smaller, less bulky molecules. The properties of the system the molecule is diffusing through also influence the diffusion coefficient value, in particular the viscosity of the system. A change in the temperature of the system can also increase the diffusion coefficient value. The diffusion coefficient is constant for a specific molecule diffusing through a specific system at a constant temperature. Diffusion coefficients are expressed as distance travelled per unit time, e.g. $m^2\ s^{-1}$.

Fick's laws of diffusion

The process of diffusion outlined above is described by Fick's 1st law of diffusion,

$$J = -D\frac{dc}{dx},$$

(3.39)

J is the flux of molecules, D the diffusion coefficient and dc/dx the concentration gradient. The negative sign in this relationship indicates that molecular flux occurs in a 'down' gradient direction, i.e. from regions of higher to regions of lower concentration.

Fick's 1st law of diffusion assumes steady-state conditions. Diffusion processes may be described as steady state and non-steady state. Steady-state diffusion takes place at a constant rate, meaning that the flux of molecules across a certain unit volume is constant with time. Non-steady-state diffusion occurs when the flux of molecules across a certain unit volume is not constant with time. So, for steady-state conditions, constant flux with time, dc/dx must also be constant with time. This means that the concentration difference in the system does not change with time. This type of steady-state situation occurs if the concentration of molecules in the region of high concentration is saturated or constantly replenished, so it remains constant and the concentration of molecules in the low region of concentration is negligible. In these circumstances the concentration gradient does not increase with time.

For non-steady-state conditions to arise, the concentration difference (dc) in the direction x must change with time. This change of concentration difference with time is described mathematically as (dc/dt). Normally concentration difference decreases with time. Consider a region of high concentration that is constantly being depleted by diffusion of molecules to the region of low concentration. If the region of high concentration is not replenished with molecules and the concentration of the region of low concentration increases with time, a decrease in concentration difference (dc) occurs as time progresses.

In order to describe non-steady-state diffusion, the changing concentration difference with time (dc/dt) has to be considered. This is described by Fick's 2nd law,

$$\frac{dC}{dt} = \frac{dJ}{dx}.$$ (3.40)

Equation 3.40 can be rewritten by replacing the term J in eqn 3.40 with the mathematical description of J in Fick's 1st law (eqn 3.39) to give

$$\frac{dC}{dt} = \frac{d\left(D\frac{dC}{dx}\right)}{dx} \quad \text{or} \quad \frac{dC}{dt} = D\frac{d^2C}{dx^2}.$$ (3.41)

Ficks' laws of diffusion describe diffusion in one dimension; however, in real pharmaceutical systems diffusion occurs in three dimensions. They can be rewritten to describe diffusion in three dimensions as observed in most drug-delivery systems, but this is outside the scope of this text. Having an understanding of Ficks' 1st and 2nd laws will assist in understanding all pharmaceutical processes that involve diffusion, dissolution, partition and gas absorption.

3.3.4 Dissolution (solid–liquid transition)

The term dissolution has already been introduced in Section 2.3. Dissolution generally refers to the phase change of a molecule from a solid phase into a liquid phase. This phase change occurs when a solid drug dissolves in a solvent system to produce a solution and this process is called dissolution. Remember that the drug molecules usually need to be in solution in the physiological fluid adjacent to their pharmacological targets if they are to produce therapeutic effects. The dissolution process is also required for the preparation of solutions for injection and oral administration. The measurement of the dissolution of a drug from a dosage form is a widely performed test in pharmaceutical laboratories to relate the performance of dosage forms *in vitro* to their performance *in vivo* and to compare batches.

Precipitation involves the phase transition of molecules from a liquid phase to a solid phase. While the process of precipitation is not dealt with in this text it can be considered the reverse of the dissolution process described.

Dissolution theory

Dissolution occurs spontaneously when a negative free-energy difference occurs between the free energy of the drug molecules in the solid phase (G_{solid}) and the free energy of the drug molecules in the liquid or solution phase (G_{liquid}). If $G_{solid} > G_{liquid}$, then the negative gradient will favour the spontaneous movement of molecules from the solid phase to the liquid phase until the free-energy gradient is eliminated and equilibrium is reached ($G_{solid} = G_{liquid}$). On the other hand, if $G_{liquid} > G_{solid}$ then the negative gradient will favour the spontaneous movement of molecules from the liquid phase to the solid phase until equilibrium is reached.

Whether dissolution occurs is highly dependent on the chemical nature of the drug and the solvent system. It is also influenced by the energy applied to the system by mixing and heat. The free-energy difference in the system is due to a combination of the enthalpy difference (ΔH) and entropy difference in the system (ΔS) and the temperature of the system, as previously described in eqn 3.15

$$\Delta G = \Delta H - T\Delta S.$$

Consider the dissolution process in three steps; (1) drug molecules break away from the solid structure, (2) a cavity suitable for the accommodation of the drug molecules is generated in the solvent medium and (3) the drug molecule locates within the solvent cavity, producing a solution. First, consider the enthalpy component that relates to the internal energy of the system. During the first two steps the enthalpy is decreased because interactions between the molecules (solute or solvent) are being broken and to break these bonds energy is required. During the third step enthalpy is increased as energy is released because of the solute–solvent interactions that are being formed.

The total enthalpy change involved in the process of a molecule moving from the solid phase to the liquid can be positive or negative, depending on the net enthalpy change over the three steps. If the solvent and drug molecules have an affinity for each other (solute–solute, solvent–solvent and solute–solvent interactions are of similar strength) a reduction in enthalpy difference between the solid and liquid states will occur ($\downarrow\Delta H$) during bond breakage and formation. If the solvent and drug do not have an affinity for each other (solute–solute and solvent–solvent interactions are stronger than solute–solvent interactions) an increase in enthalpy difference between the solid and liquid states will occur ($\uparrow\Delta H$) during bond breakage and formation.

Now consider the change in the entropy component. The dissolution process generally leads to a greater disorder in the system due to mixing of two or more components and therefore an increase in entropy ($\uparrow \Delta S$). If the increase in enthalpy ($\uparrow \Delta H$) is not sufficient to overcome the increase in entropy ($\uparrow \Delta S$) then the change in free energy of the system will be negative and dissolution takes place. Increases in temperature favour dissolution taking place.

Dissolution process

The solvent during dissolution is referred to as the *dissolution medium*. The interface where the solid molecules meet liquid molecules, can be described more accurately as an infinitesimally thin region allowing the transfer of a component of one phase to another, as shown in Fig. 3.14. In Fig. 3.14 the interfacial region is represented as two separate films but a more accurate representation would be a continuum of solid phase with increasing water content to a liquid phase with low solid content.

The movement of molecules from the solid phase to the liquid phase can be broken down into two stages. The first stage occurs within the interfacial region – the drug undergoes a solid–liquid phase transition. The solution in the interfacial region is quickly saturated with the drug as graphically represented in Fig. 3.15. C_s is the saturated drug concentration in the dissolution medium. A concentration gradient is established between the saturated concentration in the interfacial region and the bulk solution concentration, C_B. The concentration gradient ($(C_s - C_B)/x$) drives the diffusion of drug molecules away from the interfacial region in the direction of the bulk solution (x). This diffusion of drug molecules from the saturated solution at the interface is considered the second stage of dissolution. The movement of drug molecules from the interfacial region allows additional solid molecules to dissolve and replenish the solution of the interface maintaining a saturated concentration.

The movement and flow of the dissolution medium surrounding the solid material is not homogeneous and can be divided into two regions of flow. A region of low flow is attached to the interfacial region and is called the **boundary layer**. The bulk solution beyond the boundary layer in comparison is a region of high flow movement, as represented in Fig. 3.15. Before drug in solution in the interfacial region can be transferred into the bulk

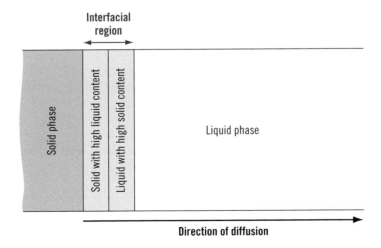

FIGURE 3.14 Schematic diagram of the interfacial region during dissolution.

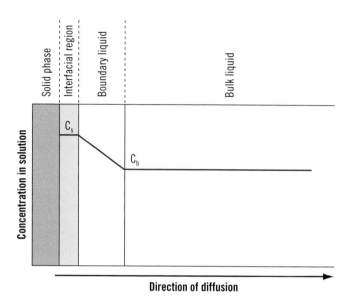

FIGURE 3.15 Concentration profile of the drug during dissolution.

solution it must diffuse through the boundary layer. It is then transferred rapidly away from the outer edge of the boundary layer by diffusion and convection currents.

Dissolution rate

Dissolution rate is the movement of mass from a solid phase into a liquid phase per unit time (dm/dt). The rate at which the dissolution process occurs is influenced by a range of factors. The driving force for dissolution is the free-energy difference between the solid and liquid phases. However, once the interfacial layer is saturated with drug in solution this energy difference is eliminated. The concentration gradient between the interfacial layer and the bulk solution is the driving force for the diffusion of molecules from the interfacial layer. In many situations it is the concentration gradient between the interfacial layer and the bulk solution that controls the rate of dissolution. When this occurs, dissolution can be said to be a diffusion-controlled process.

The Noyes–Whitney equation, eqn 3.42, is the mathematical expression used to describe the dissolution rate, (dm/dt). Upon examination of this equation it can be seen that the rate of dissolution is proportional to the concentration gradient between the interfacial layer and the bulk solution

$$\frac{dm}{dt} = -k(C_s - C_B).$$

(3.42)

The rate of dissolution is also proportional to a second factor, a constant k. The constant k is influenced by the diffusion coefficient of the drug in the dissolution medium (D), the surface area of the solid in connection with the liquid media (A) and the thickness of the boundary layer (h). As previously stated, the diffusion coefficient for a specific molecule can vary depending on the size of the diffusing molecule and the viscosity of the system the molecule is diffusing through. Increases in the viscosity of the dissolution medium can decrease the diffusion coefficient value. The thickness of the boundary layer, h, is influenced by a number of factors including the chemical composition of the drug and dissolution medium, the temperature of the systems and the degree and type of dissolution medium agitation. The thicker the boundary layer the slower the rate of dissolution.

The solid/liquid interfacial surface area, A can be increased by reducing the solid particle size, increasing particulate porosity and increasing wetting of the solid surface. As the surface area increases, the rate of dissolution increases. k can be described mathematically by

$$k = \frac{DA}{h}. \tag{3.43}$$

As dissolution proceeds the concentration gradient across the boundary layer decreases and the rate of dissolution decreases, eqn 3.42. As the concentration at the edge of the boundary layer, C_B approaches the saturated concentration, C_s, then the dissolution rate will be reduced. If the bulk dissolution medium becomes saturated with the drug, C_s equal to C_B, then the dissolution rate will become zero and no further dissolution will occur.

Dissolution conditions where the concentration of drug in the bulk solution, C_B, is negligible and there is no appreciable increase in this concentration gradient during the dissolution process are termed **sink conditions**. The concentration of the bulk solution during the dissolution process is considered negligible if it is less than 10% of the saturated solubility value, C_s. Two scenarios occur where C_B is less than 10% C_s throughout the dissolution process:

- The concentration of drug at the edge of the boundary layer is constantly depleted due to removal of the drug from the bulk solution faster than it is added. This situation can occur if a drug is absorbed rapidly from solution in the gastrointestinal tract into the intestinal membranes.
- The total quantity of a drug and the volume of dissolution medium are such that C_B cannot exceed 10% C_s even after complete dissolution of the drug.

When sink conditions exist the dissolution rate is constant with time because the term (C_s-C_B) in the Noyes–Whitney equation is constant as shown in Fig. 3.16.

If the concentration at the edge of the boundary layer, C_B, is considered negligible then it can be removed from the Noyes–Whitney equation and the equation can be further simplified to eqn 3.44 when sink conditions occur

$$\frac{dm}{dt} = -kC_s. \tag{3.44}$$

This simplified equation is useful as it shows that under sink conditions the rate of dissolution is directly proportional to the attainable saturated concentration; C_s, at the boundary layer that can be determined from the saturated solubility of the drug in the dissolution medium. So, by knowing the saturated solubility of the drug in the dissolution medium we can have a good indication whether the rate of dissolution for that drug in the medium will be fast or slow. Factors that influence the saturated solubility of drugs are dealt with in greater detail in Chapter 2.

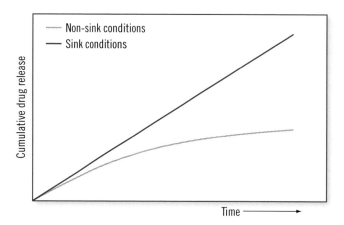

FIGURE 3.16 Cumulative drug release into dissolution media under non-sink conditions compared to sink conditions.

3.3.5 Partitioning (liquid–liquid transition)

Partitioning of a drug between two immiscible systems is the most frequent phase transition that occurs within the physiological system during drug delivery. As the physiological membranes can be considered as liquid systems, in order for a drug to reach a target receptor at least one partitioning step must occur. In most cases a number of partitioning steps occur. Examples of these steps are described in more detail in Chapter 7.

Theory of partitioning

The spontaneous movement of drug molecules between two immiscible liquid phases is due to differences in chemical potentials of the drug in both phases (μ_i). As previously described for the process of diffusion, differences in chemical potential result in differences in free energy (ΔG) and if that difference is negative then spontaneous movement of molecules can occur to eliminate this difference.

In the case of diffusion this movement is the spontaneous movement of a drug from one region of the solution to another. In the case of partitioning it is the spontaneous movement of a drug from one liquid phase to a second immiscible phase. The two immiscible liquid phases are separated by a liquid/liquid interface.

Consider a drug initially dissolved in solution A. Solution A comes into contact with a second immiscible liquid phase B. An interfacial region is formed as represented in Fig. 3.17. If the chemical potential of solution A is greater than the chemical potential of solvent B, the drug will diffuse from solution A into solution in solvent B across the interfacial region.

As for solid/liquid interfaces discussed in Section 3.3.4, the dissolution medium exists in two regions of flow. Again, the region of low flow adjacent to the interfacial region is called the boundary layer. The movement of drug molecules in the boundary layer is due to diffusion, whilst in the bulk solution it will be more rapid due to a combination of convection and diffusion. During the partitioning process two concentration gradients are established at either side of the interface, see Fig. 3.18. A difference in drug

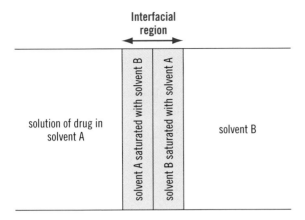

FIGURE 3.17 Schematic diagram of the interfacial region between two immiscible liquids.

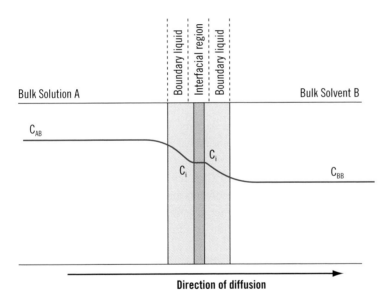

FIGURE 3.18 Schematic diagram of solute diffusion between two immiscible liquids prior to reaching equilibrium.

concentration occurs between the bulk solution concentration, C_{AB}, and the interfacial concentration, C_i and the bulk solution concentration, C_{BB}, and the interfacial concentration C_i. These concentration gradients result in two separate diffusion processes at either side of the interface.

The extent and rate at which the drug will partition between two immiscible liquid phases is influenced by the chemical compositions of the liquids involved and the drug. The addition of hydrophobic groups to the drug will increase the extent and rate of partitioning from a liquid aqueous phase to a liquid lipid phase. The rate of transfer is also influenced by the interfacial area available for molecular transfer.

Partition coefficient and distribution coefficient

Partition coefficients are a measure of how a drug will distribute between two immiscible liquid phases at equilibrium. Partition coefficient (P) values are specific to a particular drug in a particular mixture of immiscible liquids at a particular temperature. The partition coefficient of a drug between two liquid phases can be calculated from the relative concentration of drug in both liquid phases (C_A and C_B) using an equation explaining the Nernst distribution law, eqn 3.45. This equation is only application to calculate the partition coefficients for dilute solutions and ideal solvents

$$P = \frac{C_A}{C_B}. \tag{3.45}$$

The partition coefficient term P should indicate the solvents it refers to. For example, the term, $P_{oil/water}$, would represent the partition coefficient of a drug between oil and water liquid phases where the concentration in oil is C_A and the concentration in water is C_B in the calculation. $P_{water/oil}$ is the partition coefficient between water and oil and its value is the inverse of $P_{oil/water}$. In the case of poorly soluble drugs, the partition coefficient can be expressed as the ratio of the drug's solubility in the individual solvents (S_A/S_B) instead of concentration. This eliminates the need to measure concentration

$$P = \frac{S_A}{S_B}. \tag{3.46}$$

As can be seen from the equations above, the factors that affect a drug's partition coefficient between two solvent systems are the factors that affect the solubility of the drug in the respective solvent systems. These factors relate to the chemistry of the drug and solvent systems, which are described in detail in Chapter 2. For example hydrophobic drugs will have higher solubility in hydrophobic solvents and hydrophilic drugs will have lower solubility in hydrophobic solvents and higher in hydrophilic solvents.

The logarithm of the partition coefficient is log P, eqn 3.47. Log P values are often referred to instead of P values for convenience. Partition coefficient values can be very low or high and by expression as the logarithm of the values it is easier to compare between values.

$$\text{Log } P_{A/B} = \log\left(\frac{C_A}{C_B}\right). \tag{3.47}$$

3.3.6 Gas absorption (gas–liquid phase transition)

A small number of drugs are administered as gases. Anaesthetics are the main group of drugs administered as gases, for example, halothane. In order to exert a therapeutic effect they must first be absorbed across the pulmonary epithelium into the blood. The process by which they are absorbed across the epithelium is gas absorption. Gas absorption is the movement of a drug from a gas phase into a liquid phase. Gas absorption can only occur if the drug has solubility in the liquid phase.

The absorption of the drug from a gaseous phase across an interface into a liquid phase is analogous to partitioning in a liquid–liquid system. The gaseous drug is generally delivered into the lungs as a mixture of gases. The concentration of drug molecules in the gas mixture is the driving force for gas absorption into the liquid phase. The concentration of the drug in the gas mixture is described as its partial pressure (P_i). Partial pressure is described in greater detail in Section 3.2.2. The higher the concentration difference between the drug in the gas phase, the higher the driving force for gas absorption into the liquid phase. As well as the concentration or partial pressure of the drug in the gas mixture, the absorption of a drug from a gas phase into the liquid phase at a constant temperature is related to the chemical nature of the drug and liquid phase. The solubility of a gaseous drug in a liquid phase is expressed as a partition coefficient.

Coefficients to describe gas absorption

The amount of dissolved gas in a liquid phase is described by Henry's law, previously described in Section 3.2.2 (eqn 3.36)

$$P_i = k_i y_i.$$

P_i is the concentration of drug in the gas phase, y_i is the concentration of the drug in the solution phase and k_i is a constant value at equilibrium.

The constant k_i can be referred to as the Ostwald solubility coefficient, K, when measured at constant temperature. The Ostwald solubility coefficient is determined from the ratio of the concentrations of drug molecules in the liquid and gas phases at equilibrium. For convenience, the logarithm of the partition coefficient is often referred to as

$$\text{Log } K_{\text{Liquid}} = \left(\frac{\text{Concentration drug in liquid}}{\text{Concentration drug in gas}} \right). \tag{3.48}$$

The Ostwald solubility coefficient used to quantify the uptake of a gaseous drug from an inhaled gas mixture into the lung tissue is referred to as a lung/gas partition coefficient (K_{lung}). The log K_{blood} values for a range of volatile anaesthetics are given in Table 3.4. These values indicate that the extent of absorption of methoxyflurane into the lung tissue is much greater than that of sevoflurane.

TABLE 3.4 The blood/gas partition coefficients (log K_{lung}) of a range of volatile anaesthetics

Compound	log K_{lung}
Halothane	0.46
Methoxyflurane	1.20
Isoflurane	0.20
Enflurane	0.56
Sevoflurane	−0.11

Values obtained from a larger list from ref. Abraham H. A. *et al.*, Air to lung partition for volatile organic compounds and blood to lung partition coefficients for volatile organic compounds and drugs, (2008). *Eur. J. Med. Chem.* **43**, 478–485

The Ostwald solubility parameter is measured at constant temperature. Gas solubility in a liquid phase is also influenced by the temperature of the system. The Ostwald solubility parameter can be described as an equilibrium constant, as it is determined from the solubility of the drug in the liquid and gaseous phases at equilibrium. Equilibrium constants are temperature dependent. The relationship between an equilibrium constant and temperature is described by the van't Hoff equation, described previously in Section 3.1.3, eqn 3.26. In the case of gas absorption, as the temperature increases the magnitude of the Ostwald solubility parameter will increase.

◉ 3.4 SUMMARY

Medicines are systems of several components and phases existing at equilibrium. A drug that is interacting with its biochemical target to produce a physiological response is also part of a system at equilibrium. Equilibrium processes are controlled by the energetics, or thermodynamics, of the system. Energetics and equilibrium processes control the movement of the drug molecule between phases during its transport to the target receptor.

Energy is the capacity to do work. The first law of thermodynamics states that whenever a quantity of energy of one form is consumed, an exactly equivalent quantity of some other form of energy is produced. The total energy contained in a molecular assembly is called its internal energy, U. The internal energy of a system can be changed by transferring energy to or from the system either as heat, or as work, or both.

The heat energy transferred to or from a system by some process occurring at constant pressure is known as the enthalpy change, ΔH, for that process. Endothermic processes, in which heat is absorbed by the system at constant pressure, have positive ΔH. Exothermic processes, in which heat is released by the system at constant pressure, have negative ΔH.

The thermodynamic quantity that describes the capacity for a system to release or absorb heat energy is the entropy, S, of the system. The entropy is a measure of the disorder of a system. The less ordered a system, the greater its entropy. The second law of thermodynamics states that the entropy of an isolated system increases in a spontaneous change.

Systems undergo change so as to minimize enthalpy and/or maximize entropy. The thermodynamic property that takes into account both the enthalpy and entropy of the system is the free energy, G. For a process occurring at constant temperature and pressure, the change in free energy, ΔG, is given by

$$\Delta G = \Delta H - T\Delta S.$$

For complex pharmaceutical systems, the contribution of each component to the overall free energy of the system is given by its chemical potential, μ. For a pure substance, the chemical potential of the substance is its chemical potential in the standard state. When the substance is part of a mixture, its chemical potential is modified by its activity, a. Activity is related to concentration (m) by $a = \gamma m$, in which γ is known as the activity coefficient.

For a system at equilibrium, the equilibrium constant K_{eq}, is related to the change in free energy, ΔG, as follows

$$\Delta G° = -RT \ln K.$$

Many important pharmaceutical systems have many phases in which components are transferring between phases. A substance will transfer spontaneously from a phase in which it has higher chemical potential to a phase in which it has lower chemical potential.

Information on the conditions necessary for the existence of different phases of a system can be presented in phase diagrams. The phase rule tells us the number of degrees of freedom (F) we have for any particular number of components (C) and phases (P) in any system is.

$$F = C - P + 2.$$

For two-component vapour/liquid systems, Raoult's law and Henry's law both relate the vapour concentration of each component to the concentration of that component in solution. Raoult's law does so in terms of the vapour pressure of the pure component, Henry's law does so in terms of an empirical constant.

Triangular phase diagrams are convenient ways of presenting information on phase equilibria in systems of three components at constant temperature and pressure, for example an alcohol/oil/water system.

During transport of a drug from the point of delivery to the target receptor, the drug molecule has to undergo a number of phase transitions. These transitions are the result of the disruption of the equilibrium of a system and the response of the system to re-establish equilibrium. The spontaneous processes that occur in pharmaceutical systems involve mass-transfer processes. These are processes where material is moved from one location to another. Diffusion, dissolution, partitioning and gas absorption are all mass-transfer processes.

The diffusion process is driven by a difference in free energy within a single phase caused by differences in concentration, resulting in differences in chemical potential. The flux of molecules during diffusion is proportional to the concentration gradient and the diffusion coefficient. Steady-state diffusion is described by Fick's 1st law and non-steady-state diffusion is described by Fick's 2nd law.

Dissolution is the movement of molecules from a solid phase to a liquid phase or solution. This process is driven by the free energy of a molecule in the solid phase being higher than the free energy of the molecule in the solution phase. The factors that influence the rate of dissolution, are described by the Noyes–Whitney equation. The rate of dissolution is constant when sink conditions occur.

The partitioning of a molecule between two immiscible liquid systems is driven by the differences in chemical potential between the systems and the resultant free-energy difference. The relative concentration of drug in each phase when equilibrium is reached is related to the relative solubility of drug in each liquid phase. Partition coefficients (P) are used to quantify the distribution of a drug molecule between two immiscible liquid phases at equilibrium. The partition coefficient is often quoted as the logarithm of the value, log P.

Gas absorption coefficients are analogous to partition coefficients. The extent of solubility of a gaseous drug in the liquid phase is related to the solubility of the gaseous drug in the liquid phase and the concentration of drug in the gaseous phase. The partition coefficient that describes gas absorption is called the Ostwald solubility coefficient.

☰ REFERENCES

Abraham, H. A. *et al.* (2008). *Eur. J. Med. Chem.* **43**, 478–485.

Alberty, R. A., and Silbey, R. J. (1992). *Physical Chemistry.* New York: John Wiley & Sons.

Burger, A. (1982). *Acta Pharm. Tech.*, Vol. 28(1), 1–20.

Chickos, J. S., and Acree, W. E. (2003). *J. Phys. Chem. Ref. Data*, Vol. 32(2), 519–879.

📖 FURTHER READING

Alberty, R. A., and Silbey, R. J. (1992). *Physical Chemistry.* New York: John Wiley & Sons.

Amidon, G., Lee, P., and Topp, E. (2000). Transport Processes in Pharmaceutical Systems (Drugs and the Pharmaceutical Sciences), New York: Marcel Dekker.

Atkins, P. W., and de Paula, J. (1994). *Physical Chemistry* (7th edn). New York: Oxford University Press.

Aulton, M. E. (ed). (2002). *Pharmaceutics, the science of dosage form design* (2nd edn). Edinburgh: Churchill Livingstone.

Florence, A. T., and Attwood, D. (2006). *Physicochemical principles of pharmacy* (4th edn). London: Pharmaceutical Press.

❓ EXERCISES

3.1   For the reaction $2\,NO + O_2 \rightleftharpoons 2\,NO_2$, an equilibrium mixture consists of 1.2 mol dm^{-3} of NO, 1.6 mol dm^{-3} of O_2 and 8.8 mol dm^{-3} of NO_2. Write the expression for the equilibrium constant for the reaction, and calculate the equilibrium constant.

3.2   The enthalpy of fusion of paracetamol at its melting point (168 °C) is 28.1 kJ mol^{-1}. Calculate the change in entropy when one mole of paracetamol is melted.

3.3   In water, glucose exists as an equilibrium between two forms, the so-called α and β anomers. At 25 °C, the equilibrium constant for the interconversion between the α and β anomers is 1.82. Calculate the free energy change for interconversion of the α and β anomers at equilibrium at 25 °C.

3.4   The following chemical species can exist in mixtures for water and copper sulfate: H_2O, $CuSO_4$, $CuSO_4.H_2O$, $CuSO_4.3H_2O$ and $CuSO_4.5H_2O$. How many components does the water/copper sulfate system contain (using the term 'component' in the sense used in the phase rule, Section 3.2.1)?

3.5 At 335 K, the vapour pressure of pure methanol is 8.1×10^4 N m^{-2}, while that of pure ethanol is 4.5×10^4 N m^{-2}. Use Raoult's law to determine the partial pressures of methanol and ethanol over a methanol/ethanol solution of 0.6 mole fraction methanol at 335 K.

3.6 The Henry's law constant for CO_2 in water at 25 °C is 1.86×10^3 atm. Calculate the mole fraction of CO_2 in water at 25 °C under a vapour at a pressure of 1 atm containing a partial pressure of CO_2 of 0.03 atm.

THE PHARMACEUTICAL SOLID PHASE 04

Learning objectives

Having studied this chapter, you should be able to:

- appreciate the pharmaceutical importance of solid-state materials
- distinguish between crystalline and amorphous pharmaceutical solids
- understand crystallographic concepts such as crystal systems, unit cells, space groups and Miller indices as applied to pharmaceuticals
- understand what is meant by different polymorphs, hydrates or solvates of a pharmaceutical solid
- interpret powder X-ray diffraction and thermal analysis data on pharmaceutical solids
- describe the circumstances under which crystal nucleation and growth can occur
- describe particle shape, size, surface properties and mechanical strength, and outline their pharmaceutical impact
- quantify particle surface properties and their impact on solid-state dosage forms.

Previous chapters have tended to emphasize the role of drug substances as components in systems – for example, as components of medicines or as solutes in biological fluids. Chapter 2 specifically emphasized the importance of solubility in pharmaceutical activity. Why, then, devote a chapter to pharmaceuticals as solids? There are two main reasons for this. The first is that approximately 90% of drug substances exist as crystalline solids in their pure forms. The second is the importance of solid-state dosage forms, in particular of tablets. The physical properties of solids are of great importance in the pharmaceutical industry. Pharmaceutical solid-state issues have been at the centre of high-profile legal actions (Leadbeater 1991) and product failures.

By 'solid-state pharmaceutical', we could mean many things, such as a pure crystalline drug substance, a granule composed of drug and excipient particles, or a finished pharmaceutical product such as a tablet. The most fundamental of these states is the pure solid active pharmaceutical ingredient. In a finished pharmaceutical product, the interactions between individual particles of active ingredient and excipients may critically affect how the product functions as a medicine. Therefore, in this chapter, the nature of the pharmaceutical solid state will be examined, emphasizing both pure crystalline or amorphous solids and multiparticle materials.

4.1 CRYSTALLINE AND AMORPHOUS SOLIDS

As discussed in Section 1.2.1, drug substances are generally molecular in nature. The discovery of new drugs involves finding molecular structures that possess particular pharmacological properties. However, actual samples of pure drug substances are, most often, solid materials. Any one particle of these solid materials will be composed of very many drug molecules. These solid materials may fall into one of two possible types: they may be either crystalline or amorphous solids. Crystalline solids tend to have sharp melting points. When viewed under a microscope, the individual crystals may often be seen to have flat faces and straight edges. Some properties of crystals (for example the transmission of light) may be found to be different when measured in different directions of the crystal. (These properties are said to be 'anisotropic' in crystals.) A very important property of crystals, which is discussed further in Section 4.4.1, is that they diffract X-rays to give rise to a recognizable diffraction pattern. By contrast, amorphous solids tend to lack sharp melting points. Properties such as the transmission of light tend not to vary with the orientation of amorphous solids. (These properties are said to be 'isotropic' in amorphous solids.) Amorphous solids do not give rise to distinct X-ray diffraction

TABLE 4.1 Characteristics of crystalline and amorphous solids.

Crystalline solids	Amorphous solids
Melt over narrow, reproducible temperature ranges	Melt over broad, variable temperature ranges
Particles have well-defined faces and edges	Particles do not have well-defined faces and edges
Physical properties may depend on orientation of particles (anisotropy)	Physical properties do not depend on orientation of particles (isotropy)
Diffract X-rays to give well-defined and characteristic diffraction patterns	Diffraction of X-rays does not give well-defined diffraction patterns
Tend to fracture or cleave when subject to sufficient pressure	Tend to yield and flow when subject to sufficient pressure

(a)

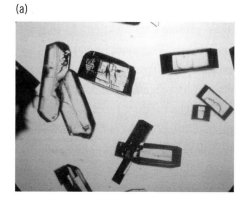

(b)

(c)

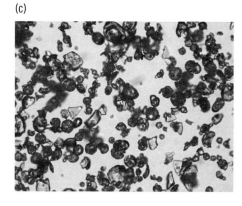

FIGURE 4.1 Crystals of (a) L-glutamic acid and (b) sulfapyridine; note the well-defined faces and edges. (c) Particles of amorphous poly(vinylpyrrolidone); note the curved surfaces and globular appearance of the particles.

patterns. Table 4.1 summarizes the differences between crystalline and amorphous forms. Figure 4.1 shows examples of crystalline and amorphous pharmaceutical materials.

The essential difference between crystalline and amorphous pharmaceutical solids lies in how molecules are ordered within the solid particle. If it was possible to shrink to the size of a molecule and travel through a crystalline solid, one would find molecules arranged in patterns that repeat in a regular manner as one travels through the crystal. If one travelled through the same crystal in a different direction, regular repetition would again be encountered, but the repeating pattern may be different. By contrast, no regular repeating pattern would be encountered upon travelling through an amorphous particle, irrespective of the direction of travel. Crystals possess regular, repeating order in three dimensions, whereas amorphous solids possess no such order. Figure 4.2 illustrates this difference.

The internal structures of crystals will be discussed in more depth in Section 4.2. However, amorphous solids are also pharmaceutically important. It is difficult to describe the structures of specific amorphous solids due to the lack of regular repeating molecular patterns. To some extent, amorphous solids can be regarded as extremely viscous, super-cooled, liquids. Amorphous solids have a general tendency for the molecules within them to gradually rearrange into ordered crystal-like patterns, so as to eventually form crystals. This means that most amorphous solids have an 'in-built' tendency to crystallize, which makes them unstable. For pharmaceutical purposes, it may be possible to slow down the crystallization of amorphous solids by adding excipients, such as polyvinylpyrrolidinone (see Section 1.2.2). This effect can be useful, as amorphous solids may sometimes have

(a)

(b)

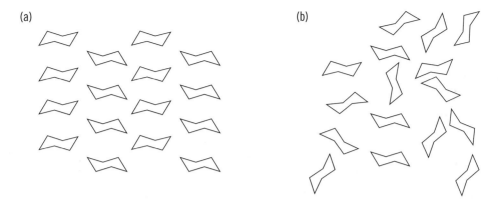

FIGURE 4.2 Two-dimensional structures for (a) a crystalline and (b) an amorphous solid. Note the regular repeating patterns in (a) and the lack of these in (b).

advantages over crystalline solids in dosage formulation. For example, amorphous solids are generally more soluble than their crystalline counterparts.

4.2 THE ESSENTIALS OF PHARMACEUTICAL CRYSTAL STRUCTURE

A critical point about pharmaceutical solids is that any particular solid drug substance may exist in several different crystalline forms. These different crystalline forms may differ in key pharmaceutical properties, such as solubility and bioavailability. For this reason, it is important to understand how crystals are 'built up' from molecules, and why different types of crystal can result. To understand this, it is necessary to grasp some of the essential principles of crystal structure. Crystal structure theory tends to appear somewhat abstract. However, what crystal structure theory makes possible is a concise way of describing the repeating three-dimensional order of crystals. This is very valuable, as it allows us to describe the different crystal structures that can form from the same pharmaceutical molecule.

4.2.1 Unit cells, crystal systems and lattices

The characteristic property of crystals is that the molecules that compose the crystals are arranged in patterns that repeat in a regular manner throughout the crystals. A specific example is shown in Fig. 4.3, which shows a view of the structure of crystals of paracetamol. The figure shows paracetamol molecules arranged in a repeating order. The actual repeat pattern depends on the direction of travel through the crystal. Figure 4.3 also shows a parallelogram superimposed on a part of the image. The parallelogram outlines one face of what is known as the **unit cell** of the crystal.

The unit cell contains the three-dimensional repeating pattern of the crystal. The crystal structure can be thought of as being composed of the unit cell being repeated in three dimensions, rather like a wall being built up by stacking identical bricks.

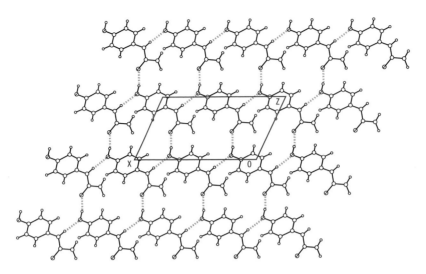

FIGURE 4.3 View of the structure of crystals of paracetamol. The parallelogram marks out one face of the unit cell of the crystal. Reprinted from Boldyrev, V. V., Vasil'chemko, M. A., Shakhtsheider, T. P., Naumov, D. Yu., Boldyreva, E. V., Etch pits morphology on p-hydroxyacetanilide single crystals, *Solid State Ionics*, 1997, **101–103**, 869–874, Copyright (1997), with permission from Elsevier.

Figure 4.3 shows an image of only a two-dimensional 'slice' through a paracetamol crystal. Actual crystals are three-dimensional, and so the unit cells that compose them are not two-dimensional parallelograms, but their three-dimensional equivalents, known as parallelepipeds. A generalized parallelepiped or unit cell is shown in Fig. 4.4. Unit cells can be characterized by three lengths, a, b and c, each measured with respect to the origin O, and the three angles, α, β and γ. As shown in Fig. 4.4, α is the angle between the sides of length b and c, β is the angle between the sides of length a and c, and γ is the angle between the sides of length a and b. Crystal structure theory shows that there are seven possible types of unit cell, defined by the lengths a, b and c, and the angles α, β and γ. (A rigorous definition also takes into account the symmetry found within the unit cell,

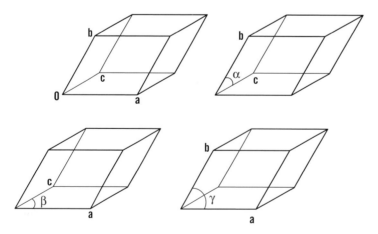

FIGURE 4.4 Generalized unit cell, showing origin **0**, sides of length a, b and c; and angles α, β and γ.

TABLE 4.2 Crystal systems and unit cell dimensions*.

Crystal system	Unit cell dimensions
Triclinic	$a \neq b \neq c; \alpha \neq \beta \neq \gamma$
Monoclinic	$a \neq b \neq c; \alpha = \gamma = 90° \neq \beta$
Orthorhombic	$a \neq b \neq c; \alpha = \beta = \gamma = 90°$
Tetragonal	$a = b \neq c; \alpha = \beta = \gamma = 90°$
Rhombohedral	$a = b = c; \alpha = \beta = \gamma \neq 90°$
Hexagonal	$a = b \neq c; \alpha = \beta = 90°; \gamma = 120°$
Cubic	$a = b = c; \alpha = \beta = \gamma = 90°$

* A rigorous definition of the crystal systems would also consider the symmetry found within the unit cells; however, this topic is beyond the scope of this book. More advanced textbooks on this topic are given in the Further Reading section at the end of the chapter.

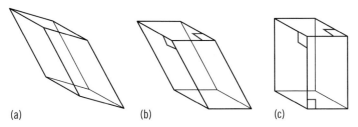

FIGURE 4.5 (a) Triclinic, (b) monoclinic and (c) orthorhombic unit cells. The 'L' symbols indicate 90° angles.

but this point is beyond the scope of this book.) Table 4.2 lists the seven types of unit cell, or crystal system. The great majority of pharmaceutical materials belong to either the monoclinic, orthorhombic or triclinic systems. Triclinic, monoclinic and orthorhombic unit cells are illustrated in Fig. 4.5.

The key point about a unit cell is that it contains the three-dimensional repeat pattern for the crystal. In Fig. 4.3, each unit cell contains four paracetamol molecules. The arrangement of these four molecules is repeated throughout each paracetamol crystal. This is equivalent to the unit cell being repeated in three dimensions like, as described earlier, bricks stacked in a wall. The three-dimensional stacking of unit cells results in a lattice, as illustrated in Fig. 4.6. This three-dimensional stacking is referred to as 'translation' of the unit cell.

In a crystal lattice such as that shown in Fig. 4.6, each unit cell contains one so-called **lattice point**. The lattice points are the arrangements of molecules or ions that are repeated in three dimensions to produce the crystal structure. A unit cell that contains only one lattice point is known as a **primitive** unit cell. In crystal-structure notation, primitive unit cells are given the symbol '*P*'. The unit cell of the paracetamol structure shown in Fig. 4.3 is a primitive unit cell. Each lattice point is the arrangement of four

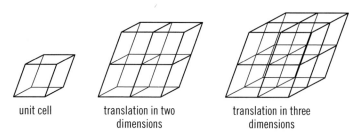

unit cell   translation in two   translation in three
dimensions   dimensions

FIGURE 4.6 A crystal lattice is composed of three-dimensional translations of the unit cell.

paracetamol molecules that is repeated in three dimensions throughout the crystal. It is possible to have unit cells with more than one lattice point per cell. However, these are relatively uncommon in pharmaceutical systems, and will not be discussed here.

4.2.2 Space groups and space-group notation

In the paracetamol structure shown in Fig. 4.3, each unit cell or lattice point contains four paracetamol molecules. The precise arrangement of these four molecules is repeated throughout the crystal lattice. Very often there is internal symmetry within the unit cell. The internal symmetry will relate two or more drug molecules, which comprise the unit cell. The symmetrical arrangement of these molecules also has to be reproduced in three dimensions by translation of the unit cell.

The types of internal symmetry that are possible depend on the crystal system. Crystallographic theory shows that there are 230 possible combinations of crystal systems and internal symmetries. These are known as the 230 space groups. A full discussion of the theory of crystallographic space groups is beyond the scope of this book. (There are many textbooks that cover this topic, see further reading at the end of the chapter.) However, to understand the basics of pharmaceutical crystal structure, it is enough to look at some of the space groups arising from the triclinic, monoclinic and orthorhombic systems (see Fig. 4.5).

Triclinic systems

The simplest triclinic space group has no internal symmetry within the crystal structure. This space group is known as '*P*1'. The only possible internal symmetry that can exist in a triclinic space group is a centre of symmetry. (A centre of symmetry is a point such that any straight line through it meets identical atoms at equal distances on either side of it.) Figure 4.7(a) shows two molecules, in this case enantiomers of the non-steroidal anti-inflammatory ibuprofen, which are related by a centre of symmetry. (In the figure, the centre of symmetry is symbolized by a dot.) Figure 4.7(b) shows a molecule, in this case of the pigment quinacridone, which possesses an internal centre of symmetry, again symbolized by a dot. In both of these examples, if any point on a structure is selected, a straight line travelling from that point to the centre of symmetry and continuing in a straight line the same distance beyond the centre of symmetry will arrive at an identical point. In space-group theory, centres of symmetry are given the symbol '$\bar{1}$' (pronounced 'bar-one'). A triclinic unit cell in which one or more molecules are related by a centre of

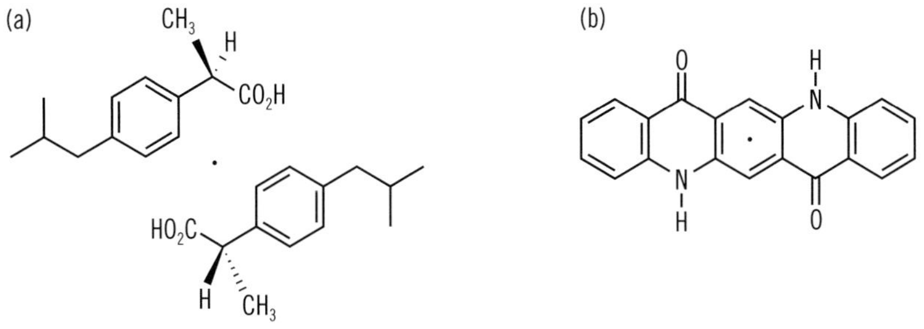

(a)

(b)

FIGURE 4.7 (a) Two enantiomers of ibuprofen related by inversion through a centre of symmetry (shown as '•'). (b) A quinacridone molecule with an internal centre of symmetry.

symmetry belongs to the space group $P\bar{1}$. The space groups $P1$ and $P\bar{1}$ are the only triclinic space groups.

Monoclinic and orthorhombic systems

Monoclinic systems are probably the most common for pharmaceutical compounds. For example, the paracetamol crystal structure shown in Fig. 4.3 is a monoclinic structure. A characteristic feature of monoclinic unit cells is that they contain internal **rotational** symmetry. Look again at the monoclinic unit cell shown in Fig. 4.5(b). In this unit cell, the angles α and γ are right angles, but not the angle β. Imagine an axis running through the centre of the unit cell parallel to the side b. If the cell were to be rotated about this axis by 180°, an identical position would be arrived at (Fig. 4.8). This type of symmetry is known as a two-fold axis of rotation, symbolized by '2'. All monoclinic space groups have a two-fold rotational symmetry element of some kind. In pharmaceutical systems, it is common for two-fold rotations to be accompanied by molecular translations to give symmetry elements known as screw axes. The effect of screw axes is to relate molecules by a combination of rotation and movement. A two-fold screw axis, given the symbol '2_1' (pronounced 'two-one'), is illustrated in Fig. 4.9.

Another type of internal symmetry element that is common in monoclinic unit cells is known as a **glide plane**. Glide planes combine reflection in a mirror plane with translation. The operation of a glide plane is illustrated in Fig. 4.10. If the unit cells shown in Figs 4.9 and 4.10 were combined, the screw axis (parallel to the b-axis) would be seen to be perpendicular to the glide plane (which is parallel to the plane containing the a- and c-axes).

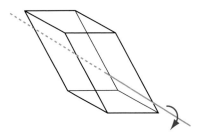

FIGURE 4.8 Rotation of a monoclinic unit cell by 180° about an axis running through the centre of the cell parallel to the b-axis produces an identical arrangement.

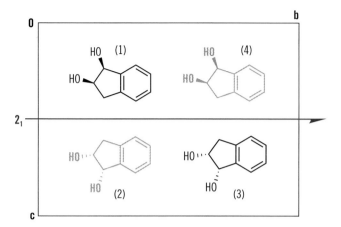

FIGURE 4.9 Operation of a 2_1 screw axis (symbolized by the half-arrow). The view shows a monoclinic unit cell viewed down the *a*-axis. (The *a*-axis is therefore pointing straight out at us and cannot be seen.) 180° rotation of the molecule in position 1 about the screw axis would generate the molecule in position 2. However, translation of the molecule by half the length of the unit cell parallel to the *b*-axis places the molecule in position 3. The 2_1 screw axis therefore relates the molecules in positions 1 and 3. We could just as well first consider the effect of translating the molecule in position 1 by half the unit cell length (which would place the molecule in position 4) and then the effect of rotation about the screw axis, which again places the molecule in position 3.

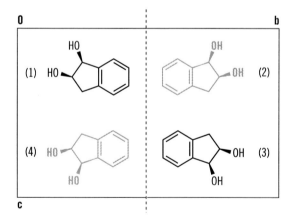

FIGURE 4.10 Operation of a glide plane. The plane, which is symbolized by the dotted line, is perpendicular to the *b*-axis and parallel to the plane of the *a*- and *c*-axes. (The view shown is that along the *a*-axis. The *a*-axis is therefore pointing straight out at us and cannot be seen.) Starting at position 1, reflection across the plane would produce the molecule in position 2. However, translation by half the length of the unit cell in the direction of the *c*-axis, leaves the molecule in position 3. The glide plane therefore relates the molecules in positions 1 and 3. (Alternatively, starting from position 1, translation half the length of the unit cell in the direction of the *c*-axis generates the molecule at position 4, after which, reflection across the plane again results in the molecule at position 3.)

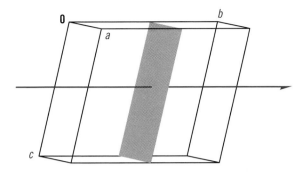

FIGURE 4.11 Combination of a screw axis perpendicular to a glide plane.

The relationship between the screw axis and the glide plane is illustrated in Fig. 4.11. The combination of the screw axis and the glide plane gives a monoclinic space group known as '$P2_1/c$' (read as 'P two-one upon c'). The '/' symbol shows that the screw axis and the glide plane are perpendicular to each other. The 'c' indicates the direction of translation in the glide plane operation (i.e. parallel to the c-axis). Many crystalline pharmaceuticals have the space group $P2_1/c$ – for example crystal forms of the antimicrobial sulfathiazole and the antipsychotics spiperone and promazine hydrochloride.

Another pharmaceutically common monoclinic space group is $P2_1/n$, in which the translation is diagonal in the direction of the ac plane, rather than parallel to the c-axis. (The 'n' in $P2_1/n$ implies that the glide plane translates molecules in a direction that is a diagonal on the plane perpendicular to the screw axis, in this case the ac-plane.) One crystal form of the antiulcer drug ranitidine hydrochloride exists in this space group. There are many other monoclinic space groups other than the two just mentioned. A common feature to all of these is the presence of a two-fold rotational symmetry element.

As shown in Fig. 4.5(c), all the angles in orthorhombic unit cells are right angles. The characteristic feature of orthorhombic space groups is the presence of three mutually perpendicular symmetry elements. For example, the orthorhombic space group $P2_12_12_1$ contains three two-fold screw axes, each one parallel to one of the edges of the unit cell. $P2_12_12_1$ is a common space group for pharmaceutical systems. For example, crystal forms of the opiate antagonists naloxone hydrochloride and naltrexone hydrochloride have the space group $P2_12_12_1$. Figure 4.12 shows a view of the crystal structure of naltrexone hydrochloride. Orthorhombic space groups also often have three mutually perpendicular glide planes. For example, under certain circumstances, it is possible to obtain crystals of paracetamol with the orthorhombic space group *Pbca*. In the space-group symbol '*Pbca*', information is conveyed both by the positions of the letters 'b', 'c' and 'a' in the space-group symbol and by the letters themselves. The first letter, 'b', refers to a glide plane that is perpendicular to the a-axis, and implies that the direction of translation is parallel to the b-axis. The second letter, 'c', refers to a glide plane that is perpendicular to the b-axis, and implies that the direction of translation is parallel to the c-axis. The last letter, 'a', refers to a glide plane that is perpendicular to the c-axis, and implies that the direction of translation is parallel to the a-axis.

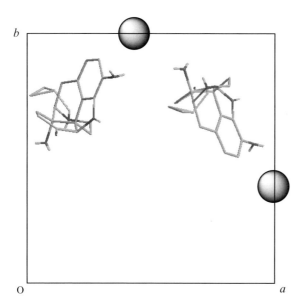

FIGURE 4.12 View along the *c*-axis of the crystal structure of the opiate antagonist naltrexone hydrochloride, which crystallizes in the orthorhombic space group $P2_12_12_1$. The large spheres represent the chloride ions. The molecules that are aligned along the direction of the *b*-axis are related by 2_1 screw axes parallel to the *c*-axis. [from K. Sugimoto, R. E. Dinnebier, M. Zakrzewski, *J. Pharm. Sci.*, Vol 96, No. 12, 2007, 3316–3323. Copyright (2007 John Wiley & Sons). Reprinted with permission of Wiley-Liss, Inc., a subsidiary of John Wiley & Sons, Inc.]

▶ **WORKED EXAMPLE 4.1**

The paracetamol crystal structure shown in part in Fig. 4.3 belongs to a monoclinic space group, $P2_1/a$. Explain what the space-group symbol means, and identify and symmetry operations that can be seen in Fig. 4.3. Also identify symmetry elements that cannot be seen in the views shown in Fig. 4.3.

The characteristic features of monoclinic space groups are unit cells with three different cell lengths, two cell angles at 90°, and one non-90° angle (that is, as shown in Fig. 4.5(b)). Monoclinic systems also have a two-fold rotational symmetry element parallel to the *b*-axis. The space group $P2_1/a$, has a primitive monoclinic unit cell with a 2_1 screw axis parallel to the *b*-axis. A glide plane is perpendicular to the screw axis, with a translation of half the unit cell length parallel to the *a*-axis. In Fig. 4.3, the structure is viewed down the *b*-axis (the sides of the unit cell shown are the *a* and *c* edges, labelled *X* and *Z* in Fig. 4.3). The operation of the screw axis cannot be seen in Fig. 4.3, as the translation direction is straight out of the plane of the diagram in the direction of the *b*-axis. However, $P2_1/a$ also has a glide plane, which is perpendicular to the *b*-axis, with the direction of translation parallel to the *a*-axis.

The operation of this glide plane can be seen in Fig. 4.13, which shows part of the view seen in Fig. 4.3 in more detail. The paracetamol molecule at position 1 appears to be sitting close to the origin 0. Half-way along the *X*-axis, at position 2, is another paracetamol molecule. If this second molecule is compared with the first one, it should

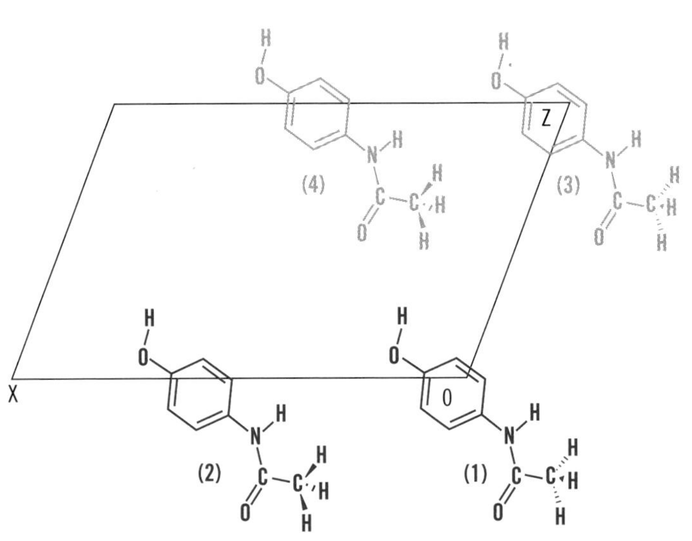

FIGURE 4.13 Operation of glide plane in the unit cell of paracetamol.

be possible to see that in the first molecule one of the CH_3 hydrogens is pointing straight out of the page, while in the second molecule one hydrogen is pointing inwards. The molecule at position 1 has undergone reflection in the XZ-plane, which inverts the hydrogens on the CH_3 group. Translation along the X-axis by half the length of the cell produces the molecule at position 2. The same glide operation relates the paracetamol molecules at positions 3 and 4. Other symmetry elements in addition to the screw axis and the glide plane are also present in the structure, specifically centres of symmetry, but these cannot be seen in the view observed in Fig. 4.3.

4.2.3 Crystal planes and faces

A crystal structure is composed of translations of the unit cell in three dimensions. If we imagine taking a 'slice' of a crystal, what we should have is a two-dimensional plane with a regular pattern of atoms. Because the crystal has a regular repeating structure, it should be possible to take identical 'slices' that are spaced apart at some constant repeating distance. These 'slices' are known as **crystal planes**. The faces of a crystal are also like such 'slices'.

It is necessary to have a system for referring to crystal planes and faces that relates to the content of the unit cells. In this way, it is possible to identify the atoms or parts of molecules that are lying on a particular plane or face. We refer to crystal planes and faces by using sets of three integers known as **Miller indices**. A Miller index labels a set of evenly spaced parallel planes. The planes will all have the same two-dimensional pattern of atoms. An important point about such a set of planes is that they will cut the edges of unit cells at identical points in each unit cell.

Figure 4.14 shows a unit cell with one plane outlined by dashed lines. Relative to the origin O, the plane cuts the a-axis half-way along the unit cell edge, that is at a/2. The plane cuts the b-axis at point b, or at b/1. The plane cuts the c-axis at a point one-third of the distance along the edge from the origin, that is, at c/3. Hence, the intercepts of the

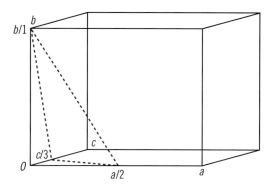

FIGURE 4.14 A plane of Miller index 213.

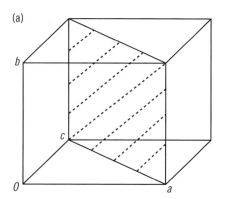
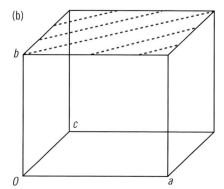

FIGURE 4.15 (a) Shaded area shows a plane of Miller index 101. (b) Shared area shows a plane of Miller index 010.

plane with the three axes a, b and c are, respectively, $a/2$, $b/1$ and $c/3$. The Miller index for this plane is said to be '213' (read as 'two-one-three'). Miller indices are therefore the reciprocals of the fractional intercepts of the plane with the three axes a, b and c. Miller indices are given the general symbol hkl. In general, a plane has Miller index hkl if the plane cuts the a-axis at a/h, the b-axis at b/k and the c-axis at c/l.

If a plane is parallel to one of the axes of the unit cell, it will not cut that axis at any point. The intercept of the plane with that axis is then thought of as being at infinity. The reciprocal of this intercept is therefore zero. Hence, a Miller index that contains a zero is the index of a plane that is parallel to a unit cell axis. A Miller index that contains two zeros is the index of a plane that is parallel to a face of the unit cell. Examples are shown in Fig. 4.15.

4.3 CRYSTAL POLYMORPHISM OF PHARMACEUTICALS

The pharmacological activity of a drug substance is a property of its molecular structure. However, actual samples of the drug substance will, most often, be crystalline solids. A

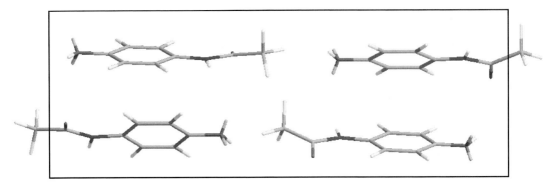

FIGURE 4.16 View of the structure of the orthorhombic crystalline form of paracetamol. Note the 'slip planes' separating the layers of paracetamol molecules [data from Haisa, M.; Kashino, S.; Maeda, H. *Acta Cryst.* **1974**, *B30*, 2510–2512, copyright permission International Union of Crystallography.]

key point is that for any one molecular structure, more than one crystal structure may be possible. Different crystal structures built up from the same molecular structure are known as crystal polymorphs. Crystal polymorphism is important pharmaceutically, as differences in the physical properties of polymorphs can impact on issues such as solubility, bioavailability, formulation, regulatory compliance, and patent protection.

For example, paracetamol crystals usually have the monoclinic structure shown in Fig. 4.3. However, as mentioned above, under some circumstances it is possible to obtain crystals of paracetamol with the orthorhombic space group *Pbca*. A view of this orthorhombic structure is shown in Fig. 4.16. The orthorhombic structure consists of ordered planes of paracetamol molecules stacked upon each other. Figure 4.16 views these planes side-on. The planes are cleanly separated by 'slip planes'. These allow the planes to slide apart easily when the paracetamol tablets are compressed into crystals (Joiris *et al.* 1998). For this reason, it is easier to produce tablets of the orthorhombic form of paracetamol than of the monoclinic form.

The effect of polymorphism on pharmaceutical activity is well demonstrated by the case of chloramphenicol palmitate. Four solid-state forms of this antibiotic are known, three crystal polymorphs and an amorphous form. Of these, the amorphous form and one of the crystal polymorphs are active when taken orally, while the remaining two crystal polymorphs are inactive when taken orally. The inactive forms are very poorly absorbed from the gastrointestinal tract, and hence are unable to exert an effect.

The legal and patent issues raised by pharmaceutical crystal polymorphism are illustrated by the case of ranitidine hydrochloride, the active pharmaceutical ingredient of the best-selling antiulcer drug Zantac. A U.S. patent was granted on the first crystal form of ranitidine hydrochloride, known as Form 1, in 1978. The manufacturers subsequently discovered a second crystal polymorph, known as Form 2. In 1985, a U.S. patent was granted on Form 2. Crystals of Form 2 belong to the monoclinic space group $P2_1/n$. These crystals have favourable processing properties. Crystals of Form 1 contain a partially disordered arrangement of the ranitidine molecules, making it difficult to obtain details of the crystal structure. Also, crystals of Form 1 were not as easy to process as those of

Form 2. For these reasons, Form 2 crystals were used in the marketed product. The original 1978 patent on Form 1 was due to expire in 1995. In the run up to this date, generic pharmaceutical manufacturers sought to market products containing crystals of Form 1. This resulted in a number of high-profile legal actions between the manufacturers of Zantac and various generic manufacturers.

4.3.1 Relative stabilities of polymorphs

One of main reasons why crystal polymorphism causes difficulties in the pharmaceutical context is that the appearance of new polymorphs is often unexpected. For example, two years after introduction onto the market, batches of the antiviral Norvir began to fail some final product tests (Chemburkar *et al.* 2000). The problem was found to be that a new poorly soluble crystal polymorph of the active pharmaceutical ingredient, ritonavir, had formed in these batches.

Unexpected formation of new crystal polymorphs is a relatively common and often costly occurrence in pharmaceutical manufacturing. In some instances, transfer of a crystallization process from one reactor vessel to another can sometimes result in a new crystal polymorph being obtained. In general, it is very difficult to be certain that all the crystal polymorphs of a pharmaceutical have been obtained and characterized. Many pharmaceutical systems are very polymorphic. For example, five crystal polymorphs of the antimicrobial sulfathiazole are known, three crystallizing in the space group $P2_1/c$ and two in the space group $P2_1/n$. Reproducibly obtaining the same crystal polymorph can be a challenge.

In all polymorphic systems, the polymorphs have an order of relative thermodynamic stability. This ordering of stabilities may be closely related to the relative solubilities of the polymorphs, with the more stable polymorphs generally being the less soluble. However, the relative order of stability of crystal polymorphs is also temperature dependent. The relative thermodynamic stability of polymorphs is quantified by their relative free energies, G. As discussed in Section 3.1.3, free energy G is related to enthalpy H, temperature T and entropy S by (eqn 3.14)

$$G = H - TS.$$

The third law of thermodynamics (see Section 3.1.2) states that at absolute zero (0 K) the entropy of a perfectly crystalline substance is zero. The enthalpy, H, and the entropy, S, of a crystalline substance increase with increasing temperature, T. These terms are plotted in Fig. 4.17, which shows that free energy G tends to decrease with increasing temperature T. This means that the energy that can be obtained from a crystalline substance decreases with increasing temperature. A plot of free energy G and enthalpy H for a polymorphic system would look similar to Fig. 4.17, except that the curves would be duplicated for each polymorph. Such a plot is shown in Fig. 4.18. Figure 4.18 shows the temperature dependence of the free energy G and the enthalpy H for two polymorphs. Also shown in Fig. 4.18 are the free energy and enthalpy of the liquid state. The more stable polymorph at any particular temperature is the one with the lower free energy G at that temperature. In Fig. 4.18, polymorph II has the lower free energy at all temperatures. In this system, polymorph II is the more stable and polymorph I the less stable at all temperatures. A system

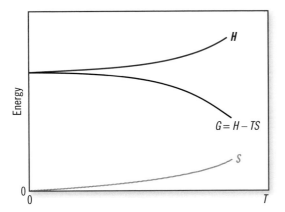

FIGURE 4.17 Variation of free energy G of a crystalline solid with increasing temperature T. Enthalpy H and entropy S increase with increasing temperature T. As the product TS increases more rapidly than H, and as $G = H - TS$, G tends to decrease with increasing temperature T.

of two polymorphs that has the same ordering of relative stabilities at all temperatures is known as a monotropic system. For example, crystal polymorphs I and V of the anti-microbial sulfapyridine are monotropic.

In Fig. 4.18, the melting point of each polymorph occurs at the temperature at which the free energy of each form equals the free energy of the liquid state. For the less stable polymorph I, this occurs at the temperature labelled 'm.p. (I)' in Fig. 4.18, while for the more stable polymorph II, melting occurs at the higher temperature 'm.p. (II)'. The enthalpy change upon melting, ΔH_f (known as the enthalpy of fusion; see Section 3.1.2) corresponds to the enthalpy difference between the polymorph and the liquid state at the melting point temperature. In Fig. 4.18, the enthalpy of fusion of the less stable polymorph, $\Delta H_{f\,(I)}$, is smaller that that of the more stable form, $\Delta H_{f\,(II)}$. This is a characteristic of **monotropic** systems.

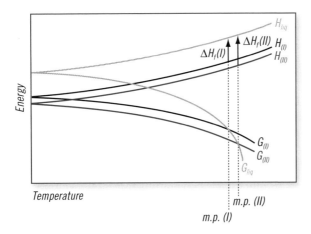

FIGURE 4.18 Energy–temperature diagram for a monotropic polymorphic system. [Reproduced with Permission from Burger, A., Zur Interpretation von Polymorphie-Untersuchungen, *Acta Pharm. Tech.*, 1982, **28**, 1–20, Copyright (1982) Wissenschaftliche Verlagsgesellschaft mbH.]

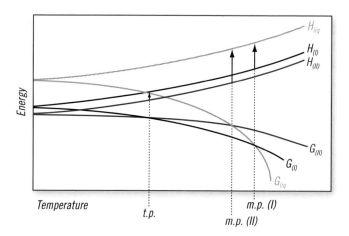

FIGURE 4.19 Energy–temperature diagram for an enantiotropic polymorphic system. [Reproduced with Permission from Burger, A., Zur Interpretation von Polymorphie-Untersuchungen, *Acta Pharm. Tech.*, 1982, **28**, 1–20, Copyright (1982) Wissenschaftliche Verlagsgesellschaft mbH.]

In a polymorphic system, at any particular temperature, one particular polymorph will be the most stable one at that temperature; it is called the thermodynamically preferred polymorph. However, kinetic effects are also very important in stabilizing crystal forms. In other words, the processes by which the crystal forms are produced may be as important as their relative energies. It can be the case that less stable forms are formed more rapidly, that is, they may be preferred kinetically under some circumstances. So for kinetic reasons, it is possible that a polymorph that is not the most thermodynamically preferred one at a specific temperature can be sufficiently stabilized by the process of formation to be isolated and used. Such a polymorph is known as a metastable form.

Polymorphic systems may also display the energy–temperature behaviour shown in Fig. 4.19. In Fig. 4.19, the stabilities of polymorphs I and II 'swap over' at the transition temperature, labelled 't.p.'. A system of two polymorphs in which the relative stabilities of the polymorphs is reversed at some temperature is known as an enantiotropic system. Crystal polymorphs IV and V of the sedative-hypnotic barbital have an enantiotropic relationship to each other. In Fig. 4.19, the melting point of polymorph I is higher than that of polymorph II. However, the enthalpy of fusion of the more stable polymorph at the melting point temperature, $\Delta H_{f\,(I)}$, is smaller that that of the more stable form, $\Delta H_{f\,(II)}$. This is a characteristic of enantiotropic systems.

Monotropic or enantiotropic relationships can only be defined between pairs of polymorphs. For systems of more than two polymorphs, the behaviour of two specific polymorphs has to be examined to see if they display monotropic or enantiotropic behaviour. Whether a polymorphic system is monotropic or enantiotropic also affects dissolution. In general, at any specific temperature, the more stable polymorphs at that temperature will be less soluble at that temperature. However, if the system is enantiotropic, the temperature dependence becomes important: the order of stabilities, and hence the solubilities, will depend on whether the temperature is above or below the transition-point temperature. This is illustrated in Fig. 4.20. In the monotropic system (Fig. 4.20(a)), polymorph II is the more stable at all temperatures, and so has the lower solubility at all temperatures. In the

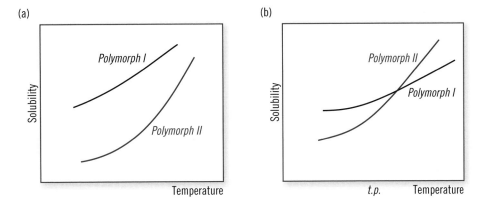

FIGURE 4.20 Variation of solubility with temperature for (a) monotropic and (b) enantiotropic polymorphic systems.

enantiotropic system (Fig. 4.20(b)) polymorph II has the lower solubility at temperatures below the transition temperature, and polymorph I has the lower solubility at temperatures above the transition temperature.

4.3.2 Solvates and hydrates

In crystallization from solution, it is often possible that the crystals that form contain both solute and solvent molecules as part of the crystal lattice. Crystals of these types are known as **solvates**. If the crystallization solvent is water, and water molecules become part of the crystal lattice, the resulting crystals are known as **hydrates**. Because solvates have different chemical composition to crystals that contain only solute molecules, they are not strictly crystal polymorphs. Solvates and hydrates are sometimes referred to as pseudo-polymorphs.

Solvate and hydrate crystal forms of pharmaceuticals are very common. An extreme example is the antimicrobial sulfathiazole. Five 'proper' crystal polymorphs of sulfathiazole are known. However, over 100 solvate forms have been reported (Bingham *et al.* 2001). Hydrates can be used pharmaceutically. For example, a hydrate crystal form of the anti-biotic doxycycline has been used, and a dihydrate crystal form of quinine sulfate has been used in tablet formulations.

4.4 METHODS OF CHARACTERIZING PHARMACEUTICAL SOLIDS

As described in Section 4.3, the crystal form of a pharmaceutical substance can impact on key issues such as solubility, bioavailability and patent protection. Obtaining the correct polymorph or hydrate of a drug substance is therefore important. We need experimental methods that can identify different crystal forms, and, if necessary, quantify them. The two most important methods for examining crystal form are X-ray diffraction (XRD) and thermal methods. Special attention will be given to these. However, other methods, especially spectroscopic and microscopic methods, are also important. These will be discussed briefly.

4.4.1 X-ray diffraction methods

Diffraction occurs when the direction of electromagnetic radiation is changed by interaction with an object. A requirement for diffraction is that the wavelength of the radiation be similar to the dimensions of the object. When diffraction occurs from a set of regularly spaced objects, the diffracted waves combine with each other to form diffraction patterns. Crystal lattices are sets of regularly spaced objects, the objects being the atoms that make up the crystal lattice. Radiation of similar wavelength to the dimensions of the bonds in molecules should form diffraction patterns upon interaction with crystals.

The spacing between the atoms in a pharmaceutical molecule could typically be about 0.1 nm (or 1 Å in the Angstrom units often used in crystallography). To give an idea of the scale of these systems, remember that 1 nm equals 10^{-9} m. The type of electromagnetic radiation corresponding to this order of wavelength is X-radiation. Hence, if we direct X-rays at a crystal sample, we should be able to observe an X-ray diffraction pattern that is characteristic of the sample.

W. L. Bragg proposed a law governing the diffraction of X-ray by crystals, based on the idea that the diffraction can be treated as reflection of the X-rays by planes of the crystal, similar to the reflection of light by a mirror. As was discussed in Section 4.2.3, crystals can be considered to contain stacks of identical, equally spaced planes. Each stack of planes can be characterized by a Miller index, *hkl*, and by the distance between the planes, d_{hkl}. If the wavelength of the X-rays used is λ, Bragg's law is as follows

$$n\lambda = 2d_{hkl} \sin \theta, \qquad (4.1)$$

in which θ is the angle that the incoming X-rays form with the set of reflecting planes *hkl*, and *n* is an integer. Given that diffraction of X-rays by crystals is a complex physical phenomenon, Bragg's law is a relatively simple equation. A short derivation of Bragg's law is given in Box 4.1.

BOX 4.1 Derivation of Bragg's law.

Bragg's law can be derived using the following construction.

Figure 4.21 shows beams of X-rays reflecting from two planes that are two of a set of identical planes of Miller indices *hkl*, each separated from the next by a distance d_{hkl}. The incoming beams of X-rays are parallel and reflect off the upper plane along the path *abc* and off the lower plane along the path *def*. The X-rays are of wavelength λ. Because the X-rays are reflecting off the planes, the incoming and the reflected X-ray beams both make the same angle, θ, with the surface of the planes. For a diffraction pattern to form, it is required that the wave travelling along *bc* be in phase with the wave travelling along *ef*. This means that both waves are at the most intense parts of their wavecycles at the same points, so that their intensities add together rather

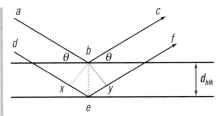

FIGURE 4.21 Construction for Bragg's law.

than cancelling out. For this to happen, the difference in the pathlengths *abc* and *def* must be an integer multiple, *n*, of the X-ray wavelength, λ. In other words

$$n\lambda = \text{pathlength } xey.$$

The distance *be* is equal to d_{hkl}. The angles *abx* and *bxe* are right angles. The angle *xbe* is therefore also equal to θ and so the distance *xe* is equal to $d_{hkl} \sin \theta$. The symmetry of the construction means that the distance *ye* is also equal to $d_{hkl} \sin \theta$. The pathlength *xey* is equal to the distances *xe* + *ey*. Hence,

$$n\lambda = xe + ey = 2d_{hkl} \sin \theta,$$

which is Bragg's law.

When a beam of X-rays strikes a crystal sample, if the wavelength of the X-rays and the orientation of the crystal sample are such that eqn 4.1 is satisfied for some set of planes *hkl*, then diffraction occurs and the diffracted beam can be observed by a detector. If the sample consists of a single well-formed crystal, the diffraction pattern that is then observed consists of a series of spots, such as that shown in Fig. 4.22.

There are two characteristics of these spots that are important in helping us to understand the structure of a crystal: their **location** and their **intensity**. The *location* of a spot in

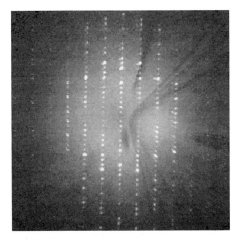

FIGURE 4.22 Example of an X-ray diffraction pattern from a single crystal.

the pattern can be related to the Miller index of the reflecting plane. The *intensity* of the spot can be related to the type and location of the atoms in the unit cell. This information can be used to determine the crystal structure, and to generate images of the crystal structures such as those shown in Fig. 4.3.

The information that is provided by a single-crystal X-ray diffraction experiment includes the space group, unit cell dimensions, the number and precise positions of all of the atoms of the molecules in the unit cell, and the shape, or conformation, of the molecules. A full explanation of how this is done is beyond the scope of this book. More detailed treatments of single-crystal X-ray diffraction are given under Further Reading at the end of the chapter.

Single-crystal X-ray diffraction provides the fullest details of crystal structures in the most accessible way. However, it is often not possible to obtain suitable single crystals of pharmaceutical materials. More importantly, a single crystal may or may not be representative of the bulk sample. Hence, we also need to record X-ray diffraction patterns from polycrystalline, or powder, samples.

In powder X-ray diffraction, the sample is not rotated around in the X-ray beam in the same manner as for single-crystal diffraction. However, it is assumed that there are very many crystals in the powder sample, and that these crystals are lying in a randomly distributed range of orientations. If this is the case, for every set of planes *hkl* in the crystal structure, there should be a sufficient number of crystals lying oriented with respect to the X-ray beam so that Bragg's law is satisfied for that set of *hkl* planes, and reflections are observed.

Each of the reflections diffracts the X-ray beam by an angle of 2θ. This is because the incoming and reflected beams both make angles of θ with the reflecting planes. For each *hkl* plane, diffractions of the X-ray beam by an angle 2θ will occur from crystals with those *hkl* planes in the correct orientation. These diffracted beams will travel in different directions due to the different orientations of the diffracting crystals. What all of the diffracted beams have in common for each set of planes *hkl*, is that the angle formed with the incoming X-ray beam is 2θ. Because there are very many crystals in the sample, the complete set of diffracted beams for each set of planes *hkl* forms a cone of diffracted X-rays at an angle 2θ with the incoming X-ray beam. This is illustrated in Fig. 4.23 for one *hkl* plane.

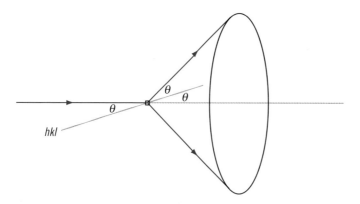

FIGURE 4.23 Formation of a cone of diffracted X-rays from a polycrystalline powder sample. (Solid State Chemistry and its Applications, A. R. West, 1984, ©John Wiley & Sons Ltd., reproduced with permission.)

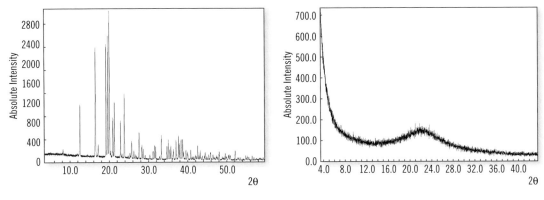

FIGURE 4.24 Powder X-ray diffraction patterns of (left) a crystalline pharmaceutical solid (lactose monohydrate) and (right) an amorphous solid (amorphous silica).

The full crystal structure would have many such planes, each giving rise to similar cones of diffracted X-rays at some angle 2θ. A detector can measure the intensity of each cone at particular 2θ values, so that a plot of the 2θ value for each diffraction as a function of its intensity can be produced. Such a plot is a powder X-ray diffraction (PXRD) pattern. Examples of PXRD patterns are shown in Fig. 4.24.

Figure 4.24 shows (on the left) the powder X-ray diffraction pattern of crystalline lactose hydrate. Each peak in the pattern corresponds to diffraction of the X-ray beam by some set of planes of Miller index *hkl*. At the most basic level, the pattern can be regarded as a 'fingerprint' of the crystal form. That is, any pure sample of lactose hydrate should give the same pattern, although there may be variations in the relative intensities of the peaks.

Each of the distinct peaks in the diffraction pattern shown on the left-hand side of Fig. 4.24 corresponds to a solution of the Bragg equation for lactose hydrate. The fact that these distinct diffraction peaks are observed means that there are distinct sets of *hkl* planes in the diffracting material, which means it is a crystalline solid. Indeed, the ability to give rise to a PXRD pattern with distinct recognizable peaks is a *characteristic* of crystalline solids.

Amorphous solids also diffract X-rays, but because the atoms and molecules in an amorphous solid are not in a regular repeating three-dimensional array, a distinct diffraction pattern does not form. Instead, amorphous solids diffract X-rays over a continuous envelope of 2θ values, giving rise to a broad 'hump' in the PXRD pattern of an amorphous solid. The right-hand side of Fig. 4.24 shows the PXRD pattern of a sample of an amorphous material. Broad continuous diffraction over a range of 2θ values can be seen, with an absence of distinct peaks such as those on the left-hand side of Fig. 4.24. A PXRD profile of this sort is characteristic of amorphous solids.

As described above, crystal polymorphs are 'built-up' from the same molecular structure, but have different crystal structures. Hence, crystal polymorphs will generally have different unit cell dimensions and atom locations. They therefore have different sets of reflecting *hkl* planes, which give rise to different PXRD patterns. PXRD patterns are therefore one of the main ways of distinguishing between crystal polymorphs of the

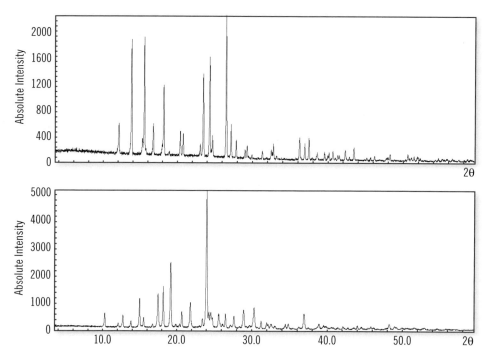

FIGURE 4.25 PXRD patterns of (top) the monoclinic and (bottom) the orthorhombic forms of paracetamol.

same molecular substance. For example, Fig. 4.25 shows the PXRD patterns of the monoclinic and orthorhombic polymorphs of paracetamol, that is, for the paracetamol crystal structures shown in Figs. 4.3 and 4.16, respectively. The two PXRD patterns are distinctly different, allowing clear identification of which polymorph is present in a particular sample.

> ▷ **WORKED EXAMPLE 4.2**
>
> Figure 4.26 shows the PXRD patterns obtained from samples of an acetonitrile solvate of quinapril hydrochloride. (Quinapril is an antihypertensive drug.) To study the physical stability of this particular solid form of quinapril hydrochloride, the samples have been ground in a mill for the time intervals shown. What do the PXRD patterns tell us about the behaviour of the substance under these conditions?
>
> The PXRD pattern obtained before any grinding (that is, at 0 seconds grinding time), shows distinct peaks corresponding to diffractions from specific *hkl* planes. The acetonitrile solvate of quinapril hydrochloride is therefore a crystalline substance that gives rise to a distinct PXRD pattern. However, as grinding of the sample proceeds, the intensities of the diffraction peaks decreases relative to the baseline noise. Eventually, after 10 minutes grinding, no distinct diffraction peaks can be seen, and the PXRD is simply a broad envelope of diffractions between approximately 12° and 40° 2θ. This is very characteristic of an amorphous material. The acetonitrile solvate of quinapril hydrochloride is therefore not stable under the grinding conditions, and transforms from a crystalline to an amorphous solid.

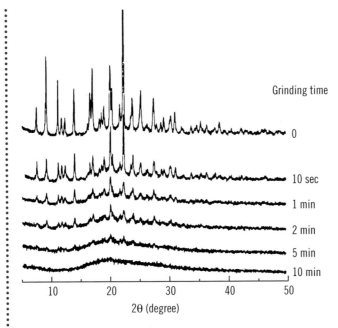

FIGURE 4.26 PXRD patterns of samples of an acetonitrile solvate of quinapril hydrochloride after grinding for various times. [from Y. Guo, S. R. Byrn, G. Zografi, *J. Pharm. Sci.*, **89**(1), 2000, 128–143; copyright 2000 John Wiley & Sons; reprinted with permission of Wiley-Liss, Inc., a subsidiary of John Wiley & Sons, Inc.]

4.4.2 Thermal methods

The heat absorbed or evolved by a process occurring under constant pressure, q_p, is equivalent to the change in enthalpy, ΔH, accompanying that process (see Section 3.1.1). Thermal methods of analysis allow for the measurement of the enthalpy change for processes by measuring the heat flow to or from the sample under investigation relative to a reference sample. For example, thermal methods can be used to measure the enthalpy of melting of crystals (also known as the enthalpy of fusion), something that varies from crystal to crystal depending on their stability. This basic principle underlies a variety of thermal analysis methods. However, for pharmaceutical analysis, the most widely used thermal method is differential scanning calorimetry (DSC).

Differential scanning calorimetry (DSC)

In DSC, the flow of heat to or from both a reference and the sample is measured as a function of temperature. During the DSC experiment both the sample and the reference are separately heated at the same rate. The DSC instrument measures the difference in the flow of heat to or from the sample as the temperature changes ($\Delta H/dT$) relative to that of the reference. If at any particular temperature, the sample undergoes a chemical or physical change, the DSC instrument detects this as a change in $\Delta H/dT$. In practice, many DSC instruments detect the difference in electrical power, ΔP, needed to maintain the sample and reference at equal temperatures. A DSC scan is therefore a plot of differential power,

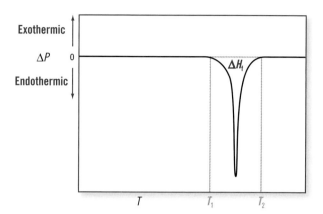

FIGURE 4.27 DSC scan for a stable crystalline solid. The melting point range is from temperature T_1 to T_2. The area bounded by the endothermic peak and the baseline is equivalent to the enthalpy of fusion, ΔH_f, for the solid.

ΔP, versus temperature T. Figure 4.27 shows an example of a DSC scan that could be obtained from a stable crystalline solid.

The scan in Fig. 4.27 is generated by heating a sample of the solid from some starting temperature at a set rate of heating, say 4 °C min^{-1}. As the heating temperature increases steadily with time, the x-axis of the scan can be given in either time or temperature. If the solid is a pure thermodynamically stable crystal form, no chemical or physical change occurs until the solid begins to melt. As there is no difference in the power supplied to the sample and the reference while no physical or chemical change is occurring, ΔP remains constant at zero. Melting of the sample commences at temperature T_1 in Fig. 4.27, and is completed by temperature T_2. Melting, or fusion, is an endothermic process – that is, heat is transferred into the system (or sample) from the surroundings. By contrast, ΔH for an exothermic process is negative.

While the melting process is occurring, the sample is absorbing heat so that additional power input to the sample is required relative to the reference. At the commencement of the melting process (at T_1), the differential power, ΔP, begins to increase, rises to a maximum, after which it drops back to the zero baseline level at T_2. The common convention in DSC is that an endothermic process is shown as a downward deviation from the baseline, and an exothermic process as an upward deviation (but some instruments use the opposite convention). The area bounded by the DSC peak and the baseline is equivalent to the change in enthalpy, ΔH, for the process. As the process corresponding to the endothermic peak in Fig. 4.27 is due to melting of the sample, the area bounded by the peak and baseline is equivalent to the enthalpy of fusion, ΔH_f, for the stable crystalline solid.

Figure 4.28 shows the DSC scan that might be obtained from a sample of a metastable crystal form. As the sample is heated, at some temperature it undergoes an exothermic transition to the thermodynamically stable form. This is indicated by the exotherm labelled 'A' in Fig. 4.28. When the melting point of the stable form is reached, endotherm B, corresponding to melting, is observed. DSC scans for metastable crystal forms may often be more complex than that shown in Fig. 4.28. For instance, the entire sample

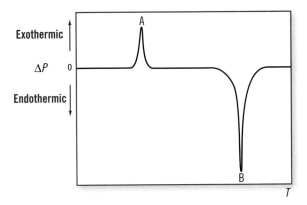

FIGURE 4.28 DSC scan of a sample of a metastable crystal form. Exotherm A corresponds to transition of the metastable form to the thermodynamically stable form. Endotherm B corresponds to melting of the thermodynamically stable form.

may not undergo transition to the more stable form, so that a peak corresponding to the melting of the metastable form may also be observed.

Figure 4.29 shows a DSC scan that might be expected from an amorphous solid. As the sample is heated, at some temperature it undergoes a transition to a more liquid-like phase, in which the molecules in the sample have increased freedom of movement. This sort of phase change is known as a **glass transition**. The glass transition is neither exothermic nor endothermic, but is marked by a change in the heat capacity of the material, which is observed in the DSC scan as a lowering of the baseline. This is the event occurring at point A in Fig. 4.29. As the sample is further heated, the increased freedom of the molecular components allows an ordered crystalline arrangement to be obtained, so that crystallization occurs. This is observed as the exothermic event B in Fig. 4.29. Finally, the now crystalline material melts, which is observed as the endothermic event C in Fig. 4.29.

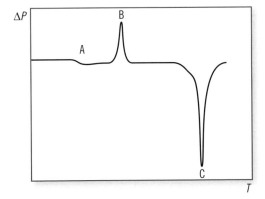

FIGURE 4.29 DSC scan of an amorphous material. The shift in baseline at A corresponds to the glass transition. Crystallization occurs at temperature B. Melting of the resulting crystals occurs at temperature C.

The DSC scans shown in Figs 4.27, 4.28 and 4.29 are somewhat idealized. DSC scans of real pharmaceutical samples may not always fit tidily into one of these patterns, or may combine features of all three. An amorphous phase may first crystallize into a metastable crystal form that undergoes partial transformation to a more stable form. Consequently, several 'thermal events' may be observed in the DSC scan, including glass transition, crystallization to a metastable form, transformation to a stable crystal form, melting of the metastable crystal form and melting of the stable crystal form. Various factors may affect whether or not all of these events are observed, and the temperatures at which they are observed. For example, variation in heating rate can produce DSC scans of very different appearance from the same sample.

Thermogravimetric analysis (TGA)

Another form of thermal analysis that is widely used pharmaceutically is thermogravimetric analysis (TGA). TGA measures changes in the mass of samples as a function of temperature. Changes in mass occur if there are volatile components in the crystal structure that can be evaporated upon heating. This makes TGA especially useful for investigating hydrates and solvates.

▶ **WORKED EXAMPLE 4.3**

Figure 4.30 shows the TGA curve and DSC scan for a sample of the pharmaceutical excipient lactose monohydrate, $C_{12}H_{22}O_{11}.H_2O$. In the TGA curve, weight loss is observed at 140–150 °C. The weight loss corresponds to 5% of the total weight of

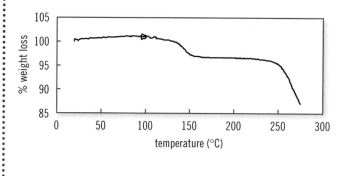

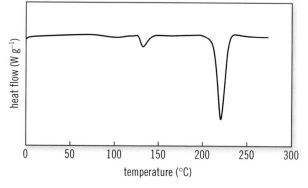

FIGURE 4.30 (Top) TGA curve and (bottom) DSC scan for a sample of lactose monohydrate.

the sample. The molecular weight of lactose monohydrate is 360 g mol⁻¹, while the molecular weight of water is 18 g mol⁻¹. The water content of lactose monohydrate makes up 5% of the mass of the compound. The weight loss observed at 140–150 °C therefore corresponds to loss of the water content. That this loss is observed confirms that the crystal form is a hydrate. The extensive weight loss occurring from 240 °C onwards corresponds to decomposition of the sample. In the DSC scan of the same sample, the loss of water at 140–150 °C is observed as an endotherm. The other event observed is a large endotherm at 215 °C. This corresponds to melting of the anhydrous crystal form of lactose.

4.4.3 Other methods of analysis

X-ray diffraction and thermal analysis are the most informative and general methods of characterizing pharmaceutical solids. However, there are many other methods. Two that can be very useful for particular systems are microscopy and vibrational spectroscopy.

Microscopic methods rely on differences in the external appearance of crystals. The external appearance of a crystal is known as its morphology or habit. Crystal morphology is discussed in more depth in Section 4.5.2. Broadly, two factors affect the morphology of crystals. The first is the internal structure of the crystal. The second is the process of crystal growth. As the crystal structure has an impact on the external appearance of crystals, examination of the shape of crystals may allow assignment of crystal form. This is really only possible in very well studied cases, in which both the crystal structure and the external appearance of the crystals are well known for all the relevant crystal forms.

Paracetamol provides such an example. Figure 4.31 shows examples of crystals of both the monoclinic and the orthorhombic polymorphs of paracetamol. Broadly speaking, the crystals of the monoclinic form (on the left in Fig. 4.31) appear 'lop-sided', while crystals

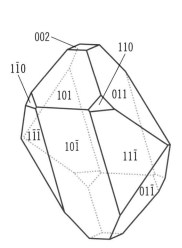
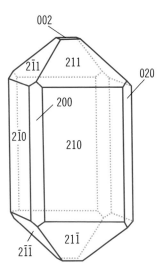

FIGURE 4.31 Morphologies of crystals of the (left) monoclinic and (right) orthorhombic polymorphs of paracetamol. [G. Nichols and C. S. Frampton, *J. Pharm. Sci.*, **87**(6), 684–693, 1998, copyright 1998 John Wiley & Sons; reprinted with permission of Wiley-Liss, Inc., a subsidiary of John Wiley & Sons, Inc.]

of the orthorhombic form (on the right in Fig. 4.31) appear 'upright'. This is a reflection of the geometries of the monoclinic and orthorhombic unit cells (see Fig. 4.5). An optical microscope is often adequate to observe crystal morphology. However, for finer crystals, electron microscopy may be required.

In the structures of molecular crystals, the molecules may often be interacting by weak intermolecular bonds, such as hydrogen bonds or dipole–dipole interactions. In different crystal polymorphs, the networks of intermolecular interactions may vary slightly

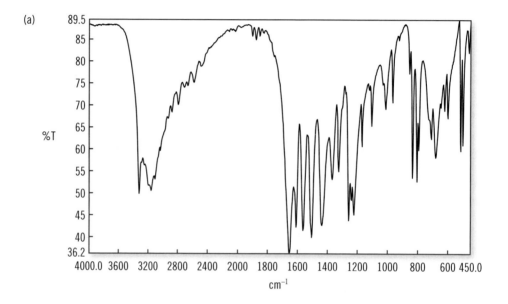

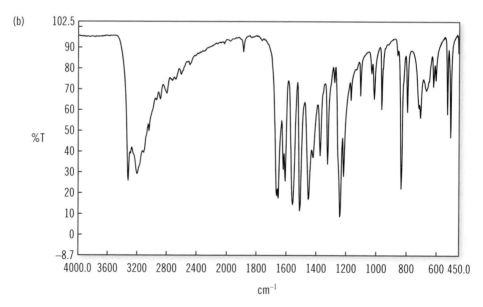

FIGURE 4.32 4000–450 cm^{-1} region of the infra-red spectra of (a) the monoclinic and (b) the orthorhombic crystal forms of paracetamol.

between polymorphs. This difference can affect the vibrational properties of the molecule so that, in some cases, small differences in the vibrational spectra of the molecules may be observed for different polymorphs. The types of vibrational spectroscopy that can be used to detect different crystal polymorphs include infra-red, near infra-red and Raman spectroscopy. Paracetamol again provides a good example. Figure 4.32 shows the region between 4000 and 450 cm^{-1} in the infra-red spectra of the monoclinic and the orthorhombic polymorphs of paracetamol. Within the region of these spectra between 1260 and 1225 cm^{-1}, the two polymorphs give rise to clearly different patterns of absorptions.

4.5 PHARMACEUTICAL CRYSTALLIZATION

Approximately 90% of active pharmaceutical ingredients are crystalline solids in their pure states. The final step in the processes of manufacturing these materials is a crystallization step. The starting point for a crystallization process will typically be a solution of the active compound in a solvent. As discussed in Section 2.1, for any combination of solute and solvent, there is a temperature-dependent equilibrium (or saturation) solubility of that solute in that solvent. If the concentration of solute is below the equilibrium solubility, the solution is stable and no crystals can form. If the concentration of the solute is at the equilibrium concentration, crystals can potentially form. However, in practice, for crystals to precipitate from solution, the concentration of solute usually needs to exceed the equilibrium concentration. When this happens, the solution is said to be 'supersaturated' with respect to the solute. Supersaturation provides the driving force for crystallization.

4.5.1 Supersaturation

Supersaturation is a necessary condition for crystallization to occur. When a solution is supersaturated with a solute, precipitation of the solute in the form of crystals becomes possible. This is shown graphically in Fig. 4.33.

Figure 4.33 describes the behaviour of a solution of a solute in a solvent as a function of temperature and solute concentration. The figure is partitioned into three regions,

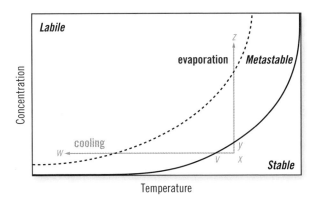

FIGURE 4.33 The stable, metastable and labile zones for a crystallization process.

described as the stable, metastable and labile zones. The curve separating the stable and metastable zones is the equilibrium solubility curve for that solute in that solvent. When the temperature and concentration are such that the system lies in the stable zone, a stable solution exists and crystallization cannot occur. The point x in Fig. 4.33 lies in the stable zone; a solution at the temperature and solute concentration occurring at point x will be stable and will not precipitate crystals.

Let us imagine that the solution is allowed to stand at a constant temperature, so that the solvent begins to evaporate. The effect of this is to increase the concentration of the solute. In terms of the description shown by Fig. 4.33, this means that the system travels from point x along the line labelled 'evaporation' in the direction of increasing solute concentration. The solution remains stable until the equilibrium concentration curve is reached (point y). Beyond point y, the solution is supersaturated with solute, and crystallization becomes possible. Further evaporation of solvent will eventually result in a concentration, say at point z, which is so high that precipitation of solute crystals is unavoidable. The solution is then said to be labile – that is, crystallization of solute occurs spontaneously.

A similar process occurs if the temperature of the solution is lowered while the concentration remains constant. Starting again at point x in Fig. 4.33, cooling of the solution at constant concentration means that the system travels along the line labelled 'cooling' in Fig. 4.33. At point v, the solution is at the equilibrium concentration of solute at that temperature. With further reduction of temperature, the system will eventually arrive at a point, say at w, at which the solution is labile and spontaneous crystallization occurs.

When the temperature and concentration are such that the system is in the stable zone, no crystallization is possible. When the temperature and concentration are such that the solution is in the labile zone, crystallization is unavoidable. However, the system does not change directly from being stable to being labile. Once the equilibrium concentration is reached, further cooling or evaporation will firstly produce a supersaturated solution. The system may remain in a supersaturated state until the temperature and concentration are such that the labile zone is entered and spontaneous crystallization occurs.

The region of the system in which the solution is supersaturated with solvent, but spontaneous crystallization does not yet occur, is known as the **metastable zone**. The metastable zone is bounded on one side by the equilibrium solubility curve, and on the other by the curve separating it from the labile zone. While the equilibrium solubility curve can be well defined by solubility studies (see Section 2.1.1), the boundary between the metastable and labile zones is less well defined and must be determined by careful studies on a case-by-case basis. Such studies allow determination of the 'metastable zone width' (MSZW) for particular systems. For manufacturing-scale crystallizations in the pharmaceutical industry, knowledge of the MSZW is essential. When the system is in the metastable zone, controlled crystallization can be carried out, for example, by seeding. Once the system enters the labile zone, controlled crystallization is no longer possible, as uncontrolled spontaneous precipitation occurs.

As Fig. 4.33 suggests, the most straightforward way of generating supersaturation is by either cooling or concentrating a stable solution. In practice, a combination of cooling and evaporation often occurs. Supersaturation can also be generated by reducing the solvating power of the solvent system. This is achieved by adding a second solvent, known as an **antisolvent**, in which the solute is poorly soluble. Crystallizations by antisolvent addition are often used in pharmaceutical manufacturing. Another method of generating supersaturation is to form the solute by reaction in a solvent medium in which the solute is poorly soluble. This is known as reactive crystallization. Crystallization by salt formation is an example of reactive crystallization.

4.5.2 Nucleation, growth and crystal morphology

The first formation of crystal particles from a supersaturated solution is known as crystal nucleation. After nucleation, further molecules of solute can add to the growing crystals (that is, crystal growth occurs). Once the process of crystal growth is believed to be complete, the crystals are collected and dried.

The crystal growth process has an effect on the external appearance of the crystals. The shape of a crystal is known as its morphology or habit. Examples of variations in crystal habit as a consequence of different growth patterns are illustrated in Fig. 4.34.

The three crystals illustrated in Fig. 4.34 have essentially the same morphology: they all have a pair of six-sided faces opposite each other, between which are six four-sided faces. The differences in their appearances are a consequence of differing growth rates in various directions. The crystal shown in Fig. 4.34(a) has grown reasonably evenly in all directions. Such a crystal can be described as being 'prismatic'. This is the preferred crystal habit, as prismatic crystals generally have good flow and processing properties. The crystal shown in Fig. 4.34(b) has grown to a much greater extent in the direction perpendicular to the six-sided faces. The result is a column-like or needle-like crystal. Such a crystal is said to have an acicular or needle-like habit. Needle-like crystals often

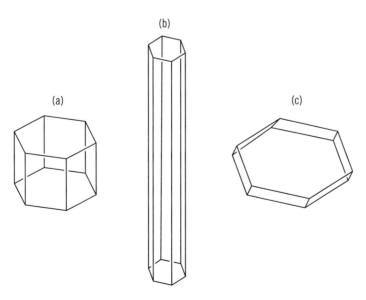

FIGURE 4.34 (a) Prismatic, (b) acicular (or needle-like) and (c) tabular (or plate-like) habits of the same crystal morphology.

form if the crystallization occurs close to the labile zone. Under the conditions within or close to the labile zone, crystals are forced to form rapidly. To allow this, solute molecules tend to add in much greater quantity to the fastest-growing crystal face, giving rise to needle-like crystals.

The crystal shown in Fig. 4.34(c) has grown relatively little in the direction perpendicular to the six-sided faces, but has grown extensively in the directions of the four-sided faces, giving a plate-like crystal. A crystal with such a habit is referred to as being tabular or plate-like. In the tabular crystal shown in Fig. 4.34(c), the six-sided faces appear very large in comparison to the four-sided faces. The six-sided faces are said to be 'well developed'. In general, the occurrence of well-developed faces is a consequence of slow crystal growth in the direction perpendicular to those faces. Faces at which rapid crystal growth occurs will generally appear as smaller, less well-developed faces, such at the four-sided faces in Fig. 4.34(c).

Acicular and tabular habits are less desirable than prismatic habits, as the resulting crystals have inferior processing properties to prismatic crystals. For example, plate-like or needle-like crystals may be more difficult to filter and dry. Unfortunately for pharmaceutical materials, acicular and tabular habits are very common. Careful generation of supersaturation and control of the crystal growth process may be able to avoid the formation of needle-like or plate-like crystals.

4.5.3 Ripening and the rule of stages

Crystal nucleation refers to the first appearance of solid particles in a crystallization process. The process of crystal growth refers to the subsequent development of these particles before they are collected and dried. While crystal growth is occurring, the crystals remain suspended in the crystallizing medium, which is usually a solvent. Often, the initially formed crystals are small particles and some of these can redissolve in the solvent. The molecules from these redissolved crystals are therefore returned to solution. Later in the crystal growth process, these molecules can be added to growing crystals that have remained in suspension. This process by which some portion of the initially formed smaller crystals is redissolved, and the redissolved molecules are subsequently added to the larger crystals remaining in suspension is known as **ripening** (or Ostwald ripening).

Ripening can occur while the initially formed crystals remain in contact with the crystallizing medium. It is usually observed as a decrease in the number of crystals accompanied by an increase in the size of the crystals. Control over ripening processes is very important in pharmaceutical manufacturing, as crystal size significantly affects the handling and formulation of the pharmaceutical product.

Ostwald ripening affects crystal population and size. A closely related phenomenon affects crystal form – that is, the particular crystal polymorph or solvate that is obtained. As discussed in Section 4.3.1, different crystal polymorphs of the same molecular substance have different relative stabilities, with one polymorph being the most stable at any particular temperature. In general, less stable or metastable forms will have higher solubilities than more stable forms. Hence, if the crystal form initially formed is a metastable one, while it remains in contact with the solvent it may redissolve. The redissolved molecule may subsequently reprecipitate in a more stable crystal form. This type of ripening is therefore observed as a tendency for less stable crystal forms to be converted into more stable forms. This effect is summarized in Ostwald's Rule of Stages, which can be stated as follows. *A crystallizing system proceeds from an unstable supersaturated solution to a stable suspension of crystals in solvent. The system does not necessarily proceed directly to the most stable equilibrium state, but tends to proceed through each stage that represents the smallest change in the free energy of the system.*

For a polymorphic system, this means that the least stable polymorph should appear first, followed by the next most stable, and continuing in this manner until the most stable form at that temperature is obtained. For example, careful study of the crystallization of the antimicrobial drug sulfathiazole from water has shown that the least stable polymorph (Form I) appears first, followed by the next least stable (Form II), followed by the second most stable form (Form III), finally giving the most stable form (Form IV) (Blagden 2001).

4.6 SOLID-STATE PROPERTIES OF POWDER PARTICLES

In the earlier sections of this chapter we focused on providing an understanding of the arrangement of drug molecules into crystalline structures, and highlighting the important properties relating to these crystalline structures. In this section the focus will move towards understanding the particulate properties of pharmaceutical solids. Firstly, consider the structure of a pharmaceutical powder: a pharmaceutical powder is composed of a number of solid particles and these particles in turn are composed of a number of molecules organized in a crystalline or amorphous structure. Solid-state properties can be divided into molecular and particulate properties. Examples of particulate and molecular solid-state properties are given in Table 4.3.

The molecular properties of the constituent molecules influence particulate properties, for example, a particle's moisture-adsorption behaviour will differ depending upon whether the constituent molecules' structure is crystalline or amorphous, as amorphous materials generally adsorb more moisture than the corresponding crystalline material.

Particulate solid-state properties can influence both the clinical and processing behaviour of drug substances. They can alter a drug's dissolution rate. This is of importance for the oral bioavailability of biopharmaceutical class II drugs (drugs that exhibit low aqueous solubility and high permeability of the intestinal membrane), which can be enhanced by increasing the drug's dissolution rate. The dissolution rate of a solid material can be increased by increasing the particle surface area exposed to the dissolution medium.

The delivery of drug particles from a dry powder inhaler to the lungs is influenced by the size and density of the drug particles. Table 7.6 gives the approximate dimensions of particles that can reach different regions of the lungs. Drug particle size and density also influences the rate of sedimentation of drug particles dispersed in pharmaceutical suspensions (explained in greater detail in Section 5.4.2). Particulate properties can also influence the degree of mixing and segregation within powder blends during production. Poor mixing and segregation within these blends can result in uneven dose uniformity in the dosage forms produced.

TABLE 4.3 Some examples of molecular and particulate solid-state properties.

Property grouping	Example of properties
Molecular properties	Crystal structure
	Polymorphism/solvates
	Crystallinity
	Dynamic and equilibrium solubility
	Melting point
Particulate properties	Particle size
	Particle shape
	Surface area
	Surface energy
	Hygroscopicity
	Wettability
	Dissolution
	Mechanical strength

For these reasons an understanding of particulate properties is required. The following sections provide a basic introduction to these properties. For more detailed information please refer to the relevant texts listed in Further Reading.

4.6.1 Particle shape

Particle shape is also referred to as particle morphology. Particles are three-dimensional structures and come in a wide range of shapes and sizes. In order to describe these shapes descriptive terms such as flakes, needles and spheres can be used. If we want to describe the shape in greater detail or compare the sphericity or flakiness of different particles, the following factors can be calculated from the dimensions of the particles.

A parameter called the **aspect ratio** can be used to compare the sphericity of particles. The aspect ratio is defined as the ratio of a particle's major axis to its minor axis, as shown in Fig. 4.35. An aspect ratio of 1 would describe a perfect sphere. The greater the deviation of the calculated aspect ratio from 1 is, the less spherical the particle.

Describing the shape of irregular-shaped particles can be very challenging. Parameters have been used to quantify the elongation of a particle by dividing the length of the particle by its breath. Particle flakiness has been quantified by dividing the particle's breadth by its thickness. Figure 4.36 shows how these calculations can be performed for irregular-shaped particles.

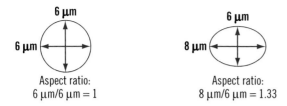

FIGURE 4.35 An example of the calculation of a particle's aspect ratio.

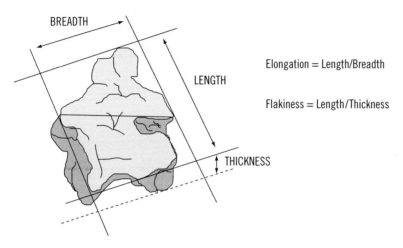

FIGURE 4.36 Schematic diagram showing particle dimensions that can be used to calculate flakiness and elongation factors. Elongation = length/breadth, flakiness = length/thickness.

4.6.2 Particle size

Particle size can be described by descriptive terms such as coarse, medium and fine particles. Table 4.4 gives details of the particle size ranges that these descriptors refer to according to the *British Pharmacopoeia*. While these descriptors can be useful, often more detail regarding particle size is required. Particle size is generally described by one length dimension, a 'diameter'. As mentioned earlier, particles are three-dimensional structures and therefore it is a challenge to describe them by a single distance measurement. A perfect sphere and cube are the only particle shapes that can be described by one dimension and, unfortunately, very few pharmaceutical powders are perfect spheres or cubes.

A range of different diameters can be used to describe particle size. These diameters are not directly measured but calculated from another measured particle property. When indirect measurements of a particle diameter are made they are referred to as an 'equivalent diameter'. These equivalent diameters are based on a property that is influenced by particle size, for example the ability of a particle to pass through a mesh or diffract light, and then calculated based on the behaviour of a hypothetical sphere. As a result, the size of particles can vary depending upon the method of measurement, see Fig. 4.37.

TABLE 4.4 Particle-size classification descriptions *(British Pharmacopoeia, 2008).*

Particle-size classification	Description of powders
Coarse powder	> 95% by weight passes through a 1400 µm sieve
	< 40% by weight passes through a 355 µm sieve
Moderately coarse powder	> 95% by weight passes through a 710 µm sieve
	< 40% by weight passes through a 250 µm sieve
Moderately fine powder	> 95% by weight passes through a 355 µm sieve
	< 40% by weight passes through a 180 µm sieve
Fine powder	> 95% by weight passes through a 180 µm sieve
	< 40% by weight passes through a 125 µm sieve
Very fine powder	> 95% by weight passes through a 125 µm sieve
	< 40% by weight passes through a 90 µm sieve
Microfine powder	> 90% by number of the particles are less than 10 µm in size

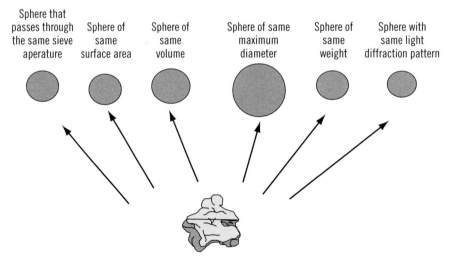

FIGURE 4.37 Examples of different types of 'equivalent diameters' that can be used to measure particle size.

4.6.3 Particle-size measurement

The particle-size distribution of a powder sample may be determined by a range of different methods, we will explain briefly two of the common methods used – sieve analysis and laser light scattering.

Sieve analysis

The sieve diameter, d_s, is the particle dimension that passes through a square aperture (opening) of a defined width. The technique is performed using a series of six to eight sieves with apertures of different sizes stacked on top of each other – the sieve with the smallest aperture at the bottom and increasing in aperture size by an order of $\sqrt{2}$ or $2 \times \sqrt{2}$. Powder is placed on the top of the sieve stack and the sieve stack is vibrated for a period of time. Powder will be retained on sieves it is too large to pass through and the powder will be separated into different size fractions, which are collected and weighted. The weight of each fraction is expressed as a percentage of the initial sample weight.

Laser light scattering

Measuring particle size using laser light scattering is based on the interaction of a laser beam with powder particles. Laser beams are used because they are a monochromatic (of a single wavelength) light source and therefore it is easier to interpret their scattering behaviour. The laser light interacts with particles as it shines through the particle sample suspended in air or liquid.

The interaction of the laser beam with the particles causes light to be diffracted, reflected and transmitted after internal refraction within the particle, as shown in Fig. 4.38. A diffraction pattern is obtained and the diffraction pattern gives the light scattering intensity and the diffraction angle α, as shown as Fig. 4.39. The diffraction pattern is composed of concentric rings and the distance between the different rings provides information regarding the particle-size distribution in the sample. Smaller particles scatter light at a wider angle than larger particles.

Various theories relating the scattering of light to particle size can be used to determine the particle-size distribution from the diffraction data obtained – Mie theory is widely used. (An explanation of the Mie theory is beyond the scope of this text.) The equivalent particle size diameters determined using this technique are expressed as area diameters d_a or volume diameters, d_v.

Particle-size measurement using laser light scattering can be described as static light scattering and dynamic light scattering. When using a static light scattering technique,

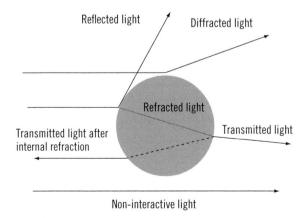

FIGURE 4.38 Illustration of the interactions of light with a particle in its path.

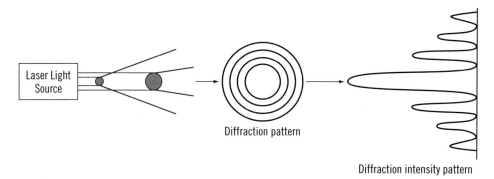

FIGURE 4.39 The diffraction pattern created when laser light interacts with particles in its path, the light is scattered at a wider angle for smaller particles.

particle-size information is calculated from the intensity of the scattering pattern at various angles. When using dynamic light scattering techniques, particle-size information is determined by correlating variations in light intensity due to the Brownian movement of the particles. Brownian motion is the random movement of small particles or macromolecules due to collisions with the molecules of the suspending liquid (this is explained in greater detail in Section 5.4.1). Only small particles that exhibit Brownian motion can be measured by dynamic light scattering.

4.6.4 Particle surface properties

In order to understand the behaviour of pharmaceutical solids, an understanding of particle surface properties is required. Critical pharmaceutical processes occur at the solid surface interfaces. For example, dissolution, crystallization and wetting occur at solid/liquid interfaces and drying and moisture adsorption occur at solid/gaseous interfaces.

Surface area

The surface area of particles can be described as the amount of surface area per volume of particles and can be referred to as the specific surface with units such as mm^2/mm^3. Surface area can also be described per unit weight and can be given units such as mm^2/g. Surface area is inversely related to particle size. As the drug particle size decreases the particle surface area increases. The surface area is also related to the shape of particles – a sphere will have the smallest possible surface area for its volume, irregular-shaped particles will have greater surface areas per unit volume compared to spheres. The surface area of porous particles is highly dependent on the porosity of the particle.

Porosity

Porosity can be expressed as the ratio of the volume of voids (space) in a particle to the total volume of the particle. It is expressed mathematically by

$$E = \frac{V_P}{V_T},$$

(4.2)

TABLE 4.5 The dimensions of the pore entrance of the pore-size classifications.

Pore-size classification	Entrance diameter
Micropores	< 2 nm
Mesopores	between 2 and 50 nm
Macropores	> 50 nm

where E is the porosity of the sample, V_p the volume of the pores in the particle and V_T the total volume of the particle occupied by pores and material.

Pores can be on the exterior or in the interior of a particle. Particulate pores can be divided into three size classifications according to the diameter of the entrance of the pore as detailed in Table 4.5. Porosity and surface area are related properties due to the fact that the amount of surface area available at a solid interface is dependent upon the location of pores on the exterior of the particle, the size of the pores and the dimensions of the entrance to the pores. For example, if the pore entrance is too small to allow the ingress of water molecules then the surface area within the pore of the particle will be unavailable for dissolution until the entrance increases to a sufficient size.

Surface energy

Solid surface energy is analogous to surface tension in liquids (surface tension is explained in greater detail in Section 5.6.1. The surface energy of a solid material is comprised of a polar surface energy component (due to polar and hydrogen bonding groups) and a dispersive surface energy component (due to van der Waals interactions). The constituent molecules at the surface of a solid particle exist in a different environment from that of the molecules in the interior of the particle. The surface energy of the material is influenced by its environmental surroundings. Interactions occur between molecules on the particle surface and the molecules in the particle's environment, for example a solid and water during dissolution.

Surface energy influences a particle's behaviour at interfaces with other solids, liquids and gases. Surface energies influence processes such as moisture adsorption and desorption, wetting, dissolution, gas absorption, cohesion and adhesion. Techniques used to determine surface energy include contact-angle measurement and inverse gas chromatography.

Gas adsorption

Gas adsorption is the phenomenon that occurs upon exposure of a solid surface to a gas. Molecules in the gas phase collide with the solid surface and gas molecules adsorb onto and desorb from the surface of the solid, as shown in Fig. 4.40(a). At equilibrium the rate of molecules adsorbing and desorbing is constant and this results in a constant solid coverage with gas molecules Fig. 4.40(b).

The potential of a solid surface to adsorb gas is related to surface structure and composition. The gas molecules adsorb due to the attractive forces that act between the exposed surface of the solid and the gas molecules. Gas adsorption onto solid surfaces

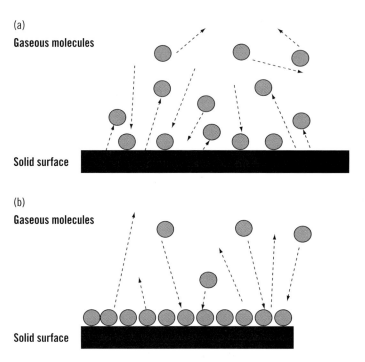

(a)
Gaseous molecules

Solid surface

(b)
Gaseous molecules

Solid surface

FIGURE 4.40 The adsorption and desorption of gaseous molecules on to a solid surface, (a) shows incomplete coverage and (b) shows a complete monolayer coverage.

can occur in two ways: 1) physical adsorption (adsorption without chemical bonds) and 2) chemical adsorption (adsorption involving chemical bonding).

The adsorbed gaseous molecules on the surface can exhibit a different behaviour depending on whether the adsorption is physical or chemical, Table 4.6.

The quantity of gas adsorbed is dependent on the temperature and pressure at which the adsorption occurs. Gas adsorption occurs over a wide range of temperatures and pressures but can normally only be measurable at very low temperature, such as the boiling temperature of liquid nitrogen at atmospheric pressure, −196 °C.

As the concentration of the adsorbing gas in the gaseous phase increases, the pressure of the gaseous phase increases. Relative pressure of the adsorbing gas (P) is a measure of the pressure of the adsorbing gas in the atmosphere compared to the saturated vapour

TABLE 4.6 Comparison of the behaviour of physically and chemically adsorbed gas.

Property	Physical adsorption	Chemical adsorption
Specificity of adsorption site	Low	High
Adsorbate structure	Mono and/or multilayer	Monolayer
Adsorption–desorption	Reversible	Irreversible
Adsorption rate	High	Depends on the activation process
Temperature dependence	Decreasing with increasing temperature	Reaction dependent

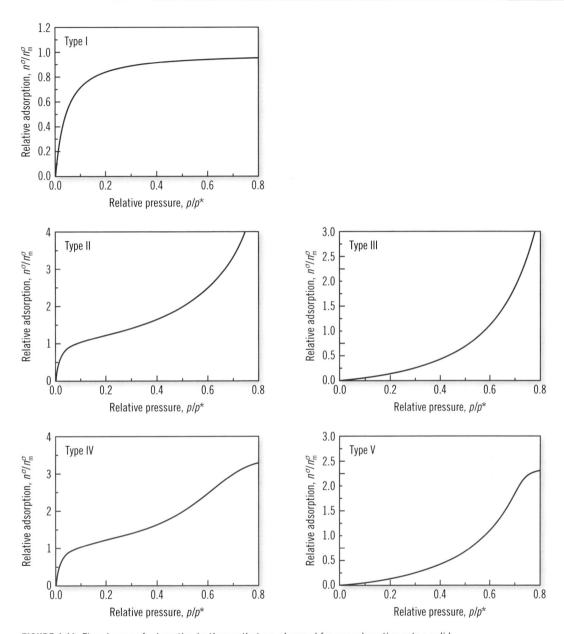

FIGURE 4.41 Five classes of adsorption isotherms that are observed for gas adsorption onto a solid surface.

pressure of the gas (P_o). The quantity of gas adsorbed per unit surface area increases as its relative pressure, in the atmosphere surrounding the solid, increases. An adsorption isotherm is a plot of the equilibrium amount of gas adsorbed as the relative pressure of the adsorbing gas increases at constant temperature. Five classes of isotherms that describe the various types of gas adsorption behaviour have been proposed, and these are shown in Fig. 4.41. Most materials exhibit type I, II and IV behaviour and these are the isotherms we will explain in a little more detail.

The type I isotherm is also referred to as the Langmuir isotherm, it is observed when the gas adsorption is limited to a monolayer, as seen when a gas is chemically adsorbed. The amount of gaseous molecules adsorbed increases as the relative pressure of the adsorbing gas increases and eventually reaches a constant value as the relative pressure of the gas approaches 1. The constant amount of gas is the amount of gas required to form monolayer coverage of the solid surface.

Type II and IV are seen where there is high affinity between the adsorbing gas and the solid surface. The type II isotherm is also referred to as the BET isotherm (BET are the initials of the scientists Brunauer, Emmet and Teller who first described this isotherm in 1938). This isotherm is observed where gas adsorption is not limited to a monolayer and multiple layers of gaseous molecules are adsorbed. It is assumed that the solid surface has a non-porous, uniform surface. Multilayer adsorption will only occur when the adsorbed gas molecules in the initial layers act as a substrate for further gas adsorption. The amount of gas adsorbed increases towards infinity as the relative pressure of the adsorbing gas approaches 1. This isotherm is used routinely to determine the surface area of solids, as described below.

Before we explain isotherm IV, we must first explain the concept of capillary condensation. A vapour can condense into a liquid in the pores of a solid even when its relative pressure is not saturated (less than the saturated vapour pressure). Adsorption will initially occur only as a thin layer along the walls of the pores but as the relative pressure of the adsorbing gas increases, at a certain pressure capillary condensation takes place. It will start initially in the pores with the narrowest diameter and as the pressure increases it occurs in wider pores until at the saturation pressure the entire system is full of condensate. Capillary condensation occurs mainly in mesoporous material. Gas molecules adsorbed and condensed in pores are not completely desorbed (the reverse of adsorption) as the relative pressure is reduced. This results in a deviation between the adsorption and the desorption isotherms, as shown in Fig. 4.42.

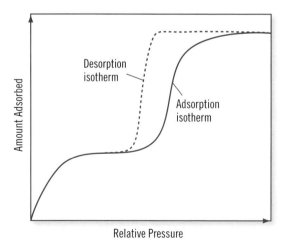

FIGURE 4.42 Adsorption and desorption curves of a mesoporous solid.

Now, if we look at the Type IV isotherm we see a second inflection in the amount of gas adsorbed due to the condensation of adsorbed gas in the mesopores of the adsorbing material.

Surface-area and particle-porosity determination using gas-adsorption techniques

By determining the amount of gas adsorbed onto the surface of powder particles at increasing relative pressures of the adsorbing gas and constant temperature, the total surface area of the sample can be determined from the BET isotherm as explained in Box 4.2. When performing the measurement the particle surface area and pores must be cleaned of contaminants by vacuum pumping and/or purging with an inert gas at elevated temperatures. Care must be taken not to alter the surface structure of the particles by melting or dehydration. A gas mixture of adsorbant gas and helium as an inert carrier gas is then passed over the sample. Nitrogen is typically used as the adsorbant gas; however, other gases can be used such as krypton.

Complete monolayer coverage of the adsorbant gas is best achieved at low temperatures and pressures. Various relative pressures of adsorbant gas are achieved by varying the concentration of adsorbant gas in the gaseous mixture. The equilibrium amount of gas adsorbed at different relative pressures can be measured gravimetrically or by a volumetric technique. When using a volumetric method the equilibrium pressure is measured directly

---

BOX 4.2 Application of BET theory to determine surface area.

The behaviour observed in the BET isotherm can be expressed mathematically by the BET equation

$$\frac{P}{V \cdot (P_o - P)} = \left(\frac{b-1}{V_m \cdot b} \right) \cdot \left(\frac{P}{P_o} \right) + \frac{1}{V_m \cdot b}.$$

P_o is the saturated vapour pressure of the adsorbant gas, V_m the volume of the adsorbant gas required to form a monolayer of gas molecules, P the relative vapour pressure of the adsorbant gas, V the volume of gas and b is a constant.

The BET equation is a linear relationship and if $\left(\dfrac{P}{P_o} \right)$ is plotted against $\dfrac{P}{V \cdot (P_o - P)}$ a linear plot will be obtained as shown in Fig. 4.43. From this linear plot the slope $\left(\dfrac{b-1}{V_m \cdot b} \right)$ and intercept $\dfrac{1}{V_m \cdot b}$ can be determined, and hence the value of V_m (the volume of the adsorbant gas required to form a monolayer) can be determined.

From V_m, the total surface area of the material (S_T) can be determined using the following equation, if the area occupied by one molecule of the adsorbing gas A, and the molar volume of the adsorbing gas M, are known. N refers to Avogadro's number in this equation.

$$S_T = \frac{A \cdot N}{M} \cdot V_m.$$

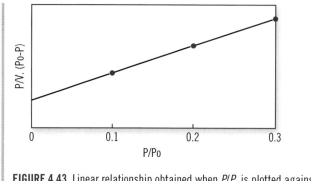

FIGURE 4.43 Linear relationship obtained when P/P_0 is plotted against $P/V \cdot (P_0 - P)$.

and the quantity adsorbed determined from pressure, volume and temperature using the gas laws.

By decreasing the relative pressure of the adsorbing gas, the volume of gas adsorbed will decrease and a desorption isotherm can be plotted, as shown in Fig. 4.42. The deviation in the adsorption and desorption isotherms would indicate a greater amount of gas is adsorbed on the solid surface during desorption compared to adsorption. This deviation is explained by capillary condensation in pores present in the sample. The pressure at which condensation initially occurs is dependent on pore diameter.

Pore-size distribution in the sample can be determined by interpreting this behaviour and the technique can measure the volume of small pores not possible by other techniques. The lower limitation of pore size measured using this technique is dependent on the molecular dimensions of the adsorbing molecule. If the molecule is too large to enter the pore then it will be unable to measure that pore.

In summary, gas adsorption is a standard technique to determine surface area. It can report a single value for surface area up to a full report on isotherm, surface area and total pore volume and pore-size distribution.

4.6.5 Moisture adsorption

Moisture adsorption is an example of the phenomena of gas adsorption onto the surface of solid particles, the gas that is adsorbed is water molecules in the gaseous phase surrounding the particles. Hygroscopic is a qualitative adjective used to describe a solid that adsorbs a relatively large amount of water vapour from the atmosphere. The rate and extent of moisture uptake by particles from the atmosphere is important to understand, as the presence of adsorbed moisture on particle surfaces can influence the drug's chemical and physical stability. Chemical degradation can occur if the drug is prone to hydrolysis. A film of adsorbed water molecules can impact upon the flow and cohesive properties of powder particles. If sufficient moisture adsorption occurs certain substances start to dissolve, this behaviour is called deliquescence.

The extent of moisture adsorption by particles exposed to moisture in the atmosphere is related to the surface area and surface energy of the particles. The greater the surface area exposed and the greater the polar component of the surface energy then the greater

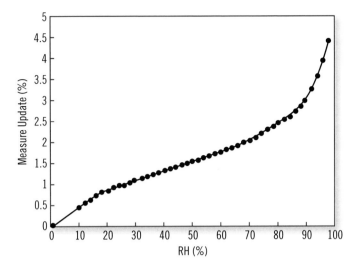

FIGURE 4.44 Example of a moisture-adsorption isotherm.

the extent of moisture adsorption. The concentration of water vapour in the atmosphere (humidity) also influences the degree of moisture adsorption. Moisture adsorption increases as humidity increases.

As temperature increases the amount of water vapour the air can hold increases. Relative humidity is a measure of the concentration of water vapour in the air divided by the maximum concentration of water vapour the air could hold at that temperature. For example a relative humidity of 80% indicates that there is 80/100ths as much water in the air as that air is capable of holding at that particular temperature.

A solid material will have equilibrium moisture content depending upon the relative humidity of the air. For example, if the relative humidity increases the equilibrium moisture content will increase and if the relative humidity of the air decreases the equilibrium moisture content will decrease. Moisture-adsorption isotherms are a plot of the water content of a sample against the relative humidity of its environment at constant temperature. They are used to determine the equilibrium moisture content of powders as the relative humidity of their surroundings are increased or decreased. An example of a moisture-adsorption isotherm is given in Fig. 4.44.

4.6.6 Particle mechanical strength

Drug particles can vary in mechanical strength. The mechanical strength of a particle refers to how the particle will deform (change shape) under applied stress. An elastic material deforms under stress but recovers when stress is removed. A rubber band or coiled spring exhibits elastic behaviour. Plastic material deforms under stress and the deformation is irreversible when the stress is removed. Playdough, putty and plasticine exhibit plastic behaviour. Brittle materials experience little or no plastic deformation but fracture after stress is applied. All materials fracture eventually but the level of stress applied to fracture the material differs for different materials.

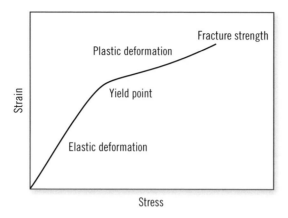

FIGURE 4.45 A typical stress–strain curve for a solid material.

Figure 4.45 is a schematic diagram of a typical stress–strain curve for a solid material. An elastic material will be able to experience a large amount of stress before its yield point is reached and then it will exhibit plastic behaviour. Consider the rubber band, if the stress applied is low it will return to its original shape when the stress is released. If the stress is very high it will not return to its original shape upon removal of the stress. This is due to the fact that it has passed its yield point and undergone plastic deformation. If the stress continues to be increased eventually the rubber band will snap (fracture). The yield point of a plastic material will be exceeded after the application of a low stress. The material will continue to deform as stress is increased and eventually fracture. A brittle material will reach the fracture point with the application of a low amount of stress.

Stress can be applied to particles during processing by compression and shearing. Particle mechanical strength is of particular importance when a milling or compression step is involved in the manufacture of a solid dosage form such as a dry powder inhaler or tablet.

We hope in this chapter we have introduced and highlighted some of the important properties of the pharmaceutical solid phase. A summary of the main points are highlighted below.

⊡ 4.7 SUMMARY

The solid state is extremely important pharmaceutically. Most drug substances are crystalline solids in their pure forms, and many finished pharmaceutical products are mixtures of solid particles. The effectiveness of these products as medicines may depend on the physical properties of these particles.

Pure pharmaceutical substances may be either **crystalline** or **amorphous** solids. Crystalline solids possess regular, repeating order in three dimensions, whereas amorphous solids possess no such order. In a crystalline solid, the three-dimensional repeating pattern of the crystal is defined by the **unit cell**. Unit cells are characterized by three lengths, a, b and c, and by three angles, α, β and γ. These dimensions can be used to assign crystals to one of seven **crystal systems**. For pharmaceuticals, the most common crystal systems are the **monoclinic** and the **orthorhombic** systems.

In a **primitive** crystal structure, each unit cell contains one lattice point. The crystal structure is generated by **translation** of the unit cell in three dimensions, to produce a lattice. Unit cells may also have internal symmetry. There are 230 possible combinations of the seven crystal systems and the possible internal symmetries. These are known as the 230 **space groups**. Many pharmaceutically important space groups contain **centres of symmetry**. Other common types of symmetry include rotation about axes and reflection in mirror planes. Combinations of these elements with translation gives **screw axes** and **glide planes**. These are common features of pharmaceutical crystal structures. Crystals can be thought of as being composed of sets of identical evenly spaced **planes**. Each plane can be characterized by **Miller indices (*hkl*)**.

For any molecular structure, more than one crystal structure may be possible. This is known as **crystal polymorphism**. Different crystal polymorphs of a pharmaceutical can have different solubility and bioavailability. Crystal polymorphs have an order of relative thermodynamic stability that is temperature dependent. Polymorphs also have temperature-dependent ordering of solubilities. Polymorphic systems in which this rank ordering remains the same over all temperatures are known as **monotropic** systems. Those that undergo a change in rank orderings are known as **enantiotropic systems**. Crystals may also form that contain water or solvent molecules in the crystal structure. These are known as **hydrates** or **solvates**.

The two most general methods of characterizing crystalline solids are by **X-ray diffraction** and by thermal analysis. X-ray diffraction can be either single crystal or powder diffraction. **Single-crystal X-ray diffraction** allows elucidation of full crystal structures. **Powder X-ray diffraction** provides characteristic patterns for specific crystal forms.

The most widely used method of thermal analysis is **differential scanning calorimetry**. This detects heat flows associated with processes such as crystallization, solid-to-solid transformation and melting. Another method of thermal analysis is **thermogravimetric analysis**. This techniques quantifies loss of mass when a sample is heated. Microscopic and spectroscopic methods are also used to characterize pharmaceutical solids.

Pharmaceutical crystals are produced by **crystallization**. Crystallization from solution requires generation of **supersaturation**, for example by cooling or concentrating a non-saturated solution. For controlled crystallization, a **metastable solution** should be generated. Metastable solutions are supersaturated, but are not so supersaturated that spontaneous uncontrolled crystallization occurs. Such a crystallization is said to be operating within the **metastable zone**.

The first formation of solid particles from a crystallization is known as **nucleation**. After nucleation, **crystal growth** occurs to give the mature crystals. Crystals that grow relatively evenly in all directions are said to have a **prismatic habit**. Those that grow most rapidly in one direction have **needle-like (acicular) habit**, while those that grow slowly in one particular direction have **plate-like (tabular) habit**.

The population distribution of crystals may often change from a large number of small crystals to a smaller number of larger crystals. Such a process is known as Ostwald **ripening**. Crystallization may also proceed so that less stable polymorphs are first formed, which subsequently transform into more stable polymorphs. Crystallizations that behave in this way are said to obey **Ostwald's Rule of Stages**.

In addition to understanding the crystalline structure of active pharmaceutical ingredients it is important also to have an understanding of the **particulate properties. Particle shape, size, surface area, surface energies, gas adsorption** and **moisture adsorption** can influence the clinical efficiency and the behaviour of powders during production of solid dosage forms. The majority of particle-size measurement methods measure particle size indirectly by measuring a property that is influenced by particle size. An '**equivalent diameter**' is determined calculated based on the behaviour of a hypothetical sphere. Two commonly used methods are **sieve analysis** and **laser light scattering**.

The properties of the particle surface influence all the pharmaceutical processes that occur at the solid interface such as dissolution, crystallization, wetting, drying and moisture adsorption. The surface area of particles is highly dependent on the **porosity** of the particle. The surface energy of a solid material is comprised of a polar surface energy component and a dispersive surface-energy component. The surface energy of a solid material influences how it interacts with its surroundings.

Gas adsorption occurs upon exposure of a solid surface to a gas. It can be due to **physical adsorption** or **chemical adsorption**. Gas-adsorption techniques can be used to measure particle surface area and pore-size distribution. Moisture adsorption is an example of solid gas adsorption.

Particle mechanical strength is of particular importance when a milling or compression step is involved in manufacture of a solid dosage. Materials can exhibit **plastic, elastic** or **brittle** behaviour.

⊜ REFERENCES

Bingham, A. L., Hughes, D. S., Hursthouse, M. B., Lancaster, R. W., Tavener, S., Threlfall, T. L. (2001) Over one hundred solvates of sulfathiazole. *Chem. Commun.*, 603–604.

Blagden, N. (2001) Crystal engineering of polymorph appearance: the case of sulfathiazole. *Powder Technology*, **121**, 46–52.

British Pharmacopoeia (2008), The Stationery Office, London.

Chemburkar S. R., Bauer J., Deming K., Spiwek H., Patel K., Morris J., Henry R., Spanton S., Dziki W., Porter W., Quick J., Bauer P., Donaubauer J., Narayanan B. A., Soldani M., Riley D., McFarland K. (2000) Dealing with the Impact of Ritonavir Polymorphs on the Late Stages of Bulk Drug Process Development. *Organic Process Research and Development*, **4**, 413–417.

Joiris, E., Di Martino, P., Berneron, C., Guyot-Hermann, A.-M., Guyot, J.-C. (1998) The compression behaviour of orthorhombic paracetamol. *Pharm. Res.*, **15**, 1122–1130.

Leadbeater, C. (1991) Glaxo fights for zantac patent, *Financial Times*, Tuesday, April 9.

📖 FURTHER READING

Aulton M. E. (ed). (2002). *Pharmaceutics, the science of dosage form design* (2nd edn). Edinburgh: Churchill Livingstone.

Barnes, G. T. and Gentle I. R. (2005). *Interfacial Science An Introduction*. Oxford: Oxford University Press.

Bernstein, J. (2002). *Polymorphism in Molecular Crystals*, Oxford: Oxford Science Publications.

Clegg, W. (1998). *Crystal Structure Determination*. Oxford: Oxford Science Publications.

Davey, R. and Garside, J. (2000). *From Molecules to Crystallizers, An Introduction to Crystallization*. Oxford: Oxford Science Publications.

Gluster, J. P., Lewis, M. and Rossi, M. (1994). *Crystal Structure Analysis for Chemists and Biologists*. New York: VCH Publishers, Inc.

Haines, P. J. (1995). *Thermal Methods of Analysis; Principles, Applications and Problems*. Glasgow: Blackie Academic and Professional.

Klobes, P., Meyer, K. and Munro R. G. (2006). *Porosity and Specific Surface Area Measurements for Solid Materials*. Washington. National Institute of Standards and Technology.

Masuda, H., Higashitani, K. and Yoshida, H. (2006). Powder Technology Handbook, 3rd edn, CRC Press, New York.

Shekunov, B. Y. *et al.* (2007). *Particle Size Analysis in Pharmaceutics: Principles, Methods and Applications*, Pharm. Res., **24**, 203–227.

❓ EXERCISES

4.1 Assign a Miller index to the family of crystal planes shown in Fig. 4.46.

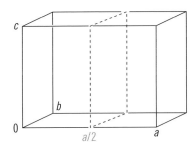

FIGURE 4.46

4.2 Copper sulfate has been used as an antifungal agent. Copper sulfate pentahydrate was prepared and analysed by TGA, which gave the data shown in Table 4.7. The mass loss at each stage shown was due to loss of water molecules. Identify the compounds formed at each stage.

TABLE 4.7

Temp/°C	% mass loss
196–200	14.4
200–210	28.8
300–360	36.0

THE THEORY OF DISPERSE SYSTEMS

05

Learning objectives

Having studied this chapter, you should be able to:

- describe the various types of pharmaceutical disperse systems
- list and describe the various classifications of pharmaceutical disperse systems
- give examples that highlight the importance of the physical stability of pharmaceutical disperse systems
- describe the types of movement (kinetic properties) exhibited by the disperse phase of disperse systems
- explain what is meant by viscosity, various viscosity terms, how viscosity is influenced by dispersed-phase concentration and how it can be used to measure the average molecular weight of dispersed species
- describe how interfacial free energy influences disperse-system stability and how surfactants can reduce interfacial tension and stabilize disperse systems
- explain how emulsions are formed and stabilized using emulsifying agents
- explain how dispersed entities acquire a surface charge and how this surface charge influences the stability of disperse systems.

Many pharmaceutical dosage forms and physiological systems cannot be described by a single phase system (liquid, solid or gas). The majority of these systems are more complex and are composed of more than one phase, such as a pharmaceutical suspension comprised of drug particles dispersed in a liquid. The term 'disperse system' or 'dispersion' is used for any system where one phase (a disperse phase) is distributed throughout a second phase (a continuous phase or dispersant). A simple illustration of a basic disperse system is shown in Fig. 5.1.

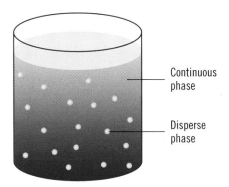

FIGURE 5.1 An illustration of a disperse system with the dispersed phase dispersed throughout a continuous phase.

The type of disperse system formed depends on the state of the disperse phase and the continuous phase, as shown in Table 5.1. The concentration of the disperse phase can also influence the type of disperse system formed. For example, when solid particles are dispersed in a liquid at relatively low concentrations a suspension may be formed, whilst at high concentrations a gel may be formed.

Disperse systems also occur in human physiology: the blood system can be considered as a disperse system of blood cells and platelets dispersed in plasma. In this chapter we will focus on pharmaceutical disperse systems. First, let us consider the various ways drug molecules can be distributed in pharmaceutical disperse systems.

TABLE 5.1 Common pharmaceutical disperse systems and their component phases.

Type of disperse phase	Continuous phase	Disperse phase
Suspension	Liquid	Solid
Gel	Liquid	Solid
Emulsion	Liquid	Liquid
Micellar system Liposomal systems	Liquid	Liquid (Molecular assemblies)
Aerosol	Gas	Solid
	Gas	Liquid
Foam	Liquid	Gas
	Solid	Gas
Solid dispersion	Solid	Solid

5.1 DRUG DISTRIBUTION IN PHARMACEUTICAL DISPERSE SYSTEMS

In some circumstances the disperse phase of the disperse system can be composed purely of drug molecules – for example, a pharmaceutical suspension where drug particles are dispersed in liquid. Alternatively, drug molecules may be a component in one phase – for example, a pharmaceutical emulsion where the drug is dissolved in the oil droplets dispersed throughout the aqueous phase. Let us consider each of these situations in turn.

One phase composed entirely of drug molecules

Upon mixing a drug phase with a second phase, one of two extreme situations can occur. On the one hand, the molecules in both phases may mix intimately with time and form a single phase (a solution) due to molecular attractions. This situation arises when particles of a water-soluble drug are dispersed in water. Alternatively, the two phases may fail to mix due to lack of molecular attraction between the phases and result in two separate immiscible phases, this arises when drug particles are dispersed in air. Both of these situations are extremes of behaviour and the actual behaviour normally lies between these two extremes.

For example, consider what happens when drug particles and water are mixed, a suspension will form if the amount of solid drug is not totally soluble in the volume of water. The suspension formed should not be thought of as drug particles dispersed in water but as drug particles dispersed in an aqueous solution of the drug. That is, the drug is present in the suspension in both the solid and liquid (solution) states. The same situation arises when two immiscible liquids are mixed. A small portion of both liquids will dissolve in each other but the majority of each liquid will not dissolve. The overall result will be two liquid phases.

When solid or liquid drug phases are dispersed in a gaseous phase, the amount of drug that dissolves into the gaseous phase can be considered negligible due to the poor solvent power of gases.

Drug molecules as a component of one phase

We also must consider situations when the drug is a component of one phase (dissolved in one phase) of a disperse system. Upon mixing, the phase containing drug molecules with a second immiscible phase, a two-phase system is formed. Drug molecules initially dissolved in one phase can partition across the interface into the second phase, as shown in Fig. 5.2. The rate and extent of partitioning between the two liquid phases is driven by the ratio of the drug's solubility in both individual phases and is quantified by its partition coefficient. (Partitioning behaviour and partition coefficients are described in greater detail in Chapters 3 and 7.) Partitioning of a drug between immiscible liquid phases is an important concept to understand as it is responsible for the distribution of a drug from the site of administration throughout the body.

In addition to drug molecules partitioning between phases, solvent molecules from the drug phase can partition into the second liquid phase and, conversely, molecules from the second phase can partition into the drug solution phase.

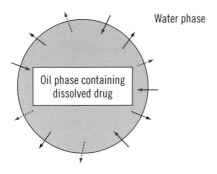

Water phase

Oil phase containing
dissolved drug

⟶ Water molecules moving to the oil phase
⋯⋯▸ Oil molecules moving to the water phase
⟶ Drug molecules moving from the oil phase to the water phase

FIGURE 5.2 Partitioning of drug, oil and water molecules between the phases of a disperse system.

This partitioning results in changes in the chemical composition of both phases that can alter the drug's solubility in both phases and hence influence the partitioning behaviour of the drug.

Now that the distribution of a drug in pharmaceutical dispersions has been explored, we will look at the ways used to classify disperse systems. Pharmaceutical disperse systems can be classified in two ways – according to either the state of the continuous or the disperse phase, as previously shown in Table 5.1, or according to the properties of the disperse system. Certain properties of disperse systems can be related to the dimensions of the dispersed entities (particles, droplets or bubbles). Based on differences in these properties, disperse systems can be classified as coarse dispersions, colloidal dispersions or molecular dispersions.

5.2 MOLECULAR, COLLOIDAL AND COARSE DISPERSE SYSTEMS

The approximate dimensions of dispersed entities in coarse dispersions, colloidal and molecular dispersions are listed in Table 5.2. Also listed are some of the differences observed between the properties of these systems. A disperse system may contain dispersed entities that fall into more than one classification, for example a suspension containing particles of a wide particle-size distribution may exhibit both colloidal and coarse properties.

Molecular dispersions

A molecular dispersion is another term for a solution: each molecule of the disperse phase is surrounded intimately by molecules of the continuous phase. Molecular dispersions are always lyophilic (solvent loving), it is their lyophilic behaviour that allows them to be surrounded intimately by the molecules of the continuous phase.

The behaviour and theory of solutions was already explained in Chapter 2. In this section we will focus on describing colloidal and coarse dispersions.

TABLE 5.2 Examples of some differences between the properties of typical molecular, colloidal and coarse disperse systems.

Type of disperse phase Approximate size	Molecular < 1 nm	Colloidal 1 nm to 1 μm	Coarse > 1 μm
Scatter light	No	Yes	Yes
Exhibit Brownian motion	Yes	Yes	No
Exhibit sedimentation	No	No	Yes
Pass through a semipermeable membrane	Yes	No	No
Diffuse	Yes	Yes	No

Colloidal dispersions

Colloidal dispersions can be described as lyophilic (solvent loving), lyophobic (solvent hating) or amphiphilic. Amphiphilic refers to the ability of a molecule to have both lyophobic and lyophilic segments within the same molecule. Molecules that exhibit amphiphilic behaviour in water contain a hydrophilic segment and hydrophobic segment within the same molecule, as shown for sodium lauryl sulfate in Fig. 5.3.

Lyophilic colloidal systems

Lyophilic colloids occur when the molecules of the dispersed phase have similar attraction for the molecules of the continuous phase as they have for themselves. When the continuous phase is water, lyophilic colloids are referred to as hydrophilic colloids. Examples of hydrophilic colloidal dispersions include water-soluble high molecular weight molecules (proteins and polymers) with molecular diameters greater than 1 nm when dispersed in water. Due to their hydrophilic nature they will disperse spontaneously in water with time, and are thermodynamically stable.

Lyophilic dispersions of proteins and water-soluble polymers have both molecular and colloidal properties. The distinction between these two systems becomes vague when the dimensions of the dispersed molecules are greater than 1 nm. Due to the size of these molecules the dispersion will exhibit colloidal properties but can still be classified as a molecular dispersion.

Lyophobic colloidal systems

Lyophobic colloidal dispersions behave very differently to lyophilic colloidal systems. The dispersed phase of a lyophobic dispersion has little or no molecular attraction for the

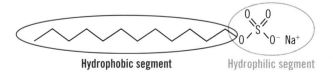

FIGURE 5.3 The chemical structure of sodium lauryl sulfate, highlighting the hydrophilic and hydrophobic segments of this amphiphilic molecule.

continuous phase. Consider oil droplets dispersed in water as an example of a lyophobic dispersion. A colloidal dispersion of oil droplets in water will not form spontaneously. In order to reduce the oil to droplets of colloidal dimensions, energy has to be added by strong agitation, such as vigorous stirring or shaking. Once the dispersion is formed and agitation stopped the oil droplets will start to aggregate and coalesce with time – that is, the colloidal dispersion will break down. Lyophobic dispersions do not form spontaneously and are thermodynamically unstable.

The instability of colloidal lyophobic dispersions drives these disperse systems to rearrange themselves to minimize the contact area between the dispersed and continuous phases. (Why this occurs is explained in greater detail in Section 5.6.2.) The rearrangement results in the aggregation of the dispersed particles or the coalescence of dispersed droplets.

Amphiphilic colloidal systems

When amphiphilic molecules are dispersed in an aqueous continuous phase, the molecules will orientate themselves so that exposure of the hydrophobic segment to the aqueous phase is minimized and exposure of the hydrophilic segment is maximized. At low concentrations the molecules will distribute themselves between the bulk solution (as individual molecules) and at the surface of the continuous phase, as shown in Fig. 5.4(a). At the surface of the continuous phase, we can imagine the hydrophilic segment of the

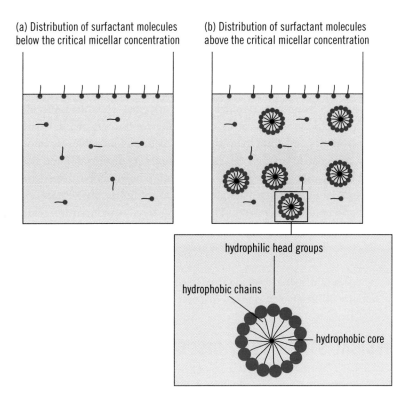

(a) Distribution of surfactant molecules below the critical micellar concentration

(b) Distribution of surfactant molecules above the critical micellar concentration

hydrophilic head groups

hydrophobic chains

hydrophobic core

FIGURE 5.4 A schematic representation of the distribution of surfactant molecules dispersed in water, (a) below the critical micellar concentration and (b) above. Note: A two-dimensional representation of a micelle is shown. Micelles are not two-dimensional structures but spherical.

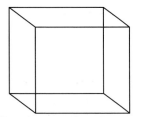

reduce the cube on the left
into 1000 smaller cubes

1000 cubes

length: 1000 μm
total area: 6×10^6 μm^2
volume: 1×10^9 μm^3
area to volume ratio: 0.006 μm^2/μm^3

length: 100 μm
total area: $1000 \times (6 \times 10^4$ μm$^2) = 6 \times 10^7$
volume: $1000 \times (1 \times 10^6$ μm$^3) = 1 \times 10^9$
area to volume ratio: 0.06 μm^2/μm^3

FIGURE 5.5 Illustration highlighting the impact of size on the surface area to volume ratio of the dispersed phase in disperse systems.

molecule immersed in the aqueous phase, with the hydrophobic segment sticking out into the air. Amphiphilic molecules are also referred to as surfactants (surface-active molecules) as they generally reduce surface tension when they orientate themselves at liquid surfaces (this is explained in greater detail in Section 5.6.3).

As the concentration of amphiphilic molecules in an aqueous system is increased, a concentration is eventually reached at which amphiphilic molecules assemble in the bulk solution with their hydrophilic segment facing the continuous phase and their hydrophobic regions adjacent to each other. The resultant structure is a spherical micelle, as illustrated in Fig. 5.4(b). The concentration at which micelles start to form is called the critical micelle concentration (CMC). Micelles exhibit colloidal properties and can be referred to as association colloids. They form spontaneously and therefore are thermodynamically stable.

Coarse dispersions

Coarse dispersions, similar to lyophobic colloidal dispersions, exist when the molecules of the disperse phase have little molecular attraction for the continuous phase. Coarse dispersions are always lyophobic, making them thermodynamically unstable. However, compared to lyophobic colloidal dispersions, coarse dispersions are more stable and have a reduced tendency for aggregation. This is because aggregation in lyophobic dispersions is a result of the system spontaneously rearranging itself to minimize the contact area between the dispersed and continuous phases. This driving force is reduced for coarse dispersions compared to colloidal dispersions because of the smaller surface area to volume ratio of the disperse phase in coarse dispersions compared to colloidal dispersions, as illustrated in Fig. 5.5.

Now that the basic types of disperse systems have been explained, let us highlight the importance of the stability of the pharmaceutical disperse systems.

5.3 THE PHYSICAL STABILITY OF DISPERSE SYSTEMS

The stability of disperse systems can be described both in terms of chemical and physical stability, however, in this book we focus on physical stability. The physical stability of a disperse system describes the ability of the disperse phase to remain uniformly (evenly) dispersed throughout the continuous phase. The uniformity of a drug substance dispersed throughout a continuous phase ensures that the patient being administered the drug receives the same dose of drug each time a volume of a dispersion is administered. Physical instability within pharmaceutical disperse systems can result in a non-uniform dispersion of the drug substance, resulting in the patient receiving too much or too little drug. Both scenarios can result in extremely serious consequences for the patient and should be avoided.

To emphasize this point, consider an oral pharmaceutical suspension in which drug particles are dispersed throughout a liquid. Physical instability due to aggregation within the suspension could result in the sedimentation of dispersed drug particles to the bottom of the container (sedimentation is explained in greater detail in Section 5.4.2). This would cause the uneven distribution of the drug throughout the suspension: there would be a higher concentration of drug particles at the bottom of the container compared to the top. If a patient pours a volume of the suspension onto a medicine spoon, without shaking, then the patient will receive less than the indicated dose of drug. If the patient continues to administer the suspension in this manner, when the patient pours out the suspension at the end of the container it will be highly concentrated with drug particles and the patient will receive more than their recommended dose of drug.

A second useful example to highlight the serious consequences of the physical instability of disperse systems is to consider an emulsion formulation designed to deliver lipids to patients intravenously, as in the case of total parenteral nutrition (TPN) (Box 5.1). Pure lipids cannot be administered directly into the blood stream as they are immiscible with blood and would not form droplets of an adequate size ($< 5\,\mu m$) to allow distribution in the blood system. Large droplets of lipid in the blood system can cause embolisms (blockages of blood vessels). Consequently, lipids must be formulated as oil-in-water emulsion systems for safe intravenous administration. The droplet size of TPN emulsions must be carefully controlled and monitored to prevent embolisms. Physical instability in a TPN emulsion on storage resulting in aggregation and droplet coalescence creating large droplets, could have serious and fatal consequences for patients.

BOX 5.1 Total parenteral nutrition (TPN).

Total parenteral nutrition is the term used to describe the delivery of glucose, amino acids, lipids, salts and vitamins intravenously. Nutrition is delivered to patients in this manner when they are unable to obtain their nutritional requirements by eating and digestion. The TPN formulation is prepared in a sterile bag and delivered to the patient intravenously at a controlled rate using an infusion pump. Formulations of TPN can contain 60 or more chemical species. Water-soluble nutrients are dissolved in the aqueous solution and lipids are present in an emulsion form. The mean lipid emulsion droplet size in TPN is typically around 300 nm. Physical instability of lipid emulsions in TPN formulations causing aggregation and coalescence of lipid droplets will increase lipid droplet size and increase the risk of the occurrence of an embolism due to the large lipid droplets.

Physical instability in disperse systems is caused by the system being in a non-equilibrium state, and it spontaneously rearranging itself to achieve an equilibrium state. After preparation, many pharmaceutical disperse systems are not at equilibrium and therefore are referred to as thermodynamically unstable. When formulating these disperse systems it is important to minimize (or, if possible, eliminate) the thermodynamic instability of the system by optimizing critical properties of the disperse system. Critical properties to be considered include (1) the movement of the disperse phase (kinetic properties), (2) the viscosity of the disperse system, (3) the surface charge on the disperse phase, and (4) the interfacial free energy of the interfacial region between the dispersed and continuous phases.

We need a basic understanding of the theory behind these properties to understand how they can be altered to optimize the stability of the pharmaceutical disperse system. The first of these properties we will explore are those that are related to the movement of the dispersed phase – the kinetic properties.

5.4 KINETIC PROPERTIES OF DISPERSE SYSTEMS

The movement of a dispersed entity (particle, droplet, bubble or micelle) in a continuous phase can be influenced by a number of factors that can be divided into three types: (1) the properties of the continuous phase (chemistry, density and viscosity), (2) the properties of the dispersed entity (chemistry, size, shape and density) and (3) external forces (heat, gravity, mechanical agitation and electrical voltage). The types of movement of a dispersed entity in a continuous phase can be described as **Brownian motion, diffusion** and **sedimentation**. Diffusion has been previously explained in Chapter 3, here we will explore Brownian motion and sedimentation.

5.4.1 Brownian motion

Brownian motion is the random movement of dispersed entities in a continuous phase due to collisions between the dispersed species and the molecules of the continuous phase. All liquids and gases are composed of atoms or molecules that are in a state of constant motion, which we call thermal motion. The speed of this movement is influenced by the temperature of the system: at higher temperatures the energy of the system is increased and this causes an increase in the movement of atoms or molecules. Due to this constant movement, molecules of the continuous phase collide with the dispersed entities and cause the dispersed entities to move.

Dispersed materials are constantly being bombarded by the atoms or molecules of the continuous phase and the resultant movement of the disperse phase is erratic and random. If you look at dust or pollen grains dispersed in water under a microscope they can be seen to move in an erratic, zigzag-like manner.

Brownian movement is exhibited by dispersed entities of molecular and colloidal dimensions. The random collisions of the continuous phase atoms or molecules will not be powerful enough to cause coarse entities to move. The viscosity of the disperse phase also influences the extent and rate of Brownian movement: more viscous continuous phases slow down the movement of dispersed entities.

5.4.2 Sedimentation

Dispersed entities of a certain mass experience movement through a continuous phase due to the force of gravity (9.8 m s^{-2}): this movement is referred to as sedimentation. If the resistance of the continuous phase (air or liquid) to this movement is negligible, then the sedimentation rate of the dispersed entity will be the same regardless of its mass. (Instinct seems to indicate that heavier particles would fall faster but this is not the case where air or liquid resistance is negligible.)

Air and liquid resistance cannot be considered negligible for pharmaceutical disperse systems, and the resistance they cause influences the movement of dispersed entities. The resistance caused by air is called aerodynamic resistance and the resistance caused by water is called hydrodynamic resistance. The resistance to sedimentation posed by a gas or liquid is called the **drag force**. Let us explore this in more detail.

Drag force

The direction of the drag force is opposite to the direction of movement of the object through air or fluid. It may be easier to think of drag force in relation to an airplane moving through the air; the airplane being the disperse phase and the air being the continuous phase as shown in Fig. 5.6.

The drag force is due to the physical interaction between the airplane and the air. For drag force to be experienced there must be a difference of velocity between the airplane and air: it will not be experienced when the airplane is stationary on a still day (no air movement). In order to experience drag force, the object must be moving and the fluid medium (air or liquid) stationary, or the fluid medium can be moving and the object stationary, or both object and fluid medium can be moving in opposite directions. An airplane sitting stationary on a windy day will experience drag force; an airplane moving through the air on a still day will experience drag force; and an airplane moving in the opposite direction to the wind's movement will also experience drag force.

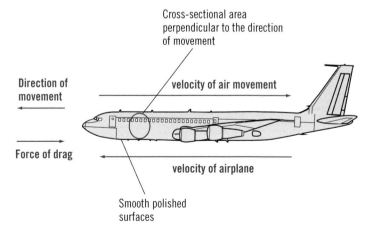

FIGURE 5.6 Factors that affect the drag force experienced by an airplane moving through the air. Understanding these can help us understand the drag force experienced during the sedimentation of the disperse phase of a dispersion.

The factors that influence the force of drag (F_D) are given in eqn 5.1, called the drag equation

$$F_D = {}^1\!/\!_2 C_D \cdot A \cdot \rho \cdot v^2, \tag{5.1}$$

where C_D is the drag coefficient, A the cross-sectional area, ρ the density of the fluid medium, and v the velocity of an object relative to the fluid medium.

Drag forces increase as the cross-sectional (A) area perpendicular to the direction of the object's movement (see Fig. 5.6) increases (this is why airplanes and formula one racing cars have a streamlined design with no unnecessary bulky projections in the direction of movement). As the size of the dispersed entity increases, the cross-sectional (A) area perpendicular to the direction of the object's movement will increase. Drag forces also increase as the viscosity of the fluid medium increases – for example, an object will move faster through air compared to water.

As the velocity of an object relative to the fluid medium increases, the drag force increases. For example, an airplane flying into a wind at constant speed moving from a region of low wind speed to high wind speed will experience an increase in drag force. Also, as the surface roughness of an object is increased, the drag force increases. Therefore, a smooth surface will produce less drag force than a rough surface. The drag coefficient (C_D) takes into account a number of variables including the viscosity of the medium, the shape of the body, and the roughness of the body's surface.

Now we must apply our understanding of drag to pharmaceutical disperse systems. Dispersed entities, such as drug particles moving through a liquid will experience a drag force. How fast a particle falls through a liquid, the sedimentation rate (V), is dependent on the combined forces of (1) acceleration due to gravity (downwards) and (2) drag (upwards).

Sedimentation rate

Factors that affect the sedimentation rate of a dispersed entity through a continuous phase are the same factors that influence drag force and acceleration due to gravity. A useful equation that summarizes these factors is called the **Stokes equation**

$$V = \frac{2r^2 \cdot (\rho - \rho_o) \cdot g}{9\eta_o}, \tag{5.2}$$

where r is the radius of the dispersed particle, ρ is the density of the disperse phase, ρ_o is the density of the continuous phase, g the acceleration due to the force of gravity and η_o is the viscosity of the continuous phase.

Looking at Stokes' equation we can see that as the size and density of the disperse phase increases the sedimentation rate increases: this explains why coarse particles sediment faster than colloidal particles. Stokes' equation also indicates that by increasing the viscosity of the continuous phase, the sedimentation rate can be decreased: this is how suspending agents work: they increase the viscosity of the continuous phase.

However, there are limitations to Stokes' equation, it is only relevant when there is parallel movement of liquid or gas relative to the movement of the falling object. This may occur in emulsions and suspensions during storage but not during the inhalation of a powdered drug – the movement of the particle in the air may be turbulent (moving in random directions and at varying velocities). Stokes' equation is also only relevant when free settling occurs – that is, when each particle sediments independently of other dispersed particles, and does not aggregate or collide with other particles during sedimentation. Free settling only occurs in dilute uniform disperse systems.

Stable dispersion can be achieved if sedimentation is counteracted by Brownian motion: this can occur in colloidal systems due to the small dimensions of the dispersed entities. Some coarse dispersions containing fine powders of very low density can act in this manner: they are referred to as diffusible powders. Other powders are referred to as non-diffusible powders.

 EXAMPLE 5.1

A pharmaceutical suspension is prepared by dispersing drug particles (radius 200 µm and density 1.6 g cm^{-2}) in a continuous phase (with a viscosity of 1 cP). What properties of the drug particles and the continuous phase could be altered to reduce the sedimentation rate of the drug particles in the suspension?

If we refer to Stokes' equation, eqn 5.2, we can see that reducing the radius and density of the drug particles and increasing the viscosity of the continuous phase will reduce the sedimentation rate of the drug particles.

Viscosity of the continuous phase was highlighted as a factor that influences the Brownian movement and the sedimentation rate of disperse systems, let us now explain viscosity in more detail.

5.5 VISCOSITY

Viscosity is a measure of the resistance of a fluid (liquid or gas) to flow: fluids with greater resistance to flow have higher viscosity values. Viscosity has already been mentioned in relation to the movement of dispersed entities in the continuous phase (see Section 5.4). In this section, we will explain the term viscosity and the factors that affect the viscosity of disperse systems.

Viscosity, shear stress and strain

A liquid flows when a stress (σ) is applied, stress being force applied per unit area. Shear stress is the type of stress that causes a liquid to flow. The resultant deformation (change of shape) that occurs when shear stress is applied is described as strain (γ); it is also referred to as the shear rate. When shear stress is applied to a liquid a strain (deformation) will result and this will be observed as liquid flow.

The extent of strain a liquid experiences increases as the level of stress applied increases. That is, the more stress that is applied to a liquid, the faster the liquid will flow (and the greater the strain). If the same shear stress is applied to two different liquids – for example, water and syrup – the extent of flow or strain experienced by each liquid will differ. The flow rate will be faster for water compared to syrup because of the differences between the viscosity of water compared to syrup. If the shear stress applied is increased, the flow of both materials will increase but the water will still flow faster than the syrup. If the shear stress is then plotted against strain (rate of flow) for both these liquids, a graph called a rheogram will be obtained, as shown in Fig. 5.7.

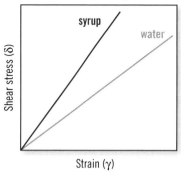

FIGURE 5.7 Shear stress plotted against strain for Newtonian fluids.

The linear relationships obtained when shear stress was plotted against strain for each liquid can be described as $\sigma = \eta\gamma$. The slope of the line η will differ for the two liquids; it is a constant value that describes the ratio of shear stress to strain (σ/γ) for each liquid. This relationship is called the viscosity coefficient (η) of the liquid, represented mathematically by

$$\eta = \frac{\sigma}{\gamma}. \tag{5.3}$$

'Viscosity coefficient' is shortened to 'viscosity' or 'dynamic viscosity' to distinguish it from other viscosity terms. Other viscosity terms of interests are described in Box 5.2. It is important to be aware of what they refer to and that they are not directly comparable.

The units commonly used to express viscosity are pascal seconds (Pa s) or centipoise (cP). 1 mPa s, the viscosity of water at 20 °C, is equivalent to 1 cP. The viscosity of a fluid is temperature dependent: the viscosity of a fluid will normally decrease as the temperature increases. When quoting a viscosity value, it is important to quote the temperature at which the viscosity of the system was measured. Table 5.3 lists the viscosity values of some familiar fluids.

The shear stress and strain relationship described in eqn 5.3 was proposed by Isaac Newton, fluids that obey this relationship are called Newtonian fluids.

TABLE 5.3 The viscosity of some familiar fluids.

Fluid	Viscosity (Pa s)	Temperature of measurement
Air	1.9×10^{-5}	18 °C
Chloroform	5.8×10^{-4}	20 °C
Water	1×10^{-3}	20 °C
Olive Oil	8.4×10^{-2}	20 °C
Castor Oil	8.9×10^{-1}	25 °C
Maple Syrup	3.2	25 °C

BOX 5.2 Various viscosity terms.

Kinematic viscosity (k): Dynamic viscosity (η) divided by the density (ρ_o) of the system

$$k = \frac{\eta}{\rho}.$$

Relative viscosity (η_{rel}): Dynamic viscosity of the solution or suspension (η) divided by the dynamic viscosity of the solvent or continuous phase (η_o)

$$\eta_{rel} = \frac{\eta}{\eta_o}.$$

Specific viscosity (η_{sp}): The difference between the dynamic viscosity of the solution or suspension (η) and the dynamic viscosity of the solvent or continuous phase (η_o), divided by the dynamic viscosity of the solvent or continuous phase (η_o).

$$\eta_{sp} = \frac{\eta - \eta_o}{\eta_o} \quad \text{or} \quad \eta_{sp} = \eta_{rel} - 1.$$

Reduced viscosity (η_{red}): The specific viscosity (η_{sp}) divided by the concentration (C) of the colloidal disperse phase expressed as % w/v,

$$\eta_{red} = \frac{\eta_{sp}}{C(\% \text{ w/v})}.$$

5.5.1 Newtonian and non-Newtonian fluids

Newtonian fluids include pure gases, liquids and solutions of low molecular weight molecules such as syrup. The majority of pharmaceutical disperse systems do not obey eqn 5.3 and these are described as non-Newtonian fluids. Instead of a straight line through the origin on the shear stress–strain plot, as shown for Newtonian fluids (Fig. 5.7), non-Newtonian fluids show a range of shear stress–strain behaviours as shown in Fig. 5.8. These are described below.

Plastic flow

The type of non-Newtonian flow displayed in Fig. 5.8(a) is described as plastic flow. From this curve we see that a certain amount of shear stress is required before the liquid will flow. This type of flow occurs when trying to get ketchup to flow from a bottle: the bottle has to be shaken to get the ketchup to flow. The shear stress that must be applied to get the liquid to flow is called the yield value (σ_y). Once shear stress above the yield value is applied, the liquid behaves like a Newtonian fluid and the strain of the liquid increases linearly with increasing shear stress. The yield value can be obtained from the plot of shear stress against strain of a material exhibiting plastic behaviour, by the extrapolation of the straight portion of the curve until it intersects the y-axis, as shown in Fig. 5.8(a).

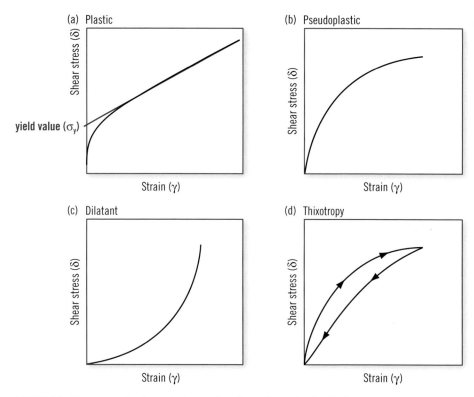

FIGURE 5.8 Rheograms showing typical examples of non-Newtonian fluid behaviour and thixotropy, (A) Plastic flow, (B) Pseudoplastic flow, (C) Dilatant flow and (D) Thixotropy.

Plastic behaviour is observed for concentrated suspensions, and disperse systems with flocculated particles. In these systems the interactions between adjacent particles must be overcome before the liquid will flow. The yield value (σ_y) gives a measure of the energy needed to break these interactions. The viscosity of liquids that exhibit plastic flow above the yield value (η_p) can be described mathematically by

$$\eta_p = \frac{(\sigma - \sigma_y)}{\gamma}. \tag{5.4}$$

▶ **EXAMPLE 5.2**

A concentrated pharmaceutical suspension has been shown to exhibit plastic flow. The yield value for the suspension is found to be 480 Pa. When shear stress is applied at levels above the yield value, the strain (rate of shear) of the sample increases linearly. At a shear stress of 620 Pa the measured strain is 200 s⁻¹.

The plastic viscosity (η_p) of this sample can be determined using eqn 5.4.

$$\eta_p = \frac{(620 - 480 \text{ Pa})}{200 \text{ s}^{-1}}$$

$$\eta_p = 0.7 \text{ Pa s}.$$

Pseudoplastic flow

The type of non-Newtonian flow displayed in Fig. 5.8(b) is described as pseudoplastic flow and is exhibited by a range of synthetic and natural polymers in solution. These disperse systems start to flow once shear stress is applied. Unlike Newtonian systems, the relationship between shear stress and strain is not linear. These polymer molecules are high molecular weight linear molecules and when dispersed in a solvent and allowed to sit for a period of time their chains may interact with each other forming a network of intermolecular bonds, and as a result these polymer solutions will have a relatively high viscosity when no stress is applied.

When shear stress is applied to the polymer solution, some of the intermolecular bonds formed may be broken and the viscosity of the solution will be reduced. If shear stress continues to increase, a minimum viscosity is eventually reached at which all intermolecular bonds are broken and no further rearrangement of the polymer chains takes place. Liquids that exhibit pseudoplastic flow are also referred to as shear-thinning liquids.

The dynamic viscosity of pseudoplastic liquids changes as the stress applied changes. Therefore, it is not possible to quote a single viscosity value for the system. An apparent viscosity (η') can be calculated at a specific point on the shear stress–strain curve. However, an apparent viscosity value is meaningless unless the shear stress value it is measured at is also quoted along with the temperature of measurement. Pseudoplastic flow can be expressed mathematically by the power law,

$$\eta' = \frac{\sigma^n}{\gamma}.$$
(5.5)

The power law is a generalized equation used to describe viscosity. The power number, n, can vary depending on the behaviour of the system. When the rheogram for the system can be described by an equation where the n value is equal to 1, then the liquid is Newtonian. If the equation that fits the rheogram for the system has an n value of < 1, then the system exhibits pseudoplastic behaviour.

Dilatant flow

Dilatant flow behaviour is the converse of pseudoplastic flow, as shown in Fig. 5.8(c). It is observed for some very concentrated suspensions with approximately 50% w/w solids or greater, such as the mixture of powder and liquid during the preparation of granules. When shear stress is applied to these systems during mixing, their viscosity can increase. This type of behaviour is called shear thickening. The increase in viscosity is due to the rearrangement of the particles in the mixture into clumps or structured arrangements of particles that exert high resistance to flow. This type of behaviour is less common in pharmaceutical systems than plastic or pseudoplastic behaviour. It can also be described by the power law eqn 5.5, with an n value > 1.

Thixotropy

The rheological behaviour of liquids can be studied by preparing shear stress–strain curves, as shown in Fig. 5.8. These are usually prepared as 'upwards curves', by increasing the shear stress and measuring the strain (or conversely increasing the strain and measuring the stress). As the shear stress applied to the system increases, changes can occur to the system due to breaking intermolecular bonds or reversible aggregates of particles (floccules). It is also interesting to look at the 'downwards curves' for these systems. Downwards curves are prepared by starting the strain measurement at high shear stress values and monitoring the strain as shear stress reduces.

When the upward and downward curves are placed on the same plot we see that they sometimes differ, as shown in Fig. 5.8(d). The difference between the upwards and downwards curves indicates that, when high shear stresses are applied, changes occur in the system that are not immediately reversible when the shear stress is reduced or removed. These changes to the system may be due to the breaking of reversible aggregates and intermolecular bonds. The time frame in which they are reversed may be too slow to be detected during the course of the measurement: the reversal of these events may occur over a time period that can extend from seconds, to minutes, to weeks depending on the structure of the system and the degree and the extent of shear stress applied. This type of behaviour is referred to as thixotropy.

Thixotropy can be a desirable property in liquid pharmaceutical suspensions, emulsions, lotions, creams and ointments. For example, the viscosity of emulsions and suspensions should be high on standing to reduce the sedimentation rate. However, viscosity should be reduced sufficiently by shaking (application of shear stress) to allow the medicine to be poured or spread on the skin.

Now that the general aspects of viscosity have been explained we will consider the viscosity of colloidal disperse systems.

5.5.2 Colloidal dispersion viscosity

The presence of dispersed entities of colloidal dimensions in a continuous phase can influence the viscosity of a dispersion. As the concentration of the colloidal phase increases, the viscosity of the dispersion increases and the difference in the viscosity of the continuous phase alone and dispersion also increases (this difference is captured in the specific viscosity (η_{sp}) value, which is described in Box 5.2). This linear relationship between the concentration of the disperse phase (C) and viscosity of the dispersion is shown by eqn 5.6, proposed to describe the viscosity of dilute dispersions of spherical particles,

$$\eta_{sp} = kC. \tag{5.6}$$

The k value is an indication of the change in viscosity of the disperse system relative to the viscosity of the continuous phase as the concentration of the disperse phase increases.

In situations where the colloidal disperse phase has an affinity for the continuous phase, there will be a large increase in viscosity as the disperse-phase concentration increases. If the colloidal disperse phase has no affinity for the continuous phase, then the increase in viscosity will be minimal as the concentration of the disperse phase increases. We can use viscosity measurements to quantify the degree of affinity of the disperse phase for the continuous phase by determining the Huggins constant, as shown in Box 5.3. An increased interaction between the continuous phase and the dispersed phase is reflected by a more positive Huggins constant. The Huggins constant can be used to compare the capability of excipients such as suspending agents to increase the viscosity of disperse systems.

BOX 5.3 Determination of the Huggins constant.

The degree of affinity of the disperse phase for the continuous phase can be measured by plotting the reduced viscosity (η_{red}) against the concentration (% w/v). The reduced viscosity is described in Box 5.2. The slope of the linear relationship, shown in Fig. 5.9, is a measure of the interaction between the continuous phase and the dispersed phase and is called the **Huggins constant**.

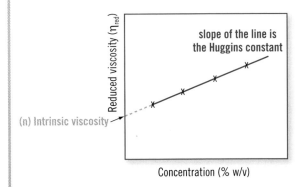

FIGURE 5.9 Plot of reduced viscosity (η_{red}) against the concentration (% w/v) indicating the intrinsic viscosity of the system (*Y*-axis intercept) and Huggins constant (the slope of the line).

The intercept of the linear relationship obtained when the reduced viscosity (η_{red}) is plotted against the concentration (% w/v) for dilute solutions of polymers is referred to as the intrinsic viscosity or limiting viscosity [η], as shown in Fig. 5.9. This value can be used to determine the molecular mass (*M*) of a dispersed polymer using the Mark-Houwink equation

$$[\eta] = K \cdot M^a. \tag{5.7}$$

K and *a* in this equation are constants relating to the shape of a particular polymer molecule when dispersed in a particular solvent at a given temperature.

▶ **EXAMPLE 5.3**

A series of dynamic viscosity values are measured for a range of dilutions of a polysaccharide solution in order to determine the average molecular weight of the polysaccharide. The intrinsic viscosity [η] was determined from the plot of reduced viscosity against concentration polysaccharide (% w/v) and found to be 1.3×10^6 mL g^{-1}.

The average molecular weight was calculated using eqn 5.7. It was assumed that the constant values in this equation are *a* equals 0.89 and *K* equals 0.4 mL g^{-1}

$$[\eta] = K \cdot M^a.$$

First, we fill in the values we know

$$1.3 \times 10^6 \text{ mL g}^{-1} = 0.4 \text{ mL g}^{-1} \cdot M^{0.89},$$

arrange the equation so the molecular weight (M) is on the left-hand side

$$M^{0.89} = \frac{1.3 \times 10^6 \text{ mL g}^{-1}}{0.4 \text{ mL g}^{-1}}$$

calculate the values on the right-hand side of the equation

$$M^{0.89} = 3.25 \times 10^6,$$

log both sides of the equation

$$0.89 \log M = 6.51 \quad \rightarrow \quad \log M = 7.32.$$

Finally, we calculate the average molecular weight of the polysaccharide (M)

$$M = 10^{7.32} \quad \rightarrow \quad M = 2.07 \times 10^7.$$

Viscoelasticity

When shear stress is applied liquids and gases flow and solids deform. Semisolid materials such as pharmaceutical gels, creams and ointments can exhibit behaviour typical of both liquid and solid materials. Materials that do not display purely viscous behaviour but viscous behaviour with an elastic component are termed viscoelastic. The deformation of viscoelastic systems cannot be described simply by the relationship between shear stress and strain. For more details on viscoelastic behaviour refer to the books suggested in the Further Reading section at the end of this chapter.

Having looked at the kinetic and viscous properties of disperse systems, let us now explore behaviour at the boundary – or interface – between two phases. The behaviour associated with the region where these two phases come into contact is important to understand as it impacts on the stability of disperse systems. The region of contact between two phases is called the interfacial region.

5.6 INTERFACIAL PROPERTIES

The various types of interface that occur in disperse systems are listed in Table 5.4. The term 'surface' is used when referring to gas/liquid or gas/solid interfaces. Interfaces are

TABLE 5.4 Various types of interfaces and associated interfacial tension.

Interface	Surface or interfacial tension
Gas/liquid	$\gamma_{v/l}$
Gas/solid	$\gamma_{v/s}$
Liquid/liquid	$\gamma_{l/l}$
Liquid/solid	$\gamma_{l/s}$
Solid/solid	$\gamma_{s/s}$

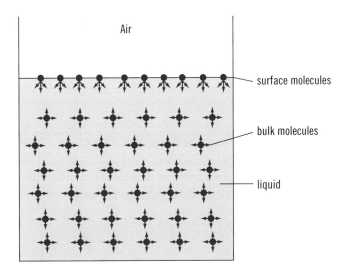

FIGURE 5.10 An illustration of the forces of attraction between the molecules in the bulk liquid and on the surface of a liquid.

not two-dimensional barriers between phases, but are three-dimensional regions of a finite thickness, often just a few molecules.

If we consider the molecules present in a single phase, each molecule is influenced by the other molecules surrounding it. Molecules in the bulk of the phase are surrounded on all sides by equivalent molecules and the forces of attraction are similar in all directions, as shown in Fig. 5.10. At the interface, where the molecules of both phases are in contact (for example a liquid in contact with air), molecules are not surrounded by equivalent molecules on all sides. Here, at the interface, molecules are surrounded by molecules of both phases, as shown in Fig. 5.10.

As a result, the behaviour of molecules at the interface will be different from molecules in the bulk phase. Interfacial behaviour influences whether two phases will form a disperse system and, once formed, whether the disperse system will be stable. There are two key properties of interfaces that influence the behaviour of disperse systems – the interfacial tension, and interfacial free energy. Let us now consider each of these in turn.

5.6.1 Interfacial tension

When molecules of two different phases are in contact, adhesive forces of attraction operate between molecules of both phases at the interface, for example between liquid and gas molecules at the surface of a liquid. Cohesive forces of attraction also operate between equivalent molecules in the bulk of each phase, for example between liquid–liquid molecules or gas–gas molecules.

If the cohesive forces of attraction in one phase are greater than the adhesive forces of attraction at the interface, then two phases will co-exist, if not both phases will disperse into each other. At a gas/liquid interface, if the cohesive forces between the liquid molecules are not greater than the adhesive forces between the liquid and gas, the liquid molecules diffuse into the gaseous air phase, resulting in a single-phase system.

In the case of two immiscible phases, the balance between the cohesive and adhesive forces of attraction at the interface results in a net cohesive force of attraction, as illustrated in Fig. 5.10. This force of attraction is called the interfacial tension (γ). (We also refer to this as 'surface tension' where gas interfaces are involved.) Interfacial or surface tensions are quantified as a force per unit length measurement. Commonly used units are dynes cm^{-1} or mN m^{-1}.

Interfacial tension can be altered by adding substances to the system causing a change in the composition of either phase. Certain substances will spontaneously accumulate at the interface and alter interfacial tension, these substances are called surfactants. If these molecules decrease the cohesive forces of attraction between molecules at the interface then they will reduce interfacial tension. Conversely, if they increase the cohesive forces of attraction they will increase interfacial tension.

5.6.2 Interfacial free energy

For a system to be thermodynamically stable, the free energy (G) of the system must be minimized, as explained in Chapter 3. The stability of a system increases as the free energy of the system approaches a minimum value. We can think of the interface as a system such that the free energy of the interface can be referred to as the interfacial free energy. Where surfaces are involved it is referred to as the surface free energy. Interfacial systems spontaneously arrange themselves to minimize interfacial free energy. The interfacial free energy is defined as the work (w) required to increase the interfacial contact area (ΔA) by 1 m^2 and can be described mathematically by

$$w = \gamma \Delta A. \tag{5.8}$$

To explain how an interface region is stabilized by its spontaneous rearrangement to minimize interfacial energy, imagine a system where 200 mL of oil is added to 800 mL of water and left to stand until it reaches equilibrium. At equilibrium, the system will separate into two layers, with the oil layer on top of the water layer. The interfacial energy of this system is minimized by this arrangement of the phases. Now, close the container of oil and water and mix vigorously. Upon mixing, oil droplets will form and disperse in the water phase. The interfacial contact area between the oil and water phases will be increased substantially due to oil-droplet formation. The more vigorously the system is shaken the smaller the droplets formed and the greater the interfacial contact area. When shaking is stopped the system will be in a non-equilibrium state, therefore the system will rearrange to achieve equilibrium: the oil droplets will coalesce and the system will return to two separate layers.

The behaviour described above can be explained in terms of the interfacial free energy. Important points to consider are: if interfacial tension increases, the interfacial free energy increases; if the interfacial contact area increases, the interfacial free energy also increases (eqn 5.8) and the free energy of a system will not increase spontaneously; it will only reduce spontaneously. For the simple oil/water system described above, the interfacial tension at the interface is relatively constant, at constant temperature and pressure. Increases in interfacial free energy will be due to increases in interfacial contact area.

When the oil/water system is shaken, work (w) is done to the system, adding energy to the system enabling the interfacial contact area to be increased. The more work done, the greater the increase in contact area, hence the smaller the droplets produced and the greater the interfacial free energy. For systems with high interfacial tension more work will be required to reduce the droplet size compared to systems with low interfacial tension. When energy is no longer added to the system, the system will be in a non-equilibrium state with a high interfacial free energy. To minimize the interfacial free energy the interfacial regions will spontaneously arrange themselves by returning to two separate layers.

The same behaviour can be observed when a poorly water-soluble solid drug is dispersed in water. The majority of the drug will exist in the solid state. If the solid particles are milled while dispersed in the liquid, the liquid/solid interfacial contact area will increase due to the work of milling. If this solid–liquid disperse system is left to stand after milling, the particles will have a tendency to aggregate to reduce the interfacial free energy of the system by reducing the interfacial contact area.

Interfacial free energy can be reduced by decreasing interfacial contact area but this is often not desirable as it can destabilize disperse systems, such as suspensions (by increasing sedimentation) and emulsions (by causing cracking of the emulsion into two phases). It can also be reduced by decreasing interfacial tension. Molecules that reduce interfacial tension stabilize disperse systems by reducing interfacial free energy. Molecules that increase interfacial tension destabilize disperse systems by increasing interfacial free energy. Surfactant molecules orientate themselves at interfaces and can alter interfacial tension. Let us how investigate them in a little more detail.

▶ **EXAMPLE 5.4**

By calculating the interfacial free energy of two different pharmaceutical suspensions, suspension A and B, the more stable suspension base can be identified. The interfacial tension between the solid particles and suspending liquid in suspensions A and B are 100 dynes cm^{-1} and 75 dynes cm^{-1}, respectively, and the interfacial contact areas are 0.6 cm^2 and 0.9 cm^2, respectively.

The interfacial free energy of suspension A is calculated using eqn 5.8 to be

100 dynes cm^{-1} × 0.6 cm^2 = 60 dynes cm or 60 erg (note: 1 dyne cm = 1 erg),

and the interfacial free energy of suspension B is

75 dynes cm^{-1} × 0.9 cm^2 = 67.5 dynes cm or 67.5 erg.

Therefore, as the interfacial free energy of suspension A is lower than suspension B, suspension A can be considered more stable than suspension B based on interfacial free energy alone. Note that other factors, such as sedimentation, may also influence the stability of a disperse system.

5.6.3 Surfactants

Molecules that adsorb at air/water interfaces and reduce surface tension are called 'surfactants', an abbreviation for surface active agents. They can be classified as amphiphiles

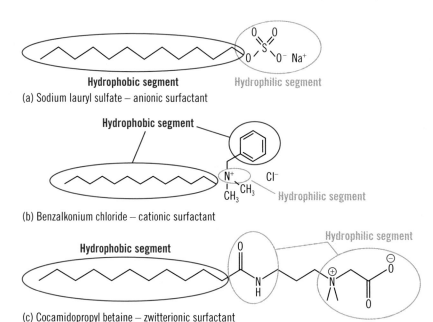

Hydrophobic segment   Hydrophilic segment

(a) Sodium lauryl sulfate – anionic surfactant

Hydrophobic segment

N⁺   Cl⁻
CH₃ CH₃
CH₃   Hydrophilic segment

(b) Benzalkonium chloride – cationic surfactant

Hydrophilic segment

Hydrophobic segment

(c) Cocamidopropyl betaine – zwitterionic surfactant

FIGURE 5.11 Examples of ionic surfactants.

$$HO-[CH_2-CH_2-O]_a-[CH_2-\overset{\overset{\displaystyle CH_3}{|}}{CH}-O]_b-[CH_2-CH_2-O]_a-H$$

polyoxyethylene oxide   polyoxypropylene   polyoxyethylene oxide
hydrophilic chain   hydrophobic chain   hydrophilic chain

FIGURE 5.12 Poloxamer copolymers are examples of non-ionic surfactants. They are composed of a hydrophobic chain with hydrophilic chains attached either side. For poloxamer copolymer the value of *a* and *b* can vary, altering the hydrophobic and hydrophilic chain lengths.

(which comes from the greek 'loves both') due to the hydrophilic and hydrophobic segments that are present in the same molecule, as illustrated for the surfactants in Figs 5.11 and 5.12.

Surfactants are classified as ionic (anionic, cationic or zwitterionic) or non-ionic. Both types are used in the formulation of drug-delivery systems. Non-ionic surfactants are preferred because they are less toxic and therefore can be used in larger quantities. The chemical structures of examples of ionic and non-ionic surfactants are shown in Figs 5.11 and 5.12.

Sodium lauryl sulfate is a commonly used anionic surfactant: the hydrophilic segment of the molecule when ionized imparts a negative charge on the molecule. An example of a cationic surfactant is benzalkonium chloride, the hydrophilic segment of benzalkonium chloride is a quaternary ammonium salt that imparts a positive change on the molecule when ionized in solution. Zwitterionic surfactants have both anionic and cationic groups on the hydrophilic segment, cocamidopropyl betaine is an example of a zwitterionic surfactant.

Non-ionic surfactants are not ionized in solution and exert their hydrophilic interaction with water through hydrogen bonding. The hydrophilic portion of non-ionic surfactants normally contains at least one free OH group or an ether group derived from ethylene or propylene oxide. Figure 5.12 shows the general chemical structure of a poloxamer copolymer, an example of a non-ionic surfactant.

Hydrophilic and lipophilic balance (HLB)

The behaviour of surfactants is influenced by the balance of hydrophilic and hydrophobic activity exhibited by the molecule. An empirical system was devised to describe the balance between hydrophilic and hydrophobic activity. It is called the hydrophilic–lipophilic balance (HLB). ('Hydrophobic' can also be referred to as 'lipophobic'.) The HLB value for some non-ionic surfactants is calculated according to an empirical formula, the % weight of hydrophile divided by five. This system of HLB vales was invented by William C. Griffin of the Atlas Powder Company to describe and compare the behaviour of a number of non-ionic surfactants, (Griffin, 1949).

 EXAMPLE 5.5

The HLB value of Tween 20 (polyoxyethylene (20) sorbitan monolaurate) can be calculated from its molecular weight. The molecular weight of Tween 20 is calculated from the following components,

> 164 Da (sorbitan) – hydrophilic
> 200 Da (lauric acid) – hydrophobic
> 880 Da (20 molelcules of ethylene oxide) – hydrophilic
> – 18 Da (water of esterification)
> 1226 Da

The molecular weight of the hydrophilic portion is 164 Da + 880 Da = 1044 Da.

The % weight of the hydrophilic portion is equal to $\dfrac{1044}{1226} \times 100 = 85\%$.

The calculated HLB value equals $\dfrac{85}{5} = 17.0$.

While the formula developed by Griffin calculates satisfactory HLB values for many non-ionic surfactants, other non-ionic surfactants and ionic surfactants exhibit hydrophilic–lipophilic behaviour that is unrelated to their composition, for example ionic surfactants do not follow this 'weight percentage' HLB basis. Consider sodium lauryl sulfate, the hydrophilic portion of this surfactant is low in molecular weight and the fact that it is ionized at certain pH values makes the surfactant more hydrophilic. Therefore, the HLB values of some non-ionic and of all ionic surfactants must be estimated by experimental methods, so that their HLB values are 'aligned' with those of the surfactants originally described by Griffin's HLB system.

The higher the HLB number the more the balance is tipped towards the surfactant being hydrophilic; the lower the HLB number the more hydrophobic the surfactant. The surfactant's HLB can indicate its suitability for certain functions, as shown in Table 5.5.

TABLE 5.5 HLB values associated with the functions of surfactants.

HLB range	Function
1–3	Antifoaming agents
3–8	Water in oil (w/o) emulsifying agents
7–9	Wetting and spreading agents
8–16	Oil in water (o/w) emulsifying agents
13–16	Detergents
> 16	Solubilizing agents

Influence of pH on hydrophilic lipophilic balance

The hydrophilic nature of ionic surfactants is dependent on the degree of ionization of the hydrophilic portion in solution, which in turn is influenced by the pH of the solution and the surfactant's dissociation constant (dissociation constants are explained in Section 2.5.1). At low pH values the hydrophilic groups on cationic surfactants will be ionized and therefore exhibit considerably greater hydrophilicity than when in an un-ionized state. Conversely, at high pH values, anionic surfactants will be ionized and exhibit considerably greater hydrophilicity than when in an un-ionized state. The hydrophilic component of zwitterionic (ampholytic) surfactants is composed of both cationic and anionic groups: it maintains hydrophilicity across a range of pH values due to the ionization of at least one group.

The hydrophilicility of non-ionic surfactants is not influenced by pH because the hydrophilic portion of non-ionic surfactants is not ionized in solution. Therefore, they retain their surfactant activity over a wide range of pH values unlike ionic surfactants.

Adsorption of surfactants at interfaces

The phenomenon by which surfactants accumulate at an interface is called 'adsorption'. Adsorption only occurs at interfaces or surfaces and it is an interfacial or surface effect. Be careful not to confuse adsorption with absorption, which refers to molecules of a material penetrating the bulk of a second material, for example water molecules taken up into the interior of a sponge is absorption and moisture on the surface of a window is adsorption.

Positive adsorption of molecules at an interface takes place when two phases are mixed and molecules in either phase have a greater affinity for the interface than for the bulk phase. This occurs when surfactant molecules are added to an aqueous solution. The driving force for adsorption is a reduction in surface or interfacial free energy.

Surfactant molecules are strongly adsorbed at interfaces as it allows the hydrophilic portion of the surfactant to be immersed in a hydrophilic medium, and the hydrophobic portion is in a hydrophobic medium – a situation that occurs at oil/water interfaces and the surfaces of aqueous solutions. Generally, surfactants decrease the forces of attraction between molecules at the interface and they reduce interfacial or surface tension. As the

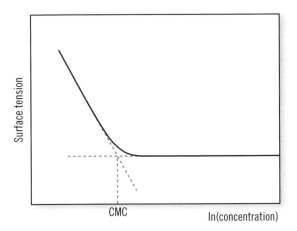

FIGURE 5.13 Surface tension decreases with increase in surfactant concentration up to the critical micelle concentration (CMC). Above the CMC the surface tension does not increase with increase in surfactant concentration. [Barnes, G.T. and Gentle, I.R. (2005). Interfacial Science. Oxford University Press, Oxford.]

concentration of surfactant in solution increases the surface tension decreases until it reaches a region of constant surface tension, shown in Fig. 5.13. The concentration at which this occurs is referred to as the critical micelle concentration (CMC), the addition of surfactant molecules to solution above the CMC results in the formation of micelles.

Pharmaceutical micellar and liposomal disperse systems are prepared using molecules with surfactant properties, these are explored in Chapter 6. Surfactants are also essential components of many emulsion systems and in the following section we will explain the role interfacial properties play in the formation and stability of emulsions.

5.6.4 Role of interfacial properties in emulsion formation

Pharmaceutical emulsions are composed of an aqueous phase and a non-aqueous phase where one phase is dispersed throughout the other as droplets. The non-aqueous phase is normally a water-insoluble organic liquid, which is generally referred to as an oil. When oil droplets are dispersed throughout an aqueous liquid the emulsion formed is referred to as an oil-in-water (o/w) emulsion. When the reverse occurs and the aqueous droplets are dispersed in a continuous oil phase then the emulsion formed is referred to as a water-in-oil (w/o) emulsion. Multiple emulsions can also occur, they feature aqueous droplets dispersed in oil droplets, which in turn are dispersed in an aqueous continuous phase. Conversely, a multiple emulsion may feature oil droplets dispersed in aqueous droplets, which themselves are present in a continuous oil phase, as shown in Fig. 5.14.

Whether an o/w or a w/o emulsion is formed depends on the ratio of oil to water, and the type of emulsifying agent used. When oil is present in a greater quantity to water, it is more likely that a w/o emulsion will be formed. When water is present in a greater quantity to oil, an o/w emulsion will be formed. The HLB values of surfactants that form w/o and o/w emulsions are listed in Table 5.5. In addition to surfactants, gums and fine solids can be used as emulsifying agents.

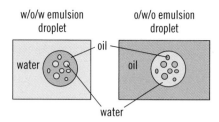

FIGURE 5.14 A simple illustration of multiple emulsions – water/oil/water (w/o/w) and oil/water/oil (o/w/o).

Emulsifying agents

Emulsions do not form spontaneously but require vigorous mixing of oil and aqueous phases. However, upon standing after preparation, the tendency of the two phases to separate into two layers is dependent on the interfacial energy between the two phases and hence the interfacial tension (as explained in Section 5.6.2). Emulsifying agents minimize the interfacial tension between the oil and water phases to close to zero, thereby increasing the stability of emulsions.

Emulsifying agents act at the interface between the oil and water phases, as shown in Fig. 5.15. Examples of commonly used emulsifying agents are listed in Table 5.6. The choice of emulsifying agent selected will influence whether an o/w or a w/o emulsion is formed. Non-ionic surfactants, sorbitan fatty acid esters (Spans) and polysorbates (Tweens) are widely used as emulsifying agents. Polysorbates are polyoxyethylene derivatives of sorbitan fatty acid derivatives. Traditionally used emulsifying agents are mixtures of surfactants and lipids, such as emulsifying wax and wools alcohol ointment. Emulsion stability can be increased by the use of a combination of emulsifying agents with both high and low hydrophile lipophile balance.

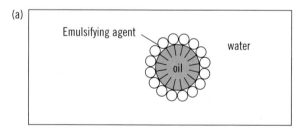

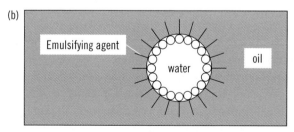

FIGURE 5.15 A simple illustration of (a) oil-in-water and (b) water-in-oil emulsions indicating the presence of surfactant type emulsifying agents at the interface.

TABLE 5.6 Examples of commonly used emulsifying agents.

Type	Name
O/W	Sodium lauryl sulfate
	Polyethylene glycol 400 monostearate
	Polyoxyethylene monostearate
	Phospholipids
	Polysorbate 80
W/O	Glyceryl monostearate
	Sorbitan monooleate
	Sorbitan monolaurate
Fine solids	Aluminium hydroxide
	Magnesium hydroxide
	Silicon dioxide
	Bentonite

Fine solids can also be used as emulsifying agents. They coat the surface of the dispersed droplets and form a layer that prevents the coalescence of droplets when they come into contact.

Pharmaceutical emulsions can be classified into coarse or macroemulsions and microemulsions according to the size of the dispersed droplets, microemulsions have droplet sizes of approximately 5–140 μm. Microemulsions exhibit behaviours different from coarse emulsions.

Microemulsions

Microemulsions form spontaneously and are thermodynamically stable when the appropriate mixture of oil, aqueous phase and emulsifying agent are mixed. This distinguishes them from macroemulsions, which require energy input (stirring or shaking) to form: macroemulsions are thermodynamically unstable.

Emulsions with very small droplet sizes have a very high interfacial surface area and will only be stabilized if the interfacial tension between the oil and water phase is minimized. To form microemulsions, very high concentrations of surfactants (15–25%) are used. As a result of these high surfactant levels the interfacial tension between the water and oil phases is lowered to a point where thermodynamic equilibrium is reached.

Microemulsions can be distinguished from macroemulsions by their optical properties. Due to their small droplet size, microemulsions scatter light weakly, or not at all. They appear transparent or translucent. By contrast, macroemulsions scatter light and appear cloudy and opaque due to their larger droplet size.

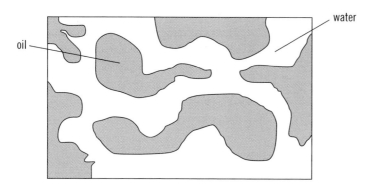

FIGURE 5.16 A simple illustration of a bicontinuous oil and water system. Surfactant molecules would be located at the interfacial regions.

The type of microemulsion formed depends on the type and concentration of surfactant used and the proportion of oil to aqueous phase. Where the proportion of oil to aqueous phase is low an o/w emulsion will form and, where the reverse occurs, w/o emulsions are formed. Where the relative amounts of oil and aqueous phase are similar a bicontinuous microemulsion can result, as shown in Fig. 5.16. The bicontinuous microemulsion consists of randomly distributed oil and water microdomains with surfactant molecules located in the interfacial regions.

Interfacial properties are one of the properties that play an important role in the stabilization of disperse systems, the next properties we will look at are the electrical properties of disperse systems.

5.7 ELECTRICAL PROPERTIES OF DISPERSE SYSTEMS

Pharmaceutical dispersions exhibit electrical properties. To help us understand the electrical properties of pharmaceutical dispersions, let us first consider what happens when two entities with the same charge are brought into close contact – they repel each other. By contrast, two entities with opposite charges attract each other. Likewise, when dispersed entities (solid particles, emulsion droplets, micelles) in an aqueous liquid acquire a surface charge they are repelled or attracted to each other depending on the type and magnitude of this charge. In this section we will highlight the role the surface charge of a dispersed entity plays in the stabilization of pharmaceutical dispersions.

So how do these dispersed entities acquire a surface charge when dispersed in an aqueous phase? There are two main mechanisms: one mechanism is the selective adsorption of ionic species present in the aqueous phase, and the other is the presence of a charge on the surface of the dispersed entity due to its chemical structure. Let's consider each of these in turn.

The aqueous phase of a dispersion normally contains ions with positive and negative charges: H_3O^+ and OH^- present in the water and ionic forms of ionizable dissolved substances added to the water. These charged ionic forms can be attracted to and adsorbed onto the surface of dispersed entities in the aqueous phase.

The second mechanism by which the surface of a dispersed entity acquires a surface charge is by the ionization of groups such as carboxylic acid and amine groups situated at the surface of the dispersed entity. The ionization of surface ionizable groups is dependent on dissociation constants and the pH of the aqueous phase, this is explained in greater detail in Section 2.5.1.

The surface charge on the dispersed entity will influence the interaction of that entity with other dispersed entities and ions in solution. Due to the surface charge it will attract oppositely charged ions to its surface and repel similarly charged ions, as shown in Fig. 5.17. The oppositely charged ions are referred to as counterions and ions of like charge are referred to as co-ions. As a result of the surface charge, a cloud of charged ions surround the dispersed entity due to the distribution of counter- and co-ions.

This electrical cloud extends from the surface of the dispersed entity, to an electroneutral region in the aqueous phase and is called the **electrical double layer** and is illustrated after point c in Fig. 5.17. ('Electroneutral' means that the charges on the positively and negatively charged ions cancel each other out and the overall net charge is neutral.)

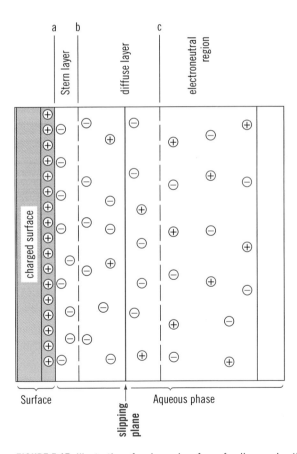

FIGURE 5.17 Illustration of a charged surface of a dispersed entity with an attached electrical double layer of ions; the attached Stern layer (a–b) and the diffuse layer (b–c). The slipping plane in diffuse layer is also illustrated. The magnitude and type of the electrical potential at the slipping plane is called the zeta potential.

Electrical double layer

The electrical double layer is so-called because it is composed of two sections. The first section is an adsorbed layer of counterions called the Stern layer and the second section is a cloud of loosely associated ions called the diffuse layer, as shown in Fig. 5.17. As the disperse species moves through the aqueous phase some of the electrical double layer will move with it – the tightly bound layer of counterions attached to the surface and a diffuse cloud of ions. The boundary at which the ions of the diffuse layer do not move when the dispersed species moves is called the hydrodynamic shear or slipping plane.

Potential energy exists when an object has the potential to cause material to move. The surface charge on the dispersed entity can attract counterions and repel co-ions, therefore it has potential energy. The potential energy due to the surface charge is referred to as the electrical potential. The greater the magnitude of the surface charge, the greater the electrical potential. The electrical potential decreases exponentially from the surface of the dispersed entity to the electroneutral region. The magnitude of the electrical potential at the slipping plane is called the zeta (ζ) potential.

Zeta potential

The zeta potential of a dispersed entity tells us whether the electrical potential is negative or positive and the magnitude of the potential. When two dispersed entities move towards each other the magnitude of the zeta (ζ) potential will determine whether they aggregate or remain dispersed as separate entities. Dispersed entities with a low zeta potential (0–5 mV) are prone to rapid aggregation, whilst disperse entities with zeta potential values of > 30 mV would remain dispersed.

The ionic composition of the aqueous phase will influence the zeta potential of the disperse system. An increase in ion concentration can result in a reduction in the zeta potential. Changes in the pH of the aqueous phase can alter the ionization of ionic species in the continuous phase and the surface charge of ionizable groups; this can also result in a change in zeta potential. A zeta potential quoted without the continuous-phase composition and pH is almost meaningless.

Colloidal dispersed entities are constantly moving in solution due to Brownian motion (explained in Section 5.4.1). When adjacent disperse entities come into contact, one of three types of association can occur. Coagulation is the permanent contact between disperse entities, and results in an increase in the size of the dispersed entity and possible sedimentation. Flocculation is a reversible contact where the contacts formed break down upon the application of energy, such as shaking. Finally, entities may simply rebound and remain free, which occurs in stable colloidal disperse systems.

Whether the dispersed entities coagulate, flocculate or rebound after contact depends on the balance between the potential energies of attraction and repulsion between particles. When dispersed in a continuous phase, dispersed entities experience potential energy of attraction (V_a) due to van der Waals forces caused by induced-dipole–induced-dipole interactions. They also experience electrostatic potential energy of repulsion (V_r) related to their surface charge and zeta potential. The DLVO theory takes into account the balance between the energies of attraction and repulsion that the dispersed entities experience.

5.7.1 Energies of repulsion and attraction – DLVO theory

The total potential energy (V_T) is the sum of the potential energies of attraction and repulsion, as shown in

$$V_T = V_a + V_r. \tag{5.8}$$

As dispersed entities move towards each other in solution the potential energies of attraction and repulsion change. The closer the entities move toward each other the higher the potential energy of repulsion and at very close proximity the potential energy of attraction increases as shown in Fig. 5.18.

It the zeta potential of dispersed entities is low, they can move close to each other and this can result in a high potential energy of attraction due to van der Waals forces at this close proximity. It the potential energy of attraction exceeds the potential energy of repulsion, coagulation (irreversible aggregation) will occur, as shown in Fig. 5.18. The potential energy of attraction decreases exponentially with increases in distances between the dispersed entities. At moderate distances there is a high potential energy of repulsion due to electrostatic repulsion. If the electrostatic repulsion is large, compared to the potential energy of attraction, the dispersed entities stay dispersed.

However, as distance continues to increase a reduced potential energy of attraction (compared to the potential energy of attraction observed at close proximity) is observed. If the potential energy of attraction can overcome the potential energy of repulsion at this distance, the dispersed species will form loose assemblies of particles or droplets – referred to as floccules. The formation of these floccules is called flocculation (reversible aggregation). The potential energy of attraction holding the floccules together is weak and can be easily broken by agitation such as shaking.

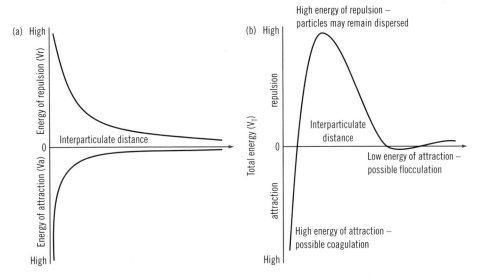

FIGURE 5.18 This diagram shows typical changes in the potential energy of repulsion and attraction between dispersed particles as the distance between dispersed particles increases. Figure (a) illustrates the changes in the separate potential energies of repulsion and attraction and figure (b) illustrates the total potential energy (the combined potential energies of attraction and repulsion).

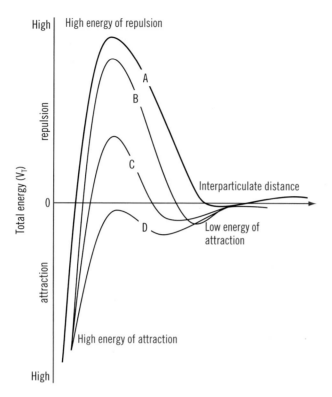

FIGURE 5.19 An illustration of the effect of electrolyte concentration on the total potential energy as the distance between particles increases. Curves A–D show typical changes in the total potential energy of dispersed particles as the electrolyte concentration increases from its lowest value A to the highest value D.

The zeta potential of dispersed entities can be altered by altering the concentration of ions in solution. The higher the ion concentration is the lower the zeta potential, as shown in Fig. 5.19. Also, at higher ion concentrations, the distance from the charged surface of the dispersed entity to the electroneutral region of the dispersion is reduced: there is compression of the electrical double layer around the dispersed entity.

The importance of zeta potential to the stability of disperse systems depends on whether the disperse system is lyophilic or lyophobic.

Optimizing dispersion stability by altering zeta potential

Lyophilic disperse systems are thermodynamically stable due both to their zeta potential and their interactions with the continuous phase. Both of these factors must be sufficiently weakened for coagulation or flocculation of the dispersed entities to occur. At low ion concentrations, changes in zeta potential do not sufficiently weaken the stability of these dispersions. At high ion concentrations, however, these lyophilic dispersed species may both lose water of hydration to the ions in solution and experience a change in zeta potential. This would result in the destabilization of lyophilic disperse systems as both mechanisms of stabilization are undermined.

Lyophobic colloids are thermodynamically unstable and primarily stabilized by electrostatic repulsion related to the magnitude of their zeta potential. They have no or minimal interaction with the continuous phase. An increase in the ion concentration can result in a reduction in zeta potential and destabilization of the disperse system. In contrast to lyophilic dispersion, low ion concentrations can affect the stability of lyophobic dispersions.

Controlled flocculation of coarse suspensions

Coarse suspensions are thermodynamically unstable because they are composed of lyophobic material. Unlike colloidal lyophobic suspensions, Brownian motion (explained in Section 5.4.1) is not observed in coarse suspensions due to the relatively large size of the dispersed particles; however, sedimentation (explained in Section 5.4.2) occurs in coarse suspensions. It is difficult to prevent sedimentation occurring during storage of coarse disperse systems because of the dispersed particle size. Therefore, a desirable pharmaceutical suspension is not a suspension free of sedimentation but rather one that can be easily redispersed after sedimentation.

Pharmaceutical suspensions can be difficult to resuspend after sedimentation due to the formation of a compacted cake during storage, a process called caking. The compacted cake forms when particles adhere to one another after sedimentation. The weight of sediment forces particles together and this can overcome the repulsive potential energy barrier. When particles are in close contact, the potential energy of attraction due to the van der Waals forces is strong, as shown in Fig. 5.18. Consequently, resuspension of these sedimented particles requires a high level of energy to overcome the attractive forces. With time, any growth or fusion of particles in the compacted aggregate will result in stronger forces of cohesion between particles, making it still harder to resuspend the sediment.

Caking is more evident in suspensions where particles do not aggregate and sediment as individual particles. Small particles fill voids created by the sedimentation of larger particles and cause the formation in a dense compacted cake. Caking can be prevented by reducing the zeta potential of the particles suspended by the addition of ions to the suspension. The reduction in zeta potential allows particles to come into sufficiently close proximity for flocculation to occur. The resultant loosely bonded floccules sediment rapidly due to their increased size. The floccules do not cake due to their low density and uniform size, and can be easily resuspended.

In addition to altering zeta potential, a dispersion can be stabilized by preventing the dispersed particles or droplets from coming into close contact. This mechanism is called steric stabilization. Steric stabilization occurs when hydrophilic polymer molecules are adsorbed onto the surface of lyophobic dispersed entities. In order to adsorb onto the surface of a dispersed particle or droplet, the polymers must have surfactant properties. Dispersions experiencing steric stabilization do not rely on zeta potential and electrostatic charge for their stability. Therefore, dispersions stabilized in this manner are more stable in the presence of added ions.

In summary, dispersed entities in aqueous dispersed systems have a surface charge that influences whether they aggregate (coagulation or flocculate) or remain dispersed. This surface charge cannot be measured but the zeta potential gives an indication of this charge. Ions in solution influence the zeta potential magnitude and their concentration can be exploited to optimize dispersion stability.

In this chapter we hope we have explained the main factors that influence the formation and stability of disperse systems: kinetic properties, viscosity, interfacial properties and surface charge.

◉ 5.8 SUMMARY

Disperse systems are composed of one phase dispersed through a second continuous phase. Various types of disperse system can be formed depending on the state of the disperse phase and the continuous phase, as shown in Table 5.1.

Drug molecules usually partition between the disperse phase and the continuous phase and can be present in both phases, but the concentration of drug is usually much greater in one phase.

Disperse systems can be classified as molecular (< 1 nm), colloidal (between 1 nm and 1 μm) and coarse dispersions (> 1 μm) – according to the properties exhibited. Some of these properties are listed in Table 5.2. Molecular dispersions are single-phase systems: the dispersed molecules are mixed intimately with the continuous-phase molecules. The disperse phase of colloidal dispersions can be lyophilic (solvent loving), lyophobic (solvent hating) or amphiphilic (both solvent loving and hating).

The physical stability of disperse systems is important to ensure uniform drug delivery. Uneven distribution of the disperse phase can cause uneven drug distribution throughout the disperse system and result in patients receiving under- and overdoses. Disperse-phase aggregation can have serious consequences for intravenous dispersion administration, as aggregates can cause embolisms.

Dispersed entities move in dispersions due to Brownian motion, diffusion and sedimentation. Only submicrometre entities experience Brownian motion, which is due to the collision of the continuous phase molecules with the dispersed entity. The rate of disperse-phase sedimentation is influenced by the force of gravity and drag force caused by air or liquid resistance to movement. Sedimentation can be reduced by decreasing the dispersed particle size and increasing disperse-phase viscosity.

Viscosity describes the resistance of a liquid to flow and the higher the viscosity of a liquid the greater the resistance of the liquid to flow when shear stress is applied. Liquids can exhibit Newtonian or non-Newtonian flow. Pharmaceutical dispersions often exhibit non-Newtonian flow such as plastic and pseudoplastic flow. Disperse systems in which changes in viscosity caused by the application of shear stress reverse over a period of time are called thixotropic. As the concentration of the disperse phase increases, the viscosity of the disperse system increases. Semisolid disperse systems can exhibit viscoelastic behaviour.

Increases in interfacial tension and interfacial contact area, increase interfacial free energy and can destabilize disperse systems. Therefore, disperse phases of coarse dimensions have greater physical stability compared to those of colloidal dimensions. The adsorption of surfactants at the interfacial region can reduce interfacial tension and thereby stabilize disperse systems.

Surfactants exhibit both hydrophilic and hydrophobic behaviour and in solution, distribute themselves at surfaces, interfaces and the bulk solution. At the critical micelle concentration (CMC) micelles form in solution, micelles are assemblies of surfactant molecules.

Emulsions can be water droplets dispersed in oil (w/o) or oil droplets dispersed in water (o/w). They are thermodynamically unstable systems and are stabilized by reducing the interfacial tension between the dispersed droplets and continuous phase by the addition of surfactants or finely divided solids. Microemulsions contain higher levels of surfactants and are thermodynamically stable systems.

The surface charge of dispersed entities influences disperse-system stability. The surface charge can be due to adsorbed ionic species from solution onto the surface of the entity or the ionization of surface groups on the dispersed entity. The kinetic dispersed entity has a tightly bound layer of ions attached to its surface and a cloud of diffuse ions surrounding it, this is called the electrical double layer. The electrical potential of the surface of the slipping plane is called the zeta potential.

The greater the zeta potential of the dispersed entity, the greater the potential energy of repulsion. When particles come into contact, if the potential energy of repulsion is not strong enough to overcome the potential energy of attraction particles will form irreversible contact (coagulation) or reversible contact (flocculation). However, if the potential energy of repulsion is greater than the potential energy of attraction particles will remain dispersed. A change in pH or the addition of ions can alter the zeta potential of dispersed particles.

⊜ REFERENCES

Barnes, G.T. and Gentle, I.R. (2005). *Interfacial Science.* Oxford University Press, Oxford.

Griffin, W.C. (1949). Classification of surface-active agents by HLB, *Journal of the Society of Cosmetic Chemists* **1**, 311–326.

ⓜ FURTHER READING

Aulton M. E. (ed). (2002). *Pharmaceutics, the science of dosage form design* (2nd edn). Edinburgh: Churchill Livingstone.

Barnes, G.T. and Gentle, I.R. (2005). *Interfacial Science.* Oxford University Press, Oxford

Florence, A.T. and Attwood, D. (2006). *Physicochemical principles of pharmacy* (4th edn). Pharmaceutical Press, London.

Lieberman, H.A., Rieger, M.M. and Banker, G.S. (1996). *Pharmaceutical dosage forms: Disperse systems,* Marcel Dekker, New York.

❓ EXERCISES

5.1 A pharmaceutical suspension is prepared by dispersing drug particles with particle radii ranging between 200 μm and 1.2 mm. Which end of the particle size range (the smaller particles or larger particles) will sediment fastest and why?

5.2   A concentrated pharmaceutical suspension has been shown to exhibit plastic flow. The plastic viscosity (η_p) of this sample was found to be 2.4 Pa s. When a shear stress of 620 Pa is applied the measured strain is 180 s^{-1}. Calculate the yield value for this suspension.

5.3   From a series of dynamic viscosity values for a range of dilutions of a polymer solution the intrinsic viscosity [η] was found to be 1.75×10^3 ml g^{-1}. Calculate the average molecular weight of this polymer (assuming that the constant values are a equals 0.95 and K equals 0.84 ml g^{-1}).

5.4   The interfacial tension between the dispersed oil drops and the aqueous continuous phase of an emulsion is reduced by the addition of surfactants A and B. Surfactant A reduces the interfacial tension of the emulsion to 12 dynes cm^{-1} and surfactant B reduces interfacial tension to 8 dynes cm^{-1}. If the oil droplets are equivalent in size in both emulsions, which surfactant will produce the most stable emulsion?

PHARMACEUTICAL DISPERSE SYSTEMS

06

Learning objectives

Having studied this chapter, you should be able to:

- describe the various types of disperse systems that can be used for drug delivery: suspensions, gels, emulsions, micellar systems, liposomal systems and foams
- describe the pharmaceutical applications of each of these disperse systems
- appreciate the factors that influence their design
- explain the factors that influence their stability.

In this chapter we will focus on the use of disperse systems as pharmaceutical drug-delivery systems (dosage forms). There are a wide range of dosage forms that can be classified as pharmaceutical disperse systems, and their properties can vary widely depending on the type of disperse system. A general introduction to disperse systems and the properties that influence their stability is given in Chapter 5. In this chapter we will describe a range of disperse systems in a little more detail and highlight their pharmaceutical applications.

6.1 PHARMACEUTICAL SUSPENSIONS

Pharmaceutical suspensions are disperse systems in which drug particles are dispersed in a continuous liquid phase. Pharmaceutical suspensions are used to deliver drug substances with poor aqueous solubility by a wide range of routes: orally, to the ear, eye and nose, topically, intramuscularly, subcutaneously, rectally and, more recently, intravenously.

Suspensions are advantageous because they can be used to deliver extended release dosage forms. These extended release dosage forms consist of particles containing a drug encased in a polymer matrix that delays drug release, such as cellulose derivatives,

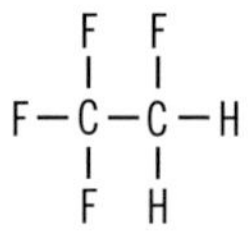

```
      F   F
      |   |
  F—C—C—H
      |   |
      F   H
```

FIGURE 6.1 The chemical structure of Norflurane.

poly lactic acid (PLA) and poly (lactic-co-glycolic acid) (PLGA). An example of these extended release suspension formulations is a prolonged-release, intramuscular injection of a GnRH agonist, leuprorelin acetate (Prostap SR®). This injection is administered monthly for the treatment of prostate cancer, endometriosis and uterine fibroids.

The continuous phase of pharmaceutical suspensions can be aqueous or non-aqueous. Aqueous suspensions of drugs are used for administration orally, rectally and parenterally (by injection). Prior to administration, some pharmaceutical aerosol formulations are stored in pressurized canisters containing a suspension of drug particles in non-aqueous volatile liquids such as norflurane (its chemical structure is shown in Fig. 6.1). Suspensions applied to the skin are often non-aqueous suspensions – for example, magnesium sulfate paste or zinc cream (containing zinc oxide).

Oral suspensions

Oral liquid dosage forms are required for patients who have difficulty swallowing tablets and capsules. Suspensions of drug particles are used as oral liquid dosage forms in situations where oral solutions are not feasible due to the drug's poor aqueous solubility. A number of poorly water-soluble antibiotic and analgesic drugs are administered as oral suspensions – for example, erythromycin and ibuprofen.

Compared to oral solutions, oral suspensions can be advantageous for bitter-tasting drugs. A patient's perception of bitter taste can be reduced by administering the drug in the form of a suspension rather than a solution. In so doing, the concentration of drug molecules in solution exposed to the patient's taste buds is reduced, and so the drug doesn't trigger the sensation of bitter taste to the same degree. For example, the antibiotic chloramphenicol has such a strong bitter taste that it is not deemed suitable for administration as an oral solution. To address this problem an inactive pro-drug chloramphenicol palmitate, which has a lower solubility than chloramphenicol, is administered as an oral suspension. (Chloramphenicol palmitate is hydrolysed after administration to form the active drug chloramphenicol.) The chloramphenicol palmitate oral suspension has a milder taste than the chloramphenicol oral solution and is better tolerated.

Injectable suspensions

A number of poorly water-soluble drugs are administered intramuscularly as suspensions. Compared to solutions, suspension formations have a longer onset of action when injected because the drug particles have to dissolve prior to exerting their pharmacological activity. Pharmaceutical suspensions injected intramuscularly and subcutaneously were amongst the earliest so-called modified release drug-delivery systems. For example, long-acting insulin injections contain suspended insulin particles. A wide range of vaccines are administered as intramuscularly injected suspensions.

The use of suspensions to deliver drugs intravenously is limited due to the risk of embolism caused by the blockage of blood vessels by dispersed particles. For intravenous administration, the size of drug particles must be reduced to submicrometre levels. However, a balance must be struck: submicrometre particles have increased surface area to mass ratios compared to coarse suspensions, and therefore have a greater tendency for aggregation.

The first approved intravenous submicrometre or nanosuspension of a drug was a paclitaxel nanosuspension (Abraxane®), for the treatment of breast cancer. It is supplied as a sterile powder for resuspension in a continuous phase of 0.9% sodium chloride prior to administration. The paclitaxel nanosuspension formulation enables larger doses to be delivered by infusion over a shorter period of time compared to a marketed paclitaxel formulation containing a non-ionic surfactant Cremophore EL® (Taxol®). The hypersensitivity reaction associated with the Cremophore EL® in the Taxol® formulation is not associated with the nanosuspension formulation.

Nanosuspensions

A nanosuspension usually refers to nanoparticles dispersed in a continuous liquid phase. In pharmaceutical terms a nanoparticle or nanocrystal refers to a drug particle with a diameter < 1000 nm (submicrometre particles). Nanosuspensions are of interest in intravenous drug delivery as mentioned above, and allow the delivery of larger doses of poorly water-soluble drugs compared to liposomal or micellar formulations. These formulations may include surfactants for stabilization but the levels used are lower than those required in liposomal and micellar formulations.

As a result of this increased contact area between drug solid particles and continuous phase, the rate of dissolution for nanoparticles is increased compared to coarse suspensions. Due to the enhanced dissolution, nanosuspensions can be used to increase the oral bioavailability of poorly water-soluble drugs with good gastrointestinal permeability.

6.1.1 Factors that influence the stability of pharmaceutical suspensions

A major disadvantage of suspensions as dosage forms is their physical instability. Physical instability problems associated with coarse dispersions are primarily caused by sedimentation leading to caking and subsequent difficulty during resuspension (covered in greater detail in Chapter 5). The rate of sedimentation can be reduced by increasing the viscosity of the continuous phase of the suspension. Commonly used excipients to increase the viscosity of suspensions are listed in Table 6.1, these are called suspending agents.

In suspensions, there is a tendency for smaller particles to dissolve and the dissolved molecules to regrow on larger particles. This phenomenon is called Ostwald ripening and is explained in greater detail in Section 4.5.3. In colloidal suspensions this process results in an increase in the size of larger particles and a reduction in the number of small particles. Why is this an issue? As particles grow they are more prone to sedimentation. The presence of large particles in topical pastes, for example, can result in a gritty texture during application. Also, the rate of dissolution is slower for larger particles compared to

TABLE 6.1 Example of commonly used suspending agents.

Type	Examples
Gums	Acacia, Xanthan, Tragacanth, Guar
Cellulose derivatives	Hydropropylmethylcellulose (HPMC), Carboxymethylcellulose (CMC), Methylcellulose (MC)
Polysaccharides	Alginates, Pectin, Gelatin, Carageenan
Colloidal particles	Silicon dioxide, Bentonite, Magnesium and Aluminium hydroxide
Synthetic polymer	Carbomer, Poloxamer, Polyvinylpyrrolidone (PVP), Polyethylene glycol (PEG)

smaller particles. So, as the number of large particles increase due to Ostwald ripening, the dissolution rate of a drug into solution can reduce.

Ostwald ripening can be minimized by minimizing the particle-size distribution in the suspension. The addition of the cellulose derivative, carboxymethylcellulose, to the suspension has been shown to prevent Ostwald ripening in some circumstances.

6.2 PHARMACEUTICAL GELS

Gels can be considered to be disperse systems in which a three-dimensional network of colloidal particles or molecules is surrounded by a liquid phase. Gels are solid-in-liquid dispersions due to the fact that the liquid phase is considerably greater than the disperse phase: in pharmaceutical gels the liquid phase normally comprises > 90% w/w of the gel. Gels are referred to as semisolid materials with properties falling between those of a liquid and a solid. They exhibit viscoelastic rheological properties.

Pharmaceutical gels are commonly used to deliver drugs locally to the skin, eye, mouth (oral mucosa) and vagina. They are also used for transdermal delivery of drugs such as analgesics and hormones. Pharmaceutical gels offer a number of advantages over other dosage forms applied topically (lotions, creams, ointments and pastes). First, the levels of surfactants and lipids required to create a semisolid texture in gels is low compared to other semisolid dosage forms, therefore, they tend to be less irritating to broken skin and mucosal tissue. Second, gels have a relatively high viscosity compared to topically applied lotions and therefore are easier to apply. Finally, the liquid phase of the majority of pharmaceutical gels is composed of water or a water–alcoholic mixture. This high water content means that they are non-greasy and absorbed into the skin or mucosal tissue where they are applied.

The viscosity behaviour of gels can be exploited to optimize drug delivery. For example, poloxamer gels have thermoresponsive properties, they are liquid when refrigerated and gel around body temperature. This property can be used to administer drugs to body surfaces such as wounds, conversion from a liquid to a gel after application facilitates their retention on the wound. The reason for their sol–gel transitions on heating is explained in Section 6.2.1.

A second example is Timoptal-LA ophthalmic gel-forming solution®, containing timolol and used to treat elevated intraocular pressure. This product exploits the thixotropic nature of liquids (explained in Section 5.5.1): when shaken it has the viscosity of a liquid and can be applied to the eye as an eye drop. The eye drop is then transformed to a gel after administration into the eye. Because of the viscous nature of the gel, the timolol is released slowly from the dosage form to the eye. Compared to the standard Timoptol ophthalmic solution®, which is administered twice daily, Timoptal-LA ophthalmic gel-forming solution® acts for a longer period of time and only requires once-a-day administration.

6.2.1 Gel formation

Gels can be formed from dispersed lyophobic or lypophilic colloidal material, the concentration of dispersed colloidal particles or molecules required to form a gel depends on their ability to form a three-dimensional network. The structure and strength of the network formed depends on the degree of interaction between the colloidal particles or molecules in the network, and the nature of bonds formed.

Lyophobic gelling agents

Lyophobic colloids that form gels include insoluble clay particles (naturally occurring aluminium magnesium silicate and aluminium silicate) and magnesium and aluminium hydroxide. The network of clay particles is formed by an electrostatic attraction between positively charged ions (Mg^{2+} and Al^{3+}) and negatively charged ions (O^-) on the clay surface in suspension. Magnesium and aluminium hydroxide gels form due to weak van der Waals interactions between particles, and by the formation of floccules in suspension. Relatively high concentrations of lyophobic particles (approximately 10% w/v or greater) are required to form gels.

The interactions formed between lyophobic gelling agents are relatively weak. As a result, these networks are easily destroyed by agitation such as simple shaking. After shaking, and upon standing, the network will reform and viscosity increases. These types of changes in viscosity due to stress are called gel–sol–gel transformations.

Gel networks formed due to electrostatic interactions and flocculation can be influenced by pH and agitation. The viscosity of these gels can change in response to changes in pH. For example, bentonite (aluminium silicate) gel is viscous above pH 6 but below pH 6 the bentonite particles precipitate and the gel is destroyed.

Lyophilic gelling agents

Lyophilic colloidal materials form gels by the formation of a network of solvated macromolecules in the liquid phase. These gel-forming macromolecules are naturally occurring polysaccharides and gums, and synthetic polymers. They disperse in the liquid phase and form an irreversible molecular network (type I gel) or a reversible molecular network (type II gel). Type II gels are the commonest form of gels used in drug delivery.

Type I gels are formed by the covalent crosslinking of polymers. When water is added to dry crosslinked polymers, a gel is formed and, as more water is added, the gel swells. Gels of this type are used as expandable implants (to fill body cavities) and as contact lenses. They are also of interest for the sustained delivery of drugs: the gel network can prolong

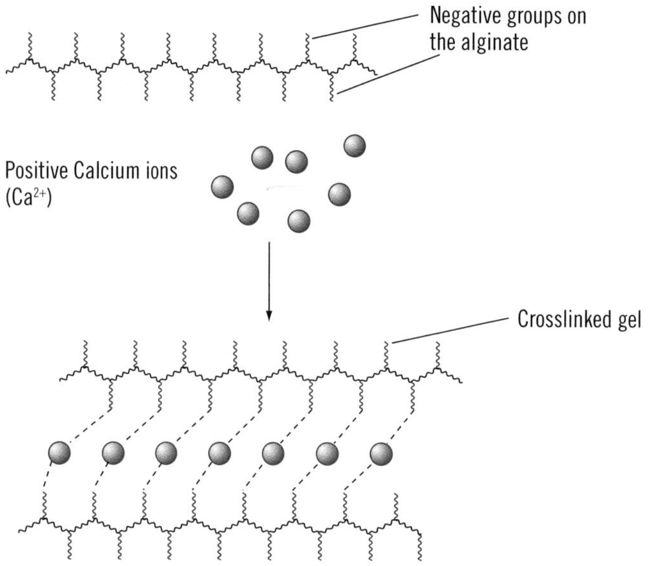

FIGURE 6.2 Schematic diagram showing alginate crosslinked with Ca^{2+} ions.

drug release. However, their drug-delivery use is restricted due to the irreversible nature of the polymer network that results in a non-biodegradable system.

Type II gels are formed by weaker molecular interactions compared to type I gels. The macromolecular network of type II gels is formed primarily by hydrogen bonding. It can also be formed by electrostatic interactions due to the addition of a divalent or trivalent crosslinking molecule, as shown in Fig. 6.2. Temperature changes can reduce the viscosity of type II gels. A common example of this is the preparation of edible jelly: a low viscosity jelly solution is prepared by heating the jelly in warm water. However, upon cooling, the solution sets and forms a rigid jelly. The point at which it sets is called the gel point.

Thermoresponsive gels

Polymers that form gels upon heating are thermoresponsive gels. The temperature at which the gel forms is called the lower critical solution temperature (LCST). This unexpected behaviour is due to the surfactant nature of these polymers. At high temperatures (above the LCST) non-ionic surfactants lose the water of hydration, which results in a reduction in hydrogen bonding associated with the hydrophilic segment of the non-ionic surfactant and solutions are converted from a solution to a gel.

Examples of these types of polymers are poloxamers, which are block copolymers with surfactant properties. Block copolymers are comprised of two different polymer chains repeated in long sequences, or blocks, for example -A-A-A-A-A-B-B-B-B-B-B-A-A-A-A- is an AB diblock copolymer. Poloxamers, illustrated in Fig. 5.12, are non-ionic surfactants. They form micelles in solution above their critical micelle concentration (CMC). At low temperatures in aqueous solutions, the polyoxyethylene hydrophilic portion of the molecule is hydrated. When the temperature is raised above the LCST, there is a reduction in the hydrogen bonding associated with the hydrophilic segment of the molecule and the hydrophilic chains become dehydrated. Because of this dehydration, the hydroxyl groups on adjacent chains become more accessible and interact with one another. This type of gel network is micellar in nature.

The sol–gel transition that occurs as the temperature increases is reversible on cooling. The LCST for poloxamers is around body temperature such that these gels undergo a sol–gel phase conversion at body temperature. This behaviour makes them interesting excipients for drug-delivery systems.

The viscosity of lyophilic gels can also be altered by changing the pH or by adding electrolytes. Some gelling agents (certain grades of carbomers and gelatin) require pH adjustment to create the gel. Changes in pH can alter the viscosity of gels after preparation. Gels formed from sodium alginate are most stable above pH 4. Below pH 4, sodium alginate is converted to alginic acid and it precipitates. The viscosity of gels formed from sodium carboxymethylcellulose is markedly reduced below pH 5 and above pH 10.

Non-aqueous gels

The continuous phase of the majority of pharmaceutical gels is composed of water. These are referred to as hydrogels. Organogels are a class of gels where the continuous phase is an organic liquid immobilized by a three-dimensional network of polymers or low molecular weight fibres. They are of interest in the area of drug delivery but are outside the scope of this text.

6.2.2 Drug-loaded gel systems

Drug substances can be incorporated into gels either dissolved or suspended in the liquid phase. The viscous nature of a gel brings with it the advantage of stabilizing the dispersed drug particles. Drugs are usually released from gel systems by molecular diffusion from solution in the gel, to the surrounding liquid. However, the suspended drug must first dissolve in the liquid phase prior to release.

The rate of drug diffusion through the gel is influenced by the dimensions and molecular weight of the drug, and the macromolecular or particulate network within the gel, which in turn influences the viscosity of the gel and the degree of swelling due to liquid uptake. As the degree of swelling increases, the rate of drug release from the gel increases. However, as the number of interactions between molecules or particles in the gel increases the rate of drug release decreases because the pathway of drug release becomes more indirect.

Hydrogels are an attractive means of delivering protein and peptide molecules due to their high water content, which helps to maintain the structural stability of these molecules. The rate of release of protein and peptide drugs from hydrogel systems can be prolonged by increasing the number of interactions between the macromolecular chains in the gel structure. This mechanism is used to control the release of drugs from many controlled release dosage forms. For controlled-release tablets, a gelling agent can be added to the tablet blend or coated onto the surface of the tablet. Upon hydration of the tablet, a gel forms on the exterior of the tablet and slows down the rate of drug release.

6.2.3 Stability of pharmaceutical gels

The main factors that influence the stability of pharmaceutical gels are microbial growth, dehydration, changes in pH, and temperature. Due to their high water content, hydrogels can support microbial growth that can adversely affect the patient and cause a loss in gel structure. Preservatives such as methylparahydroxybenzoate and propylparahydroxybenzoate are commonly included in pharmaceutical hydrogel formulations.

Changes in temperature and pH can affect the viscosity of gels, as explained in Section 6.2.1. Therefore, many gel formulations contain salts that buffer the pH of the gel. The semisolid structure of gels is the result of both the molecular or particulate network and the restricted liquid phase. If the liquid phase is lost through evaporation, the gel will become brittle and hard. Excipients such as glycerol and sorbitol are added to gel formulations to prevent evaporation of water. Molecules that prevent the evaporation of water in this way are called humectants.

The interaction between the dispersed molecular or particulate network and the restricted liquid phase influences the behaviour of gel systems. When placed in a humid or aqueous environment, hydrogel systems can take up considerable amounts of water. If the uptake of water is accompanied by a change in volume, this process is called swelling. When a change in volume does not occur with water uptake, the process is called imbibition. Gels can also shrink and squeeze out the liquid phase. When this occurs, the liquid phase can be seen on the surface of the gel as a layer of liquid or as droplets. This process it called syneresis.

6.3 PHARMACEUTICAL EMULSIONS

Pharmaceutical emulsions are used to deliver drugs locally to the skin, eye, ear, nose, vagina or rectum. The viscosity of pharmaceutical emulsion systems can vary from low viscosity lotions to viscoelastic creams. Emulsions are used for the delivery of drugs such as anti-inflammatory, antimicrobial and anaesthetic agents. Oil-in-water (o/w) emulsions can be used for the oral delivery of oils such as castor and cod liver oil. Oral emulsions can also be used as an alternative to oral suspensions for the delivery of poorly water-soluble drugs.

The drug can be either dissolved in the aqueous or oil phases of the emulsion or partitioned between the two phases. Where the concentration of a drug present in the emulsion exceeds its solubility, the drug will be suspended as drug particles in addition to being present in solution. Therefore, the drug can be present in three phases: as drug particles, as a drug in solution in the continuous phase, and as a drug in solution in the dispersed droplets.

The formulation of oral emulsions should be designed with flavourings to minimize the oily taste and feel of the oil component of the emulsion in the mouth. O/w emulsions are also used to deliver poorly water-soluble drugs by injection. Examples of this are the intravenous delivery of the poorly water-soluble anaesthetic, propofol and the delivery of nutritional lipids intravenously, as explained in Box 5.1.

Use of emulsions as moisturizers and emollients

O/w and w/o emulsions are widely used to treat dry skin conditions by rehydrating the skin. These products are classified as emollients and moisturizers. Emollients smooth and soften the skin while moisturizers increase the water content of the skin (which also has the effect of smoothening and softening the skin). Products often have both moisturizing and emollient effects.

Due to their aqueous continuous phase, w/o emulsions are absorbed into the skin, adding water to the skin and leaving a thin layer of oil on the surface. They are called non-greasy creams. The layer of oil, which is hydrophobic, prevents the loss of water from the skin and therefore increases the levels of water in the skin. The covering of skin with a barrier that prevents water loss in this way is called occlusion.

Topical w/o emulsions are referred to as greasy creams or ointments. They are not absorbed into the skin due to the oily nature of the continuous phase. The layer of oil deposited on the surface of skin is thicker than that deposited by non-greasy creams. Patients find w/o emulsions greasy to use.

Emollient and moisturizing creams can be rubbed directly onto the skin. They can also be applied by dispersing in warm water (for example while taking a bath). In the presence of water, emulsions can disperse due to the surfactant nature of the emulsifying agents present in emulsions. Emulsions therefore can be used as bath additives or as cleansers for patients with dry-skin conditions.

Factors that influence the stability of emulsions

Emulsions are thermodynamically unstable even with the inclusion of emulsifying agents. The types of instabilities observed in emulsions are cracking of the emulsion into separate oil and water phases, creaming, and phase inversion. Any factor that weakens the interfacial film at the oil/water interface will affect emulsion stability.

The addition of chemicals that are incompatible with the emulsifying agent, or the addition of an additive with an opposite charge to an ionic emulsifying agent, will cause cracking. High temperatures can affect the solubility characteristics of non-ionic surfactants. For example, proteins that act as emulsifying agents are destroyed at high temperatures. Indeed, most emulsions crack above 70 °C. Freezing will destroy most emulsions due to disruption of the interfacial film by ice formation. The interfacial films can also be destroyed by bacterial growth, which is supported by non-ionic surfactants and proteins, again leading to the cracking of emulsions. Droplets in emulsion can coagulate and flocculate, as is observed in pharmaceutical suspensions, due to their close proximity. Therefore, any weakness in the interfacial film will result in the coalescence of droplets and consequent cracking of the emulsion.

Phase inversion is the conversion of an o/w emulsion to a w/o emulsion or vice versa – it changes the properties and behaviour of the emulsion. Attempts to incorporate excess disperse phase can result in phase inversion. Emulsions with > 70% dispersed phase tend to crack or phase-invert. Phase inversion can also be caused by adding additives to alter hydrophilic–lipophilic balance (HLB) of the emulsifying agents. Heating a non-ionic surfactant reduces its hydrophilic nature; this in turn reduces the HLB, which can cause the emulsion to invert.

Creaming occurs in o/w emulsions, it involves the disperse phase rising to the top with time due to the different relative densities of disperse and continuous phases. However, if the droplets do not coalesce they can be redistributed on shaking. Creaming is not a serious stability issue but can result in an inelegant product and can cause dosage errors. Homogenization reduces the extent of creaming in emulsions by reducing the size of the dispersed droplets and increasing the viscosity of the emulsion.

6.4 PHARMACEUTICAL MICROEMULSIONS

There is major interest in microemulsions as vehicles of oral delivery of poorly water-soluble drugs. Microemulsions increase the solubility of poorly water-soluble drugs and have also been shown to increase their oral bioavailability. They can be formed in the gastrointestinal tract when a mixture of surfactants and lipids is delivered to a patient. This mixture is called a self-microemulsifying drug-delivery system (SMEDDS). These SMEDDS form o/w microemulsions in the presence of water with mild agitation. Neoral® containing cyclosporin was the first SMEDDS approved for oral delivery and has been available to patients since 1995. Cyclosporin is an immunosuppressant drug indicated in the treatment of transplant patients. The improved bioavailability of the Neoral® formulation compared to an earlier marketed formulation of cyclosporine, Sandimmun®, has been attributed to the microemulsion formed after administration.

The use of microemulsions for intravenous drug delivery is limited due to the requirement for high levels of biocompatible surfactants in their formulation and associated risks of toxicity. In some circumstances, when used to deliver drugs topically, microemulsions have shown greater bioavailability compared to emulsion, solution and gel formulations. They are also of interest for the delivery of drugs to the eye due to their optical properties and transparent appearance.

6.5 PHARMACEUTICAL MICELLAR SYSTEMS

Surfactant molecules can assemble in aqueous systems and form micelles above the critical micelle concentration, as explained in Section 5.6.3 and shown in Fig. 5.4. Micelles are three-dimensional spherical structures composed of between 20 and 50 molecules, which are dynamic rather than static. In micellar systems surfactant molecules are present in three states: as individual molecules in bulk solution, adsorbed at the surface, and as micelles, as shown previously in Fig. 5.4. At equilibrium, these molecules are constantly moving from one state to another while the portion of molecules in each state remains constant.

Due to their amphiphilic nature surfactants can disperse in both hydrophilic and hydrophobic solvents. When dispersed in hydrophobic solvents the hydrophobic portion of the molecule is on the outside and the hydrophilic portion faces inwards. These are referred to as reverse or inverted micelles.

The main pharmaceutical interest in micelles is their ability to solubilize poorly water-soluble drugs. The interior core of the micelle is a hydrophobic environment, which is

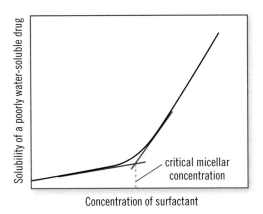

FIGURE 6.3 A plot showing the increase in the aqueous solubility of poorly water-soluble drug above the surfactant critical micellar concentration.

capable of dissolving poorly water-soluble drugs, as shown in Fig. 6.3. Solubilization is the process by which poorly water-soluble substances are brought into solution by incorporation into micelles.

Influence of temperature on micelle stability

The temperature of a surfactant solution can influence whether micelles form in solution due to the relationship between solubility and temperature, as explained in Chapter 2. To form micelles, the solubility of the surfactant molecules in solution must be above the critical micelle concentration (CMC): if the surfactant solubility is reduced below the CMC due to a decrease in solution temperature, micelles cannot form. The temperature at which the solubility of the surfactant equals the CMC value is called the **Krafft point**.

The characteristics of micelles formed from non-ionic surfactants can differ from those formed from ionic surfactants. While the hydrophobic portions of both molecules are similar, there are differences in the properties of the hydrophilic portions. The exterior of ionic surfactant micelles in water (the hydrophilic surface of the micelle) can be thought of as a compact layer of charged chemical groups. On the other hand, the exterior of non-ionic surfactant micelles in water can be thought of as a region of hydrophilic chains with hydrogen-bonded water attached to the chains and mechanically trapped water molecules within the chains. This region is referred to as the **palisade layer**. A schematic diagram of a non-ionic micelle indicating the palisade layer is shown in Fig. 6.4.

For most solutions, as temperature increases, solubility normally increases, but for non-ionic surfactants above a certain temperature this relationship is no longer true. Non-ionic surfactants, unlike ionic surfactants, are much more sensitive to increases in temperature. Heating a non-ionic surfactant solution to high temperatures can turn the solution from clear to cloudy. The cloudy appearance is due to the precipitation of the surfactant from solution. The change in the surfactant solubility that occurs on heating is reversible and if the solution is cooled it will become clear again. The temperature at which this clear to cloudy change occurs is called the **cloud point**. The precipitation of the surfactant from the solution at the cloud point is due to a loss of water of hydration and a reduction in hydrogen bonding associated with the hydrophilic segment of the non-ionic

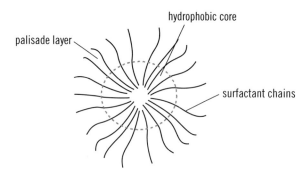

FIGURE 6.4 A schematic diagram of a micelle formed from non-ionic surfactants, indicating the hydrophobic core and the hydrophilic palisade layer.

surfactant. Concentrated solutions of non-ionic surfactants containing micelles convert from a solution to a gel when heated above the cloud point temperature. This behaviour is observed in thermoresponsive gels, as explained in Section 6.2.1.

Pharmaceutical applications

Micellar systems containing poorly water-soluble drugs can be used for drug delivery via a range of routes. One of their main applications is the solubilization of cancer drugs for delivery by intravenous infusion. Drug loading in polymeric micelles has been shown to prolong the presence of the drug in the blood stream (systemic circulation). The small size of the drug-loaded micelles prevents their uptake by phagocytes, yet they are large enough to slow down renal excretion. The micelle can also protect the loaded drugs from biological agents that could inactivate the drug.

Micelles can be designed to deliver drugs specifically to the tumour tissues. Due to their small size, micelle carriers accumulate in regions of the body with leaky blood vessels such as tumours. Functional chemical groups attached to the exterior of the micelles can be designed to target specific markers at the site of action.

Care must be taken in the choice of surfactant selected to form micelles, particularly for intravenous delivery. For example, Cremophor EL®, the surfactant used in Taxol® (a paclitaxel formulation), is associated with hypersensitivity reactions in some patients. Further, dilution of micellar drug formulations can result in the precipitation of the drug if the concentration of surfactant drops below the critical micellar concentration. Micelle instability can result due to elevated temperatures, changes of pH and addition of additives that affect the hydrophilic–lipophilic balance (HLB) of the constituent surfactants, see Section 5.6.3.

6.6 PHARMACEUTICAL LIPOSOMES

Liposomes, similar to micelles, are molecular assemblies of amphiphilic molecules. However, the liposomal structures are more complex structures than micelles. Phospholipids, the main component of liposomes, contain hydrophilic and hydrophobic regions within the same molecule (they are amphiphilic) and therefore exhibit surfactant behaviour. When

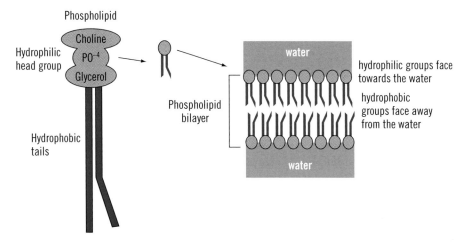

FIGURE 6.5 A basic diagram of the hydrophobic and hydrophilic regions of a phospholipid molecule and its arrangement into a bilayer structure.

a suspension of phospholipids is agitated in an aqueous medium the phospholipid molecules self-assemble into bilayer sheets that bend to form enclosed vesicles. The basic structure of phospholipids and the phospholipid bilayer are represented in Figs 6.5 and 6.6. The structure of liposomes is similar to cell membranes: both are bilayer structures composed of phospholipids. Due to the structural similarities between liposomes and cell membranes they can readily interact with each other. (Cell membranes are described in greater detail in Chapter 7.)

The ability of liposomes to interact with cell membranes makes them attractive drug-delivery vesicles. Also attractive is the ability of liposomes to encapsulate both hydrophilic and hydrophobic drugs within the same vesicular structure due to the presence of hydrophilic and hydrophobic regions. Molecules encapsulated in the core of liposomes cannot easily diffuse through the lipid bilayer and are trapped within the structure. Drugs encapsulated in the liposome can be protected for enzymatic metabolism, prolonging drug activity. Delivering drugs encapsulated in liposomes can also minimize the side effects associated with the drug by reducing the dose of drug exposed to non-target biological tissue. Liposomes can be designed to direct the encapsulated drug to specific biological targets by attaching specific chemical groups that have affinity for the biological target to the surface of the liposome.

Liposomal properties

Liposomes can be classified according to the number of bilayers present (lamellarity): multilamellar vesicles (MLV), oligolamellar vesicles (OLV), small unilamellar vesicles (SUV) and large unilamellar vesicles (LUV) as shown in Fig. 6.6. They range in size from 20 nm to 10 μm.

The composition of liposomes influences their resultant properties. Specifically, the type of phospholipid comprising the liposome or the addition of cholesterol to the liposome structure can influence the rigidity and surface charge. The majority of liposomal preparations approved for human use feature phosphatidylcholine (which possesses fatty

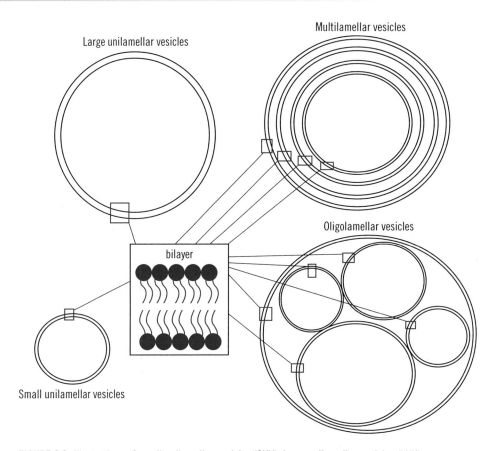

FIGURE 6.6 Illustrations of small unilamellar vesicles (SUV), large unilamellar vesicles (LUV), multilamellar vesicles (MLV) and oligolamellar vesicles (OLV). Note these are two-dimensional representations of three-dimensional structures.

acyl side chains of various lengths, and various degrees of saturation) as the main component, together with cholesterol. Figure 6.7 shows the chemical structures of phosphatidylcholine and cholesterol.

The charge on the surface of liposomes can be negative, positive or neutral. Phosphatidylcholine gives liposomes a neutral charge, stearylamine groups a positive charge and phosphatidic acid groups a negative charge. The surface chemistry of liposomes can be readily modified by attaching molecules to the surface. Modifying the size, lamellarity, charge or surface functional groups influences the distribution and fate of the liposomes in the body.

Pharmaceutical applications of liposomes

Liposomes can be used to administer drugs orally, topically, transdermally and parenterally. After intravenous injection, liposomes can disintegrate in the systemic circulation or be cleared by uptake by macrophages or escape from the circulation via leaky blood vessels such as those found at sites of inflammation and tumour sites. The leaking of liposomes from the systemic circulation at these sites is advantageous for the delivery

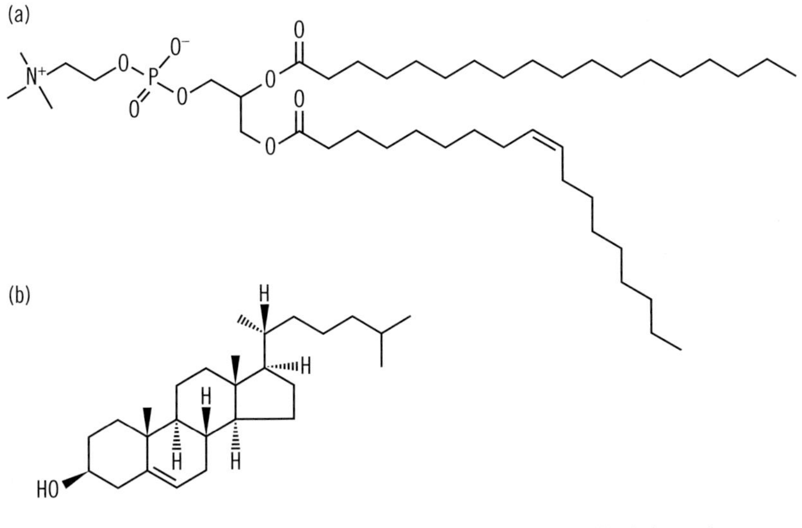

FIGURE 6.7 The chemical structure of (a) phosphatidylcholine and (b) cholesterol.

of drugs to treat inflammation and cancer. A number of pharmaceutical liposomal products are currently marketed; details of some of these are given in Table 6.2. To understand how liposomes are distributed in the body, liposomes can be also classified according to their surface properties into (1) conventional (neutral or negatively charged), (2) cationic, (3) stealth and (4) antibody-targeted liposomes.

Conventional liposomes after intravenous administration usually become coated with serum proteins and are taken up by macrophages and eliminated. Interaction of liposomes with these serum proteins can destabilize the liposomal structure, cause release of the loaded drug and disintegration of the liposome. The rate and extent of uptake of liposomes is dependent on their rigidity and size: small rigid liposomes are retained in the blood system for hours; however, most conventional liposomes are removed from the circulation too fast to accumulate in adequate amounts in the sites of leaky blood vessels.

The uptake of liposomes by macrophages can be exploited for the delivery of drugs into the interior of cells. Antibiotic delivery to the interior of macrophages can be improved by its incorporation in liposomes.

Hydrophilic polymers, such as polyethylene glycol (PEG), coated on the surface of liposomes, reduce clearance by macrophages and prolong duration in the systemic circulation. Liposomes coated with polymers in this manner are referred to as 'stealth' liposomes. Stealth is a description of something designed to avoid detection, for example stealth aircrafts are designed to avoid detection by radar systems. Stealth liposomes are designed to avoid detection by macrophages. The prolonged duration of stealth liposomes in circulation enables adequate levels to escape from the circulation via leaky blood vessels and accumulate at inflamed tissue and tumour sites. The Doxil® and Caelyx® products listed in Table 6.2 contain stealth liposomes.

Liposomes can be designed to target specific cells or tissues by covalently attaching antibodies or specific 'homing' molecules, with affinity for specific features of the biological

TABLE 6.2 Examples of drugs delivered intravenously using liposomal delivery systems.

Name	Drug	Liposomal structure	Indications
Depocyte™	Cytarabine	dioleylphosphatidylcholine (DOPC), dipalmitoylphosphatidylglycerol (DPPG), cholesterol, triolein	Lymphomatous meningitis (a complication of cancer of the lymphatic system)
Doxil™ Caelyx™	Doxorubicin	N-(carbonyl-methoxypolyethylene glycol 2000)-1,2-distearoyl-*sn*-glycero-3 phosphoethanolamine sodium salt (MPEG-DSPE), hydrogenated soya phosphatidylcholine (HSPC), cholesterol	A number of types of cancer
DuanoXome™	Daunorubicin	Distearoylphosphatidylglycerol (DSPC), cholesterol	AIDS-associated Kaposi's sarcoma
Amibisome™	Amphotericin B	Hydrogenated soy phosphatidylcholine (HSPC), Distearoylphosphatidylglycerol (DSPG)	Fungal infections

target, to the liposome surface. For example, plasminogen has an affinity for fibrin, fibrinolytic drugs loaded into plasminogen-coated liposomes will target fibrin clots.

Deformable elastic liposomes have been developed to penetrate intact skin. It is thought that they can penetrate the skin by squeezing themselves along the intercellular sealing lipids of the stratum corneum due to their deformability. Their deformable nature is due to the mixture of phospholipids and membrane softeners they contain.

6.7 PHARMACEUTICAL AEROSOLS

Aerosols are dispersions of particles or liquid in gas. The particles or liquids dispersed are of colloidal dimensions, for example a mist can be considered to be an aerosol with liquid droplets dispersed in air. Smoke is another example of a commonly observed aerosol, this time with solid particles dispersed in air. Aerosols are primarily used as drug-delivery systems to deliver drugs to the lungs, though they can also be used to administer drugs sublingually and locally to the skin or throat. The drugs can be dispersed in the gaseous carrier as a fine powder, or can be delivered as droplets of a drug solution.

Aerosol generation and delivery devices

Aerosol drug-delivery systems are usually formed immediately prior to administration. The drug can be dissolved or suspended in a volatile liquid or propellant in a pressurized container. A solvent may be added to increase solubility. The mixture in the pressurized canister is released when the valve on the container is opened. Upon entering the ambient atmosphere the volatile liquid expands and a dispersion of particles is produced in the air.

The pressurized device that produces the aerosol can vary. Metered dose inhalers (MDIs) are commonly used for pulmonary delivery. An alternative to the metered dose inhalers are the dry powder inhalers (DPIs). The aerosol, powder in air, is generated by the patient breathing in: it does not contain a liquid propellant. The performance of DPIs

TABLE 6.3 Examples of pharmaceutical dry-powder inhaler devices.

Name	Product	Drug
SpinHaler®	Intal	Sodium cromoglicate
Diskhaler®	Flixotide Relenza	Fluticasone propionate Zanamivir
Accuhaler®/Diskus®	Advair Serevent	Fluticasone and salmeterol Salmeterol
Turbohaler®	Pulmicort Symbicort	Budesonide Budesonide and formoterol
HandiHaler®	Spiriva	Tiotropium
Aerolizer®	Foradil	Formoterol

is influenced greatly by the physical properties of the powder, the patient's inhalation technique, and also by the design of the DPI device. Various designs of DPI devices are listed in Table 6.3.

Nebulizers are an alternative inhaled drug-delivery device to MDIs and DPIs and can deliver larger doses of drug. Unlike MDIs and DPIs where the efficiency of the device is dependent on the patient's breath intake, nebulizers allow the drug to be inhaled during normal tidal breathing. For this reason they are particularly useful for paediatric and geriatric patients.

Nebulizer devices create an aerosol from a liquid containing a drug either in solution or in suspension. They can be divided into two categories: jet nebulizers and ultrasonic nebulizers. Jet nebulizers create an aerosol of liquid droplets by passing a gas at high speed through a narrow nozzle. The nozzle is designed so that when the gas expands through the nozzle it results in a negative pressure at the outlet of the nozzle compared to the inlet. This pressure difference results in liquid containing the drug being drawn up the nozzle, which is then atomized by the high-speed air. Fine atomized droplets containing the drug travel in the air stream to the lungs.

The ultrasonic nebulizer atomizes liquid into a mist or droplets by supplying energy to the liquid at a high frequency from a piezoelectric crystal. The ceramic piezo element is attached to a fine sheet of metal that vibrates in response to an alternating voltage. Liquid in contact with this metal sheet is atomized into droplets by surface vibration and cavitation.

Pharmaceutical application

Drug delivery by aerosols is ideal for delivery of drugs to the airways for the treatment of diseases local to the airways such as asthma. Local delivery by aerosols avoids unwanted systemic effects, increases the onset of action of the drug, and reduces the required therapeutic dose. For example, oral doses of 2–4 mg of albuterol are equivalent to inhaled doses of approximately 0.2 µg. That said, the fraction of the inhaled drug that reaches the lungs is low: 10–15% in older devices and 30–50% in newer devices, with the fraction of the inhaler dose from dry-powder inhalers reaching the lungs being influenced by particle size. Any cohesion between particles prior to administration can reduce the fraction of the dose delivered because of the increase in particle size.

6.8 PHARMACEUTICAL FOAMS

Foams can be considered as a disperse phase of a three-dimensional liquid network surrounded by an entrapped continuous gaseous phase, or a disperse phase of gas bubbles in a continuous liquid phase. The liquid phase is generally a surfactant solution with a low surface tension that can spread as a molecularly thin layer on air bubbles. Foams are formed by passing the gas into surfactant solutions; the surfactant films formed create a network of liquid films surrounding the dispersed bubbles.

Foams are used as drug-delivery systems to deliver drugs topically, rectally, vaginally and as burn dressings. The foams are prepared by releasing a compressed gas and surfactant solution from a pressurized container. Upon release, the compressed gas expands and the foam is formed. As for the other pharmaceutical dispersions discussed, the physical stability of pharmaceutical foams influences their use as a dosage form. Pharmaceutical foams are designed with optimal stability for the required period after release from the pressurized container. Physical instability results in collapse of the foam. Physical stability is of particular importance for rectal foams, which spread through the rectal cavity by the expansion. Predfoam®, containing prednisolone, is an example of a rectal foam. It is used in the treatment of ulcerative colitis.

Mixtures of propane, iso-butane and/or n-butane are some of the volatile gases that can be used to expand pharmaceutical foams. For topical foams, the ability of the foam to be easily destroyed and rubbed into the skin at low shear is important. Bettamousse® is a pharmaceutical foam used to administer betamethasone topically to the scalp for the treatment of dermatoses, such as psoriasis. It has a similar texture to, and is applied as, hair mousse.

We hope we have highlighted some of the main pharmaceutical applications of disperse systems and some of the main points of consideration during their use.

◉ 6.9 SUMMARY

Oral pharmaceutical suspensions can be used for the delivery of drugs with poor aqueous solubility. Pharmaceutical suspensions injected intramuscularly and subcutaneously can be designed with extended drug-release characteristics. Nanosuspensions are of interest for intravenous drug delivery. Physical instability problems associated with suspensions are primarily caused by sedimentation leading to caking.

The viscosity behaviour of gels can be exploited for drug delivery. Gels can be formed from dispersed lyophobic or lypophilic colloidal material. The structure and strength of the gel network formed depends on the degree of interaction between the colloidal particles or molecules in the network, and the nature of bonds formed. The main factors that influence the stability of pharmaceutical gels are microbial growth, dehydration, changes in pH, and temperature.

Oil-in-water (o/w) emulsions can be used for the oral delivery of oils and poorly water-soluble drugs. O/w and w/o emulsions can be used to treat dry skin conditions by rehydrating the skin. The types of instabilities observed in emulsions are cracking, creaming, and phase inversion. Microemulsions can increase the solubility of poorly water-soluble drugs and have also been shown to increase their oral bioavailability. The use of microemulsions for intravenous drug delivery is limited.

The interior core of micelles is a hydrophobic environment, which is capable of dissolving poorly water-soluble drugs. One of their main applications is the solubilization of cancer drugs for intravenous infusion.

Liposomes are vesicles formed mainly from phospholipids. Molecules encapsulated in the core of liposomes can be protected for enzymatic metabolism, prolonging drug activity. The uptake of liposomes by macrophages can be exploited for the delivery of drugs into the interior of cells. Hydrophilic polymers, such as polyethylene glycol (PEG), coated on the surface of liposomes, reduce clearance by macrophages and prolong duration in the systemic circulation.

Foams are used as drug-delivery systems to deliver drugs topically, rectally, vaginally and as burn dressings. The foams are prepared by releasing a compressed gas and surfactant solution from a pressurized container. Mixtures of propane, iso-butane and/or n-butane are some of the volatile gases that can be used to expand pharmaceutical foams.

📖 FURTHER READING

Bonacucina, G. *et al.* (2008), *Colloidal soft matter as drug delivery system*, Journal of Pharmaceutical Sciences, Published online May 2008

Lawrence, M.J. and Rees G. D. (2000), *Microemulsion-based media as novel drug delivery systems*, Advanced Drug Delivery Reviews, 45, 89–121

Lian, T. and Ho, R. (2001), *Trends and developments in liposome drug delivery systems*, Journal of Pharmaceutical Sciences, 90, 667–680

Lieberman, H.A., Rieger, M.M. and Banker, G.S. (1996), *Pharmaceutical dosage forms: Disperse systems*, Marcel Dekker, New York

Rau, J.I. (2005), *The inhalation of drugs: Advantages and Problems*, Respiratory care, 50, 367–382

Wong, J. *et al.* (2008), *Suspensions for intravenous (IV) injection: A review of development, preclinical and clinical aspects*, Advanced Drug Delivery Reviews, 60, 939–954

DRUG PARTITIONING AND TRANSPORT ACROSS BIOLOGICAL BARRIERS

07

Learning objectives

Having studied this chapter, you should be able to:

- explain the differences between selective and non-selective drugs, agonists and antagonists and the various types of biological targets drugs interact with

- explain why drug partitioning is important for pharmaceutical action and how partition coefficients are determined and calculated

- describe how the partitioning of acidic, basic and zwitterionic drugs is influenced by pH and the differences between partition coefficients and distribution coefficients

- describe with the structural and biochemical features of epithelium cells that affect drug transport

- explain how drugs are transported paracellularly and transcellularly

- understand how the structural features of the blood/brain barrier create a barrier to drug transport

- list and describe the factors that influence drug absorption from the GI tract, skin and other routes of drug administration highlighted.

In order to exert a therapeutic response, drug molecules must interact with target receptors. Drugs are normally administered to a patient via a drug-delivery system, such as a tablet, inhaler, cream, injection or liquid. When a patient is administered a delivery system the drug molecules are released from the drug-delivery system into solution in the surrounding biological fluids, and subsequently transported to the desired target receptor.

The transport of drug molecules to their biological targets can be thought of as a journey. The type of journey the drug molecules undertake is variable. The distance the drug molecules must travel, the biochemical challenges they encounter, the biological barriers they must cross and the probability of reaching their biological target can all vary. Drug transport across biological barriers is influenced by the partitioning of the drug molecule between aqueous and lipid phases. Factors that influence the type of journey the drug molecules experience include: (1) the ability of drug molecules to partition into biological fluids and tissues, (2) the site of the target receptor, (3) the route of administration and (4) the delivery system used to deliver the drug.

The earlier chapters of this book have focused on explaining the physicochemical properties of drug substances and some drug-delivery systems. This chapter aims to give a brief introduction to drug–target receptor interactions, to highlight factors influencing the partitioning of drug molecules and to describe the biological barriers drug molecules encounter during delivery. During this chapter we will explore the main transport mechanisms by which drug molecules are transported across these biological barriers. Let's start this chapter with a brief overview of drug–receptor interactions.

7.1 DRUG–RECEPTOR INTERACTIONS

The interaction of a drug molecule with its target receptor initiates a chain of biochemical events that result in a desired physiological response, such as a reduction of inflammation or an inhibition of bacterial growth. When a drug–target interaction occurs, the drug molecule has done the job it was designed to do. Understanding drug–target interactions is the subject of pharmacology and medicinal chemistry. These are vast subject areas, well beyond the scope of this book. However, suggestions for relevant further reading are given at the end of this chapter. Very broadly, drugs are designed to interact with biological targets that may be part of the patient's physiology or present in an infectious agent such as a bacterium or virus.

A drug's molecular structure is what gives the drug molecule the ability to interact with its biological target. However, a drug can only interact with its target if it arrives at the right location in the first place. The physicochemical issues discussed earlier in this book can have a significant effect on drug–target interactions by impacting upon a drug's journey to its target. For example, physicochemical properties such as drug solubility in biological fluids can affect drug absorption and distribution, and hence the ability of the drug molecules to arrive at their target receptor.

7.1.1 Types of drug receptors

It is a great challenge to design a drug molecule that causes a single specific physiological effect. Drug molecules can be described as selective or non-selective. Drugs that interact with a range of receptor subtypes are non-selective and can cause a range of physiological effects. For example, salbutamol (a β-2 adrenergic receptor agonist) produces relaxation of bronchial muscle fibres through stimulation of β-2 adrenergic receptors in bronchial smooth muscle. It also acts on β-1 adrenergic receptors in cardiac muscle to a lesser extent. Some patients may experience cardiac effects – increased pulse rate or high blood pressure – at high doses of salbutamol. Such unwanted physiological effects are called adverse effects or side effects.

A drug can exert an effect by interacting with one subtype of receptor. However, that receptor may be present in a range of tissues and this can result in a range of different physiological effects depending on the locations where the receptor is activated. For example, tricyclic antidepressants act by interacting with amine reuptake transporters in the brain. However, amine reuptake transporters are also found in the mouth, bowel and bladder. Drug molecules in the systemic system that reach these receptors can cause unwanted side effects such as dry mouth, urinary retention and constipation.

Agonists and antagonists

Drugs that act by binding to receptors fall into one of two very broad categories. In the first category are the agonists. The binding of agonists to receptors causes a recognized physiological effect. For example, many bronchodilators, such as salbutamol, are selective agonists of adrenergic receptors in bronchiolar tissue. In these cases, the physiological effect is the relaxation of the bronchial smooth muscle. The second category is that of the antagonists. Antagonists act by blocking or reversing the effects of agonists. For example, β-blockers are the selective antagonists of noradrenaline at certain adrenergic receptors in vascular and cardiac tissue. One of the physiological effects caused by noradrenaline interacting with adrenergic receptors in the cardiac tissue is an increased heart rate. Blocking this interaction with an antagonist, such as a β-blocker, reverses the effect and causes a reduction in heart rate.

The biological target of a drug can be a membrane-bound receptor or an enzyme or nucleic acid. First, let us consider the membrane-bound receptors.

Membrane-bound receptors

Membrane-bound drug receptors are usually proteins that are embedded in the membranes of particular cell types. Part of the protein will be 'exposed' on the surface of the cell, and so will be accessible to the drug molecule, as shown in Fig. 7.1. Examples of these receptors include the adrenergic receptors that occur in a range of biological tissues including blood vessels and the pulmonary bronchiole. These receptors bind noradrenaline, and related compounds such as β-blockers and certain bronchodilators. Classes of receptors may be composed of several different types and subtypes. Often, a key challenge for pharmacologists and medicinal chemists is to design drug compounds that are selective for a specific receptor subtype class.

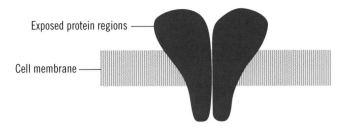

FIGURE 7.1 Exposed regions of a protein embedded in a cell membrane.

Enzymes, nucleic acids and other targets

Approximately one-third of drugs act by inhibiting specific enzymes. Examples include antihypertension drugs (such as lisinopril and perindopril), which act by inhibiting an enzyme known as an angiotensin-converting enzyme, and antiviral drugs (acyclovir and zidovudine) that act by inhibiting DNA-replicating enzymes in viruses. The issues that affect the pharmacodynamics and pharmacokinetics of drugs that act at receptors also affect drugs that act by inhibiting enzymes: the drug still has to reach the enzyme to have its effect, just as it must also reach a receptor (if the receptor is its intended target).

Nucleic acids, and DNA in particular, are another important biological target for drug binding. As DNA molecules have a phosphate backbone, they are highly negatively charged. For this reason, DNA-binding compounds (such as mitomycin C) are often polybasic – that is, they are capable of existing in forms with multiple positively charged sites. (If they were negatively charged, the DNA-binding compounds would be repelled by the phosphate backbone, making binding impossible.)

It is not necessarily the case that all drugs act by interacting with specific biological targets. For example, inhalation anaesthetics such as halothane, diethyl ether and cyclopropane do not share any obvious structural similarities, and do not appear to act by binding to a common receptor. However, what they do have in common are certain physicochemical characteristics, namely lipophilicity and volatility. It has been proposed that it is these physicochemical characteristics, rather than ability to interact with a specific target, that results in their anaesthetic properties.

7.1.2 Thermodynamics of drug–receptor interactions

The binding of a drug to a receptor is an equilibrium process. In other words, it is under thermodynamic control. The free energy of binding of a drug to a receptor is given by eqn 7.1, previously described in Chapter 3:

$$\Delta G = \Delta H - T\Delta S. \tag{7.1}$$

In the context of drug–receptor binding, the enthalpy term in eqn 7.1, ΔH, relates to the strength of the interactions between the drug and the receptor. These interactions are usually non-covalent interactions such as ion–ion bonds, hydrogen bonds, dipole–dipole interactions or hydrophobic binding. However, the entropy term, $-T\Delta S$, is often important. If binding of the drug to the receptor requires the drug to adopt a specific conformation, and so to lose degrees of conformational freedom, this results in an unfavourable reduction in entropy upon binding. Alternatively, binding of a drug to a receptor is often accompanied by release of water molecules that were solvating the drug molecule. This release of solvating water molecules results in an overall increase in the disorder of the system, and is favourable in terms of entropy.

Issues concerning the interaction of the drug and its target are often referred to as the 'pharmacodynamics' of the drug, whereas issues relating to its absorption, distribution, metabolism and excretion are referred to as its 'pharmacokinetics'. More information relating to pharmacokinetics and pharmacodynamics is provided in Chapter 8. The following sections of this chapter describe how drug molecules reach their biological targets and the barriers they encounter on the way. The movement of a drug across biological barriers involves the partitioning of drug molecules between lipid membranes and aqueous systems. Therefore, we need to first explain the factors that affect the partitioning of drug molecules.

7.2 PARTITIONING AND PARTITION COEFFICIENTS

Partitioning is the spontaneous movement of molecules between two immiscible liquid phases. The extent to which partitioning occurs is determined by the difference in the chemical potentials, μ, of the molecules in the two immiscible liquids. Partitioning processes are vital for the pharmaceutical action of drug molecules, as they are required for the diffusion of drug molecules from their site of administration to their biological target.

The theoretical basis for partitioning is described in Section 3.3.5. The partitioning of a drug molecule between two phases can be quantified by a **partition coefficient**, P. For most cases, P can be defined as given in eqn 7.2, that is

$$P = \frac{S_A}{S_B},$$ (7.2)

in which S_A is the solubility of the compound in phase A, and S_B is the solubility of the compound in phase B. In most circumstances, one of these phases will be an aqueous phase, such as an aqueous biological fluid – for example, gastric fluid or blood. The other phase is usually a non-polar, organic or lipid phase, such as an oil, or the lipid bilayer of a cell membrane.

The partitioning of a drug molecule between an aqueous and an organic phase can be described by the following equilibrium

$$(\text{Drug})_{\text{aqueous phase}} \rightleftharpoons (\text{Drug})_{\text{organic phase}}.$$ (7.3)

The partition coefficient, P, for this process is just the equilibrium constant for eqn 7.3 – that is

$$P = \frac{[\text{Drug}]_{\text{organic phase}}}{[\text{Drug}]_{\text{aqueous phase}}}.$$ (7.4)

The partition coefficient, P, for a drug can therefore be described in terms of the concentration of the drug in both the aqueous and organic phases, for instance in a two-phase mixture such as water and chloroform.

The convention for describing the partition coefficients for pharmaceutical drug molecules is that the partitioning process fits the pattern shown in eqn 7.3, that is, with the drug in aqueous solution on the left-hand side of the equilibrium, and the drug in organic solution on the right-hand side. The partition coefficient, P, is then defined by the concentration of the drug in the organic phase divided by the concentration of the drug in the aqueous phase.

To complete the description of the partition coefficient, the exact aqueous and organic phases should be specified. The aqueous phase is usually just water. The choice of organic phase is more variable. Specific oils or organic solvents may be used – for example chloroform. A common choice of organic phase for measuring partition coefficients is 1-octanol. 1-Octanol is an organic liquid that is immiscible with water. However, it is not a pure hydrocarbon, as it does have a hydroxyl (alcohol) group. 1-Octanol is therefore believed to be a good mimic of the polarity of a biological membrane lipid, as it is largely hydrocarbon, but also has some polar functionality. Unless stated otherwise, quoted partition coefficient values are usually based on the water/1-octanol system. To complete the definition of a partition coefficient, the temperature should also be specified. For example the 1-octanol–water partition coefficient for indomethacin at 25 °C is 3237. A partition coefficient of this magnitude indicates that the drug is substantially more concentrated in the organic phase than the aqueous phase.

A partition coefficient provides a measure of the relative solubility of a compound in aqueous and organic phases. Diffusion of a drug from the site of administration to the site of action requires some degree of solubility in both aqueous and lipid-rich media. Unless there is a 'balance' of solubility in both of these environments the drug will fail to move from one phase to another towards its intended target. Partition coefficients provide a measure of this balance. As a result, partition coefficients provide valuable information for drug design.

7.2.1 Log *P* values

Partition coefficients are effectively the equilibrium constants for the processes shown in eqn 7.3. As is often the case with equilibrium constants, the range of values is very large, including values much smaller than 1, as well as very large values. For this reason, the partition coefficients themselves are inconvenient to use. Instead, we tend to use the logarithms of the partition coefficients – that is, log P values – as these give a narrower range of values. For example, the partition coefficient (P) of indomethacin is 3236 and the logarithm of this partition coefficient is 3.51. It is important to determine the log P of any new chemical entity that is being considered as a candidate drug. Table 7.1 gives log P values for a selection of drug substances.

The range of log P values given in Table 7.1 is from −1.12 to 6.50, a range within which a majority of drug substances fall. Compounds with negative log P values are compounds with greater solubility in water than in non-polar organic phases. Compounds in this range tend to be polar compounds that are well solvated in aqueous media, but diffuse poorly into lipid-rich media such as cell membranes. These compounds tend to have poor absorption from the gastrointestinal tract. In practice, compounds with log P values between 0 and 1.0 also tend to be poorly absorbed into lipophilic media.

TABLE 7.1 Log *P* values for selected pharmaceuticals[a].

Compound	log *P*
Oxytetracycline	−1.12
Sulfadiazine	0.12
Aspirin	1.19
Benzylpenicillin	1.83
Temazepam	2.19
Lidocaine	2.26
Atrazine	2.75
Oxadiazon	4.09
Permethrin	6.50

[a] Data reprinted with permission from A. L. Leo, Comprehensive Medicinal Chemistry, Vol. 4, C. Hansch, P. G. G. Sammes and J. B. Taylor (eds), C. A. Ramsden (Vol. Ed.), 1990, Vol. 4., pp. 295–319; Copyright Elsevier (1990); and with permission from B. J. Herbert and J. G. Dorsey, Anal. Chem., 1995, **67**, 744–749, Copyright (1995) American Chemical Society.

In the other extreme, compounds with log *P* values greater than approximately 4.5 tend to be compounds with poor aqueous solubility and considerable solubility in non-polar environments, such as membrane lipids. Compounds with log *P* values in this range are often difficult to formulate into medicines, due to their low aqueous solubility. They also tend to be strongly absorbed into lipid-rich tissue, so that their distribution to the required site of action and subsequent excretion may be poor. They may also have inadequate solubility in gastrointestinal fluids after oral administration to allow sufficient amounts to dissolve prior to absorption.

log *P*	−1.0	0	1.0	2.0	3.0	4.0	5.0	6.0
	Polar compounds			Compounds of intermediate polarity			Non-polar compounds	
	Good aqueous solubility			Good balance between aqueous and lipid solubility			Poor aqueous solubility	
	Poor lipid solubility						Good lipid solubility	
	Poor absorption and distribution			Good absorption and distribution			Accumulation in lipid-rich environments	
							Slow excretion	

FIGURE 7.2 Effect of log *P* values on solubility, absorption and distribution of drug substances.

Compounds with log P values lying between these two extremes are preferred, as these compounds have the required balance between aqueous and lipid solubility to allow for good absorption and distribution. These trends in log P value versus absorption and distribution are summarized in Fig. 7.2.

▷ **EXAMPLE 7.1**

Thiopental and pentobarbital are very similar in terms of their molecular structures (shown in Fig. 7.3). However, under certain conditions thiopental is well distributed into the central nervous system whereas pentobarbital is not. Their log P values are 3.50 and 0.05, respectively. Are these values consistent with what is known about the absorption and distribution of these compounds?

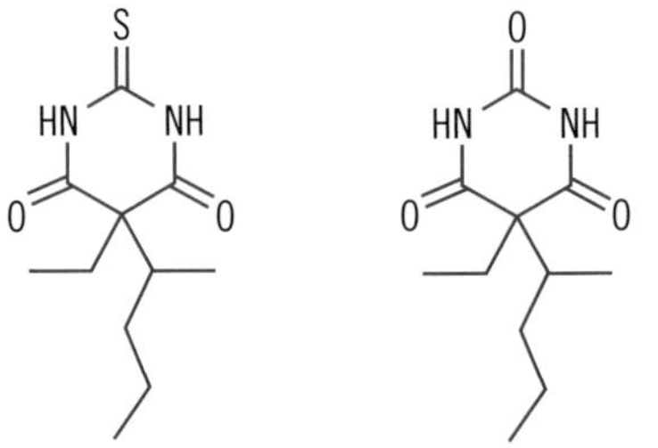

Thiopental   Pentobarbital

FIGURE 7.3 Molecular structures of thiopental and pentobarbital.

The central nervous system is protected by a complex and very lipophilic membrane system known as the blood/brain barrier (see Section 7.7.2). Only significantly lipophilic compounds can cross the blood/brain barrier, and so exert effects on the central nervous system. Thiopental has a log P of 3.5, which implies that it has a good balance between aqueous and lipid solubility, and is a reasonably lipophilic compound. This is consistent with thiopental being capable of crossing the blood/brain barrier and exerting effects on the central nervous system.

By contrast, pentobarbital has a log P of 0.05. This value implies that pentobarbital is really quite a polar compound with good aqueous solubility, but poor solubility in lipid media. It is not surprising that pentobarbital is unable to cross the blood/brain barrier, as it is just too poorly soluble in lipid-rich environments for this to happen. Although the difference in the molecular structures of these two compounds appears to be minor, examination of their log P values shows that there are considerable differences in their solubility in aqueous and lipid media. These differences have a very significant impact on their respective biological activities.

▷ **EXAMPLE 7.2**

A new therapeutic molecule, drug X, is identified. Its 1-octanol–water partition coefficient is determined experimentally at 25 °C. The concentration of drug X in 1-octanol is 0.1 mg mL^{-1} and in water 2.3 mg mL^{-1}. Calculate the partition coefficient and log P of drug X? Comment on the ability of drug X to move from aqueous medium into lipid-rich media, such as cell membranes?

Answer

The partition coefficient of drug X can be calculated using eqn 7.4 as shown below

$$P = \frac{[\text{Drug}]_{\text{organic phase}}}{[\text{Drug}]_{\text{aqueous phase}}} \rightarrow P = \frac{0.1 \text{ mg mL}^{-1}}{2.3 \text{ mg mL}^{-1}} = 4.35 \times 10^{-2}.$$

The logarithm of the partition coefficient equals $\log(4.35 \times 10^{-2}) = -1.36$

$\text{Log } P = -1.36.$

Drugs with negative log P values will have high aqueous solubility and poor solubility in lipid-rich media such as cell membranes. Therefore, the drug will be mainly concentrated in the aqueous phase, with very little drug in lipid-rich media.

7.3 DETERMINATION OF PARTITION COEFFICIENTS

Partition coefficients provide valuable information about potential drug candidates, therefore methods for determining log P values are important. Ultimately, only an experimental measurement can provide a fully certain log P value. However, software for calculating log P values based on molecular structure is now very sophisticated and widely used. In this section we discuss the basis of both experimental and computational approaches.

7.3.1 Experimental determination of log P

The most direct method of experimentally determining the log P of a compound is the so-called 'shake-flask' method. This method is illustrated schematically in Fig. 7.4. In the shake-flask method, solutions of the compound in both water and 1-octanol are added to a vessel similar to a separating funnel. The water solvent must be saturated in 1-octanol, and the 1-octanol solvent saturated in water, otherwise dissolution of the solvents in each other may complicate the measurement of drug concentrations.

The two solutions are intimately mixed by, for example, vigorous shaking, after which the water and 1-octanol phases are allowed to settle for at least 24 hours. The two phases are separated, and the concentration of the compound in each layer determined by a suitable analytical method. The resulting concentrations can be inserted directly into eqn 7.4 to give the partition coefficient, P, which can then be converted into a log P value, as shown in Example 7.2.

In principle, the shake-flask method is very straightforward. However, it can be difficult in practice to obtain reproducible data using this method. One challenge is that temperature must be constant throughout the entire process. The mixing and separating stages must also allow the equilibrium shown in eqn 7.3 to be established.

An alternative to the shake-flask method is log P determination using high-performance liquid chromatography (HPLC). Separation of compounds by HPLC is based upon differing rates of migration through a medium. In HPLC, the medium is a solid stationary phase. Compounds are brought into contact with the stationary phase by a liquid mobile phase. In standard (reverse-phase) HPLC, the stationary phase is most

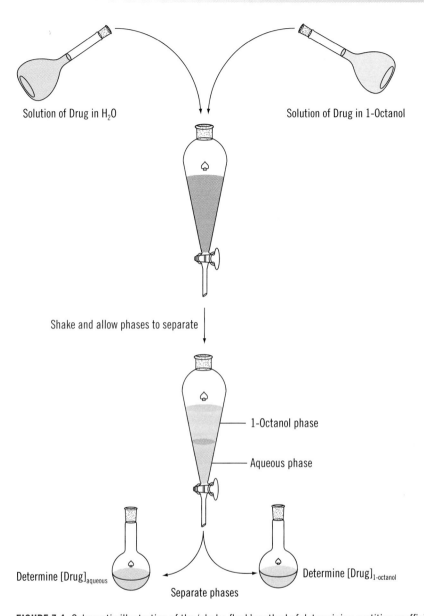

Solution of Drug in H$_2$O

Solution of Drug in 1-Octanol

Shake and allow phases to separate

1-Octanol phase

Aqueous phase

Determine [Drug]$_{aqueous}$

Determine [Drug]$_{1\text{-octanol}}$

Separate phases

FIGURE 7.4 Schematic illustration of the 'shake-flask' method of determining partition coefficients.

commonly an octyl- (C8) or octadecyl- (C18) functionalized silica, while the mobile phase is an aqueous solution. While compounds migrate through the HPLC column, they undergo partitioning between the hydrocarbon-rich stationary phase and the aqueous mobile phase. This partitioning is conceptually the same as partitioning between biological aqueous and lipid phases. The behaviour of a compound in reverse-phase HPLC can therefore be used to provide an estimate of its log P.

To estimate the log P of a compound by HPLC, the behaviour of a series of structurally similar compounds needs to be determined in the same HPLC system. The most

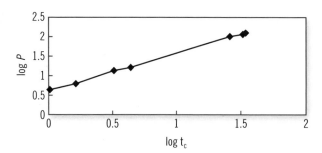

FIGURE 7.5 Plot of log P versus log of the elution time, t_c, for the following series of compounds: pyridine, practolol, acetanilide, β-picoline, quinoline, isoquinoline, anisole. Data reproduced with permission from Mirrlees, M. S., Moulton, S. J., Murphy, C. T., Taylor, P. J. (1976) Direct Measurement of Octanol-Water Partition Coefficients by High-Pressure Liquid Chromatography, *J. Med. Chem.*, **19**, 615–619. Copyright (1976) American Chemical Society.

straightforward approach (Mirrlees *et al.* 1976) is to determine the elution time t_c for each compound from the following equation

$$t_c = (t_R - t_o),\tag{7.5}$$

in which t_R is the HPLC retention time of each compound, while t_o is the retention time of a compound that is not retained by the stationary phase. From this data, a calibration plot is established using the equation

$$\log P = \log t_c + \text{constant.}\tag{7.6}$$

Figure 7.5 gives an example of such a plot. From the calibration curve in Fig. 7.5, the log P of a low molecular weight drug molecule with log t_c between 0 and 1.5 could be read off directly. For example, the log P of phenol (log $t_c = 0.81$) would be estimated at 1.45. Since this method was first employed, many more sophisticated treatments for determining log P values from chromatographic data have been developed (Demare *et al.* 2007).

The measured log P values for one water/organic system, for example water/1-octanol, can be related to those for other water/organic systems using the equation

$$\log P_{\text{solvent}} = a \log P_{\text{1-octanol}} + b,\tag{7.7}$$

in which P_{solvent} is the partition coefficient in the water/solvent system to be determined, $P_{\text{1-octanol}}$ is the partition coefficient in the water/1-octanol system, and a and b are experimentally determined constants. Some values for a and b are given in Table 7.2. For example, if the log P of a compound in the 1-octanol/water system is 2.45, an estimate of its log P in a chloroform/water system could be determined using the values of a and b from chloroform from Table 7.2. These would give

$$\log P_{\text{chloroform/water}} = (1.126 \times 2.45) - 1.343 = 1.42.$$

7.3.2 Computational determination of log P

Considerable effort has gone into developing software that can calculate log P values based on molecular structures. Log P predicting software is now common and widely used.

TABLE 7.2 Selected constants a and b for use in the relationship $\log P_{solvent} = a \log P_{1\text{-octanol}} + b$.

Solvent	a	b
1-Octanol	1.000	0
Chloroform	1.126	−1.343
Diethyl ether	1.130	−0.170
Oils	1.099	−1.310

Data reproduced with permission from Mirrlees, M. S., Moulton, S. J., Murphy, C. T., Taylor, P. J. (1976) Direct Measurement of Octanol-Water Partition Coefficients by High-Pressure Liquid Chromatography, *J. Med. Chem.*, **19**, 615 – 619. Copyright (1976) American Chemical Society.

Log P calculating commands can also often be found on molecular-structure-drawing software. The methods used by these software packages are generally based on the hydrophobic fragmental constants developed by Rekker, Hansch and Leo (Leo 1990).

The fragmental constant approach assigns a hydrophobic constant, f, to each particular molecular fragment that may go into 'building up' a pharmaceutical molecule. The fragmental constants are based on the analysis of the log P values of very many compounds. They can also be calculated using quantum-mechanical calculations of charge densities. Given a molecular structure, the log P for that structure can then, in principle, be determined by adding up the hydrophobic constants for each fragment of the molecule. Allowance can be made for structural features such as conformational freedom, branching, and interactions between groups by using correction factors. This concept can be presented mathematically as follows

$$\log P = \Sigma a_n f_n + \Sigma b_m F_m. \tag{7.7}$$

in which a is the number of fragments of type n and fragmental constant f_n, and b is the number of corrections of type m and correction factor F_m. This approach allows the log P of a compound to be calculated by notionally breaking the compound up into a series of known fragments, adding in the fragmental constant for each of these fragments, and allowing for any necessary corrections. This approach is very amenable to software development.

The approach can be demonstrated for simple compounds, such as the anaesthetic halothane (Leo 1990), the structure of which is shown in Fig. 7.6. We can estimate the log P for halothane by adding up the fragmental constants for each of the constituent atoms, and adding on any necessary correction factors. Each halogen is regarded as a branch off the basis structure. So, for each halogen one correction factor for branching is added. A correction factor is included for each halogen that is sharing a carbon with at least one other halogen (known as geminal halogens XCX), and also for each halogen involved

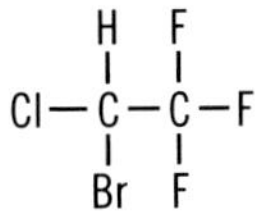

FIGURE 7.6 Molecular structure of halothane.

in a vicinal (XCCX) interaction with at least one other halothane. Putting these together gives

$$\log P \text{ (halothane)} = f_{Ce} + f_{Br} + 3f_F + 2f_C + f_H + 5F_{branching} + 5F_{XCX} + 5F_{XCCX}.$$

The accuracy of the log P value obtained would depend on the accuracy of the fragmental constants and correction factors used.

Computational determination of log P is very convenient (minutes or seconds versus hours or days for experimental measurements) and, with continuous improvements in software and databases of fragmental constants, calculated log P values have become more reliable and accurate. However, care has to be exercised in using calculated log P values, as the calculated value may differ to some extent from the true log P. This is especially the case with complex structures that contain groups, or combinations of groups, for which the fragmental constants are not available, or for which necessary correction factors have not been determined.

7.4 IONIZATION AND DISTRIBUTION COEFFICIENTS

As discussed in Section 2.5, many active pharmaceutical compounds are acidic or basic to some extent, and so can exist in both neutral and ionized forms in solution. This has a direct impact on their partitioning. An ionized molecule would not be expected to have the same solubility in any solvent as a neutral molecule. Likewise, the transfer of molecules between two immiscible solvents will differ if the molecules are ionized rather than neutral. Generally, we would expect ionization to enhance solubility in water and impede solubility in lipid-rich media. The effect of pH on the extent of ionization of an acid or base also has to be taken into consideration. To accommodate the additional complication caused by ionization, in the case of acids and bases the partition coefficient, P, is replaced by the distribution coefficient, D.

The concept of the distribution coefficient, D, can best be understood by thinking about the effect of pH on the acid-dissociation process. An acidic compound undergoes the following dissociation in solution.

$$HX \rightleftharpoons X^- + H^+. \tag{7.8}$$

If compound HX is a pharmaceutical, both the neutral form HX and the ionized form X^- have to be taken into consideration because both may have pharmacological activity. As they are related by an equilibrium process, the concentrations of both forms contribute to the overall concentration of the active pharmaceutical component. For example, in theory, if the compound undergoes dissociation as per eqn 7.8, and both HX and X^- are active, the total concentration of the active drug would be given by $[HX] + [X^-]$. However, the neutral form, HX and the ionized form, X^-, would not be expected to have the same solubility. In water, the ionized form, X^-, would be expected to have greater solubility than the neutral form, HX. By contrast, it is generally assumed that the ionized form, X^-, would be effectively insoluble in non-polar lipophilic media and that only the neutral form, HX, could exist in non-polar phases.

The total concentration of pharmaceutically relevant forms in aqueous media is therefore equal to

$$[HX]_{aqueous} + [X^-]_{aqueous},$$

while the total concentration of pharmaceutically active forms in lipid-rich media is just $[HX]_{lipid}$.

The distribution coefficient, D, attempts to quantify the effect of drug ionization, on drug partitioning between phases. It is defined, similar to the partition coefficient, as the total concentration of relevant forms in the lipid phase divided by the total concentration of relevant forms in the aqueous phase – that is,

$$D = \frac{[HX]_{lipid}}{[HX]_{aqueous} + [X^-]_{aqueous}}. \tag{7.9}$$

Equation 7.9 allows the impact of pH and ionization on the partitioning of an acid pharmaceutical to be understood. In a strongly acidic environment, the equilibrium shown in eqn 7.8 above lies entirely to the left. The compound is completely undissociated and $[X^-]_{aqueous}$ equals zero. In that case, eqn 7.9 becomes the same as the equation for the partition coefficient, P. In other words, at sufficiently low pH the distribution coefficient, D, for an acidic drug becomes equal to the partition coefficient, P, of the neutral form for the drug.

If the pH increases and the medium becomes less acidic, the equilibrium shown in eqn 7.8 becomes more significant, and increased concentrations of the ionized form, $[X^-]_{aqueous}$, develop. Increasing $[X^-]_{aqueous}$ means that the sum below the line in eqn 7.9 increases, so that the value of the distribution coefficient, D, decreases. Therefore, the value of D decreases with increasing pH, as shown in Fig. 7.7.

Note that Fig. 7.7 shows the logarithm of the distribution coefficient, log D, plotted as a function of pH, rather than the distribution coefficient itself. As with partition coefficients, the logarithms of the distribution coefficients are generally used. The point at which the plot intercepts the x-axis gives the value of the log P of the compound when it exists solely in the neutral form. The contribution of the $[X^-]_{aqueous}$ term becomes significant at pH values above the pK_a of the compound. From that point on, a steady decrease in log D is observed.

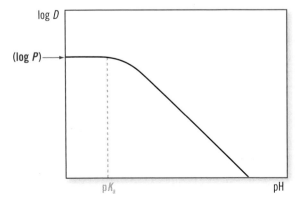

FIGURE 7.7 Variation of the distribution coefficient, D, with pH for an acidic drug.

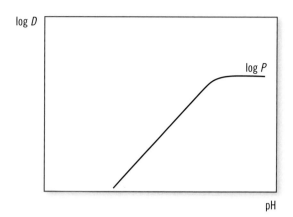

FIGURE 7.8 Variation of the distribution coefficient, D, with pH for a basic drug.

The explanation of how the distribution coefficient, D, varies with pH for a basic pharmaceutical following similar reasoning. For a basic compound, the following equilibrium can exist in solution

$$B + H^+ \rightleftharpoons BH^+. \tag{7.10}$$

Both the species B and BH^+ are pharmaceutically relevant. The total concentration of pharmaceutically relevant forms in the aqueous phase is therefore equal to

$$[B]_{\text{aqueous}} + [BH^+]_{\text{aqueous}}.$$

The only pharmaceutically relevant species existing in the lipid phase is the free base, B, as it is assumed that the ionized BH^+ form is insoluble in lipid media. The distribution coefficient, D, for a basic compound is therefore given by

$$D = \frac{[B]_{\text{lipid}}}{[B]_{\text{aqueous}} + [BH^+]_{\text{aqueous}}}. \tag{7.11}$$

In a strongly basic medium, the equilibrium shown in eqn 7.10 lies entirely to the left-hand side, and the term $[BH^+]_{\text{aqueous}}$ equals zero. Under these conditions, the distribution coefficient, D, is at a maximum and equals the partition coefficient, P, for the compound in the absence of ionization. As the basicity of the medium decreases towards lower pH, significant concentrations of the protonated form, BH^+, come into existence. The value of the sum below the line in eqn 7.11 begins to increase, so that the distribution coefficient, D, begins to decrease. This trend is shown graphically in Fig. 7.8.

▶ **EXAMPLE 7.3**

Figure 7.9 shows the variation of $\log D$ with pH for a new chemical entity that is a candidate for development as a pharmaceutical. What information does the profile shown in Fig. 7.9 provide about this compound?

Above pH ~ 5, the distribution coefficient of this compound is maximized. However, below pH ~ 5, the distribution coefficient drops off sharply. The implication is that the compound is basic; however, above pH ~ 5, there is no significant concentration of

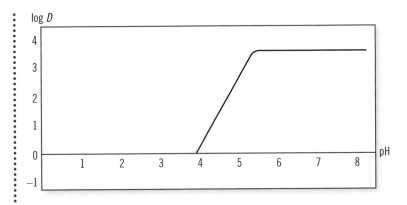

FIGURE 7.9 log D versus pH profile for a candidate pharmaceutical.

the protonated form, BH^+. This suggests that the compound is only a moderately strong base, as it is fully deprotonated at intermediate pH. Above pH ~ 5, the log D value remains constant at ~ 3.5. Therefore, 3.5 is the log P of the free base form, B. This suggests that the free base form is relatively lipophilic. Hence, the profile is that of a lipophilic drug of moderate basicity. Such a compound might be poorly absorbed in a strongly acidic medium, for example from the stomach in a fasting state at pH ~ 2. However, it should be quite well absorbed and distributed under neutral or basic conditions.

7.4.1 Distribution coefficients of zwitterionic compounds

Zwitterions are compounds that are overall neutral but that possess internal localized positive and negative charges. Examples include the neurotransmitter L-DOPA and the metabolic intermediate *para*-aminobenzoic acid (PABA). In Fig. 7.10, the structures of both of these compounds are given. Both are shown as existing in an equilibrium between a zwitterionic form, possessing a positively charged ammonium group ($-NH_3^+$) and a negatively charged carboxylate group ($-CO_2^-$), and a form with no localized charges. In many cases, the zwitterionic form is the dominant form in aqueous solution at neutral pH.

Zwitterionic compounds are relatively common pharmaceutically. Because zwitterions are also amphoteric – that is, they are capable of acting both as acids and as bases – the variation of their distribution coefficients with pH is more complex than that of purely acidic or basic compounds.

FIGURE 7.10 Compounds capable of existing in zwitterionic form: (top) L-DOPA and (bottom) *para*-aminobenzoic acid (PABA).

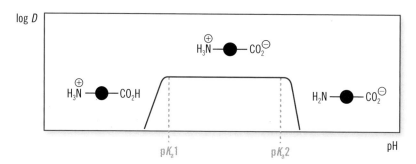

FIGURE 7.11 log D versus pH profile for a zwitterionic compound.

The variation of the distribution coefficient with pH for a zwitterionic compound is shown in Fig. 7.11.

In the range between pK_a1 and pK_a2, the log D of the compound stays constant at its optimum value. Within these pH values, the compound is overall neutral and so it is in its most lipophilic form. If the compound is existing primarily as a zwitterion, even in its most lipophilic form it will still be relatively hydrophilic. However, if the medium becomes acidic such that the pH drops below pK_a1, the distribution of the compound falls off sharply.

Under acidic conditions, the carboxylate (CO_2^-) group is protonated, so that the predominant form of the compound is the positively charged form. This is a very polar form with very low lipophilicity. At pH values above pK_a2, the ammonium (NH_3^+) group is deprotonated, and the compound exists as a negatively charged form. Again, this is an ionized form that is highly polar and lipophobic, so that its distribution coefficient drops off sharply. The value 'pK_a1' corresponds to the pK_a of the carboxylic acid group of the compound. The value 'pK_a2' corresponds to the pK_a of the ammonium group of the compound. The plateau between these values corresponds to the pH range in which the compound exists in its most lipophilic form.

Now that the importance of drug partitioning to drug delivery and the theory of drug partitioning coefficients has been explained, we can move on to explore the types of biological barriers drug molecules encounter en route to their target receptors. In this following section, we will now focus on how these biological barriers influence drug transport.

7.5 CELLULAR AND EPITHELIAL BARRIERS

Before considering the cellular barriers that drug molecules must cross to reach their target receptors, we must first understand the structure of the cells that comprise these barriers.

7.5.1 Cellular membrane structure

Cells in the body are surrounded by an exterior membrane structure called the 'cell membrane'. The primary function of the cell membrane is to maintain the structure

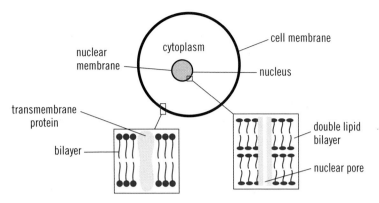

FIGURE 7.12 Simplistic representation of a cell structure, highlighting the lipid bilayer structure of the cell membrane and double-bilayer structure of the nuclear membrane.

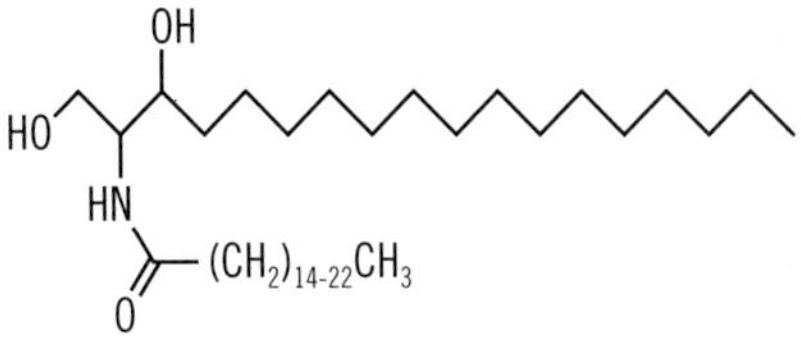

FIGURE 7.13 A representative chemical structure of the family of ceramide sphingolipids.

of the cell and to protect the contents of the cell from potential hazards present in its surrounding environment. The cell membrane is composed primarily of a thin layer of amphipathic phospholipids that are arranged in a bilayer structure, as described in Section 6.6 and shown in Fig. 7.12. The hydrophobic tail segments of these molecules face each other to minimize their exposure to aqueous environments. By contrast, the hydrophilic head segments of the molecules face the aqueous fluid in the interior of the cell (the cytosol) and the aqueous fluid surrounding the outside of the cells (extracellular fluid). In addition to phospholipids, sphingolipids (such as ceramides (Fig. 7.13)) and sterol lipids (such as cholesterol) are also involved in the cell membrane lipid bilayer structure.

The bilayer structure prevents polar solutes, such as amino acids, nucleic acids, carbohydrates, proteins, ions and drugs, from diffusing across the cellular membrane. The transport of polar molecules in and out of the cell is controlled by the presence of complexes such as pores and gates within the cellular membrane. These pores or gates comprise proteins that span the width of the membrane (transmembrane proteins) dispersed throughout the cell membrane. The bilayer structure allows the diffusion of some hydrophobic molecules into and out of the cell through its structure.

Nuclear membrane structure

The nucleus of the cell is surrounded by a membrane called the 'nuclear membrane'. The nuclear membrane separates the nuclear genetic material from the components of the cytoplasm. The nuclear membrane has a double-bilayer structure: it is composed of one bilayer surrounded by a second bilayer. A large number of pore structures are present in the nuclear membrane: they allow the transport of molecules between the cytoplasm and the nucleus. The two bilayers are fused at the point of these nuclear pores. Materials that travel in and out of these pores include proteins such as transcription factors, and RNA.

Most drug molecules pass through the cell: they travel across the cell membrane into the cell and then travel from the interior of the cell back out through the cell membrane. The transport of drug molecules through a cell is called transcellular transport. However, genomics and biotechnology discoveries have led to the design of a number of therapeutic and diagnostic agents that target components within the cellular cytoplasm and nucleus.

7.5.2 Epithelial barriers

Cells that line the surface of the body are called epithelial cells and they form a tissue called the epithelium. The inside of body cavities, the outer surface of many internal organs, the inner surfaces of all glands, cavities, tubes and ducts of the body are all lined by a layer of epithelium. The epithelium that lines the outer surface of the body is called the 'epidermis' and the epithelium that lines the inside of blood vessels is called the 'endothelium'. The function of the epithelium is to act as a barrier and to maintain structure. Drug molecules that are delivered by a number of routes are required to cross these epithelial barriers to reach their target receptors.

Epithelium structure

The structure and morphology of epithelial tissue can vary depending on its location in the body. However, all epithelia are designed to act as a barrier and therefore all epithelial cells have structural similarities. For example, all epithelia have a free surface, a surface to which no other cellular materials are attached. This free surface is called the apical surface. All epithelial tissue is separated from the underlying connective tissue by a basal membrane. The layer of cells attached to the basal membrane is called the basal layer. Epithelium structure can be described by the number of layers of cells present – a single layer of cells is described as simple and a number of layers as stratified. Some simplistic representations of basic cell shapes are given in Fig. 7.14.

Epithelial permeability

The protective barrier formed by epithelial cells is facilitated by their close proximity to one another. The space between epithelial cells (extracellular space) is limited due to the

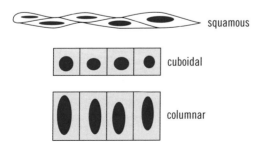

FIGURE 7.14 Simplistic representations of basic cell shapes.

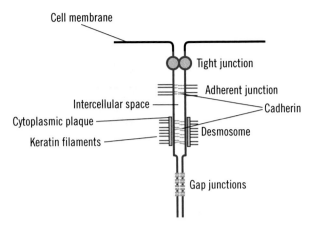

FIGURE 7.15 Simplistic representations of junctions that can be found between adjacent epithelium cells.

presence of several different cellular junctions between cells – tight junctions, adherent junctions, desmosomes and gap junctions – as shown in Fig. 7.15. The degree of space and the type of junction formed between epithelium cells depends on the location and function of the epithelial tissue. Some epithelium barriers are more impermeable than others. For example, while the epidermis of the skin is a very impermeable barrier, the epithelium lining of the intestine is permeable to some molecules. Its permeability enables nutrients and drug molecules to be absorbed across it.

Tight junctions are regions where the outer surfaces of adjacent cells fuse together, a process mediated by proteins on the cell membrane surface. Proteins found to be responsible for tight junction structure include claudin and occludin. These proteins cause cell membranes to adhere tightly together and prevent the movement of molecules between cells.

Desmosomes are the commonest type of cell junctions and are found not only in epithelial cells but also in cardiac muscle and other tissues. At the desmosomal junction, filaments protrude from the cell surface into the extracellular space between the cells. These filaments are composed of a protein called cadherin.

The adherent junctions are similar to desmosomes, in that the adhesion between cells is due to the protein cadherin. Adherent junctions and desmosomes are both involved in cell–cell adhesion that provides mechanical strength to the epithelium tissues.

Gap junctions contain transmembrane proteins that allow the transport of small molecules between cells. The transport of small intracellular signalling molecules and ions facilitates communication between adjacent cells. Gap junctions are therefore referred to as communicating junctions.

In addition to the physical barrier posed by the epithelial cell layer and the junctions between cells, enzymes present at the epithelial membrane also pose a barrier to drug transport by catalysing the degradation of drugs.

Biochemical barriers at epithelial tissue

Enzymes can be present within the epithelial cells, on the surface of the epithelium, or in the mucus covering the epithelial surface. The levels of enzyme activity and the

TABLE 7.3 Examples of substances that act as inhibitors and inducers of cytochrome P450 3A4.

	Examples
Inhibitors	indinavir, nelfinavir, ritonavir, saquinavir, clarithromycin, erythromycin, fluconazol, diltiazem, cimetidine, grapefruit juice
Inducers	carbamazepine, dexamethasone, phenobarbital, phenytoin, ethosuximide, rifabutin, rifampin, primodone

Data obtained from Michalets, E. L. (1998) Update: Clinically significant cytochrome P-450 drug interactions, Pharmacotherapy, **18**, 86–112.

types of enzymes present can vary depending on the location of the epithelium. For example, enzyme activity in the oral cavity is lower than enzyme activity further down the gastrointestinal tract. Therefore, the oral cavity may be more suitable for the delivery of some drug molecules that would be readily degraded by the enzymes in the small and large intestines.

The cytochrome P450 (CYP) family of enzymes oxidize drug molecules during their elimination from the body. The oxidization process improves the water solubility of drug molecules, facilitating their removal from the body. There are around 50 different variants (isoforms) of cytochrome P450 in humans and these are mainly membrane-bound proteins. A number of drug molecules are metabolized by cytochrome P450 3A4 (a variant present in the intestinal tissue) during their transport across the intestinal wall. Certain drug molecules and foods can act as inhibitors and inducers of different cytochrome 450 isoforms; they can increase or decrease the rate of metabolism of drug molecules and thereby influence their absorbance, metabolism and elimination. Examples of some cytochrome P450 3A4 inducers and inhibitors are shown in Table 7.3.

Now that the structural and biochemical challenges posed by cellular barriers have been highlighted, let us now proceed to examining how drug molecules cross these barriers.

7.6 TRANSPORT OF DRUG MOLECULES ACROSS CELLULAR BARRIERS

Drug molecules can travel across cellular barriers by two main routes: the paracellular route (between adjacent cells) and the transcellular route (through a cell, in one side of the cell and out the other side).

7.6.1 Paracellular transport

Paracellular transport takes place when a concentration gradient exists across the cellular barrier, such that the concentration of the drug molecule is higher at one side of the cellular barrier compared to the other side. Molecules move between the cells from areas of high concentration to areas of low concentration by diffusion, as shown in Fig. 7.16. (The process of diffusion is described in greater detail in Section 3.3.3.) Small

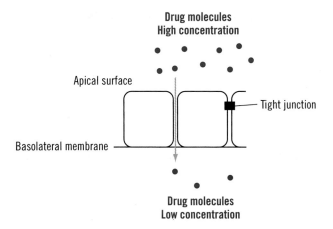

FIGURE 7.16 Simplistic diagram of paracellular transport.

molecules diffuse faster by the paracellular route than larger molecules. Indeed, there is an upper size limit at which molecules are too large to travel through the space between adjacent cells.

In addition to molecular size, the ability of drug molecules to be transported paracellularly is dependent on their chemistry and the hydrophilic or lipophilic nature of the extracellular fluid. For example, lipophilic molecules can diffuse via the paracellular route through the epidermis due to the lipophilic nature of the extracellular fluid in the epidermis. The presence of tight junctions in intestinal epithelia restricts the paracellular transport of molecules between cells.

7.6.2 Transcellular transport

Transcellular transport can be described as passive or active. Low molecular weight molecules, which are lipophilic in nature, can be transported passively across cell membranes. Passive transport is driven by the partitioning of drug molecules between lipophilic and aqueous phases, as shown in Fig. 7.17a. If the drug molecule is present at a high

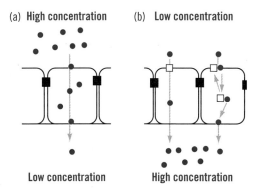

FIGURE 7.17 Simplistic diagram of transcellular transport (a) passive and (b) active (carrier-mediated transport).

concentration on the apical surface of the cell membrane it will partition from the aqueous extracellular fluid into the membrane lipid bilayer. The drug molecule will then diffuse through the lipid bilayer and partition from the lipid bilayer into the aqueous cytosol at the other side of the bilayer membrane. As the drug concentration in the cytosol increases, drug molecules will partition into the basal cellular membrane, diffuse through this lipid bilayer and finally partition into the aqueous fluid surrounding the basal cell membrane.

In addition to the passive transport, drug molecules can be actively transported transcellularly by carrier molecules. This type of transport is called carrier-mediated transport. The drug molecule can be transported via a protein molecule spanning the cell membrane. Drug molecules can also bind to a carrier molecule in the cell membrane and form a carrier–drug complex. The carrier–drug complex can then move through the cell membrane. On reaching the other side of the membrane, the carrier–drug complex is broken and the drug is released. The carrier then returns to the cell membrane surface where it is available to transport more drug molecules through the cell membrane, shown in Fig. 7.17b.

In addition to the carrier-mediated transport of drug molecules into the cell, there are also substrates bound to the cell membrane that may also act as carriers to transport drug molecules back out of the cell. These are called efflux transporters and their role is to protect the body from potentially harmful substances. P-glycoprotein, a 170-kDa membrane-bound protein, is an example of this type of efflux transporter. P-glycoprotein is found in a number of cells including the apical surface of intestinal epithelium and the cell membrane of astrocytes at the blood/brain barrier.

Unlike diffusional processes such as paracellular transport and passive transcellular transport, carrier-mediated transport can transport drug molecules from regions of low concentration to regions of high concentration (that is, against the concentration gradient). This type of transport requires a lot of energy and is therefore called active transport. Carrier-mediated transport can also be in the direction of the concentration gradient (from regions of high concentration to low concentration): this type of transport is called facilitated diffusion.

It is difficult to transport large molecules paracellularly or by carrier-mediated transport. Instead, large molecules can be transported into cells by the process of endocytosis.

Endocytosis

Endocytosis is a process by which cells absorb materials and macromolecules by engulfing them with their cell membrane. The cell membrane invaginates and is pinched off, as shown in Fig. 7.18. The pinched-off section forms a vesicle in the interior of the cell. The absorption of material into cells by endocytosis can be divided into two: phagocytosis and pinocytosis.

Phagocytosis refers to the absorption of solid materials, and pinocytosis refers to the absorption of liquids. Phagocytosis can be thought of as cell eating and pinocytosis can be thought of as cell drinking. The solid material ingested into the cells during endocytosis can be bacteria, viruses or other particulate matter. Pinocytosis involves the uptake of solutes that can include macromolecules such as proteins dissolved in solution. Particles taken up into the cells can be secreted from the cells by a reverse process, exocytosis.

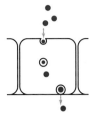

FIGURE 7.18 Simplistic diagram of endocytosis.

Pore transport also occurs across cellular barriers. The aqueous pores present in cellular membranes are approximately 0.4 nm in width and allow low molecular weight hydrophilic molecules to diffuse across these pores.

Now that the general processes involved in the transport of drug molecules across cell membranes have been explained, we can consider the structure of specific biological barriers involved in drug delivery and the transport of drug molecules across these barriers. The first important barrier to consider is the systemic circulation: the major system for drug distribution in the body.

7.7 TRANSPORT OF DRUG MOLECULES IN THE SYSTEMIC CIRCULATION

Many drug molecules reach their target receptor by distribution to body tissues via the systemic circulation. The systemic circulation is the series of blood vessels that carry oxygenated blood away from the heart to the rest of the body, and return deoxygenated blood back to the heart. The systemic circulation accesses most of the body's tissues, and it is therefore a useful transport system for the delivery of drug molecules. Drug molecules can be administered directly into the systemic circulation by intravenous injection or infusion. Drug molecules administered by other routes (such as the gastrointestinal tract, skin, lungs and rectum) travel into the systemic circulation after being absorbed into the tissue at the site of administration.

In order to interact with some target receptors, many drug molecules must travel back out of the systemic circulation once they have undergone a period of transport. To understand how drug molecules travel into and out of blood vessels, the structure of the walls of blood vessels (in particular capillaries) must be understood.

7.7.1 The structure of blood vessels

The structure of the blood vessels varies depending on their type. The walls of veins and arteries have three layers, while the walls of capillaries have only one layer of endothelial cells. Endothelium is the name given to the systemic circulation epithelium. The capillary endothelium is designed to exchange fluid, water and gas by osmotic and hydrostatic gradients. The ability of drug molecules to travel through a capillary endothelium depends on the physicochemical properties of the drug molecules – size and polarity – and the type of capillary. There are three types of capillaries: fenestrated, sinusoidal and continuous.

Fenestrated capillaries have endothelial cells with small openings about 80–100 nm in diameter. The presence of these openings called 'fenestrations' allows the rapid transport of some macromolecules through the capillary wall. Fenestrated capillaries are found in the kidney, small intestine and endocrine glands.

The interior of the sinsuoidal capillaries are lined with phagocytic cells, and have a large number of fenestrations. Sinusoidal capillaries are found in the liver, bone marrow and spleen. The fenestrations in these capillaries are larger than those found in the fenestrated capillaries and allow relatively large red blood cells and plasma protein to pass through the capillary wall.

Continuous capillaries are sealed, having no openings (fenestrations) in their endothelium. Paracellular transport between adjacent endothelial cells is restricted by tight junctions. They can be characterized by a large number of small invaginations on the endothelial cell membrane and numerous micropinocytotic vesicles in the cytoplasm of these cells. Materials can only be transported across this cellular barrier by pinocytotic transcellular transport. Continuous capillaries are present in organs where the transport of molecules from capillaries needs to be carefully controlled, such as in the central nervous system. Continuous capillaries contribute to the blood/brain barrier.

The barrier to drug transport created by the endothelial cells of the central nervous system (CNS) is a major component of the blood/brain barrier that warrants special attention.

7.7.2 Drug transport across the blood/brain barrier

The blood/brain barrier is a system of vascular structures, enzymes, receptors and transporters designed to prevent the transport of potentially hazardous molecules into the CNS and to enable nutrients – such as glucose – to pass from the blood into the brain.

Drug molecules are required to access the CNS to treat a wide range of medical conditions and illnesses including Alzheimer's disease, pain, cancer, psychosis and depression. Access of drug molecules to the extracellular fluid in the brain is restricted due to the reduced transport of drug molecules across the blood/brain barrier.

Molecules are restricted from passing between the adjacent cells in the capillaries of the CNS due to the presence of tight junctions. Pinocytosis is also limited across the capillaries of the CNS. The main mechanism by which drug molecules pass through the capillaries of the CNS into the brain is passive transcellular diffusion. Drug molecules transported by passive transcellular diffusion are low molecular weight lipophilic molecules. The permeability of the blood/brain barrier increases, as the lipophilicity of the low molecular weight molecules increases. However, above a certain molecular weight, the permeability of lipophilic molecules across the blood/brain barrier is reduced substantially due to their increased molecular size – that is, the molecular weight offsets the positive effect of lipophilicity upon diffusion across the blood/brain barrier. Some molecules that are transported by passive diffusion into the brain may be pumped back out into the systemic circulation by efflux carriers such as P-glycoprotein, as mentioned above in Section 7.6.2.

Carrier-mediated influx may transport hydrophilic nutrient molecules such as glucose and basic amino acids across the blood/brain barrier. These influx carriers may be responsible for the transport of some drugs. Endocytic transport is responsible for the transport of macromolecules such as insulin into the brain. Drug transport across the blood/brain barrier can be increased by designing drug molecules to mimic the inter-action of nutrients with endogenous carriers present at the blood/brain barrier.

For example, increasing the levels of dopamine in the brain can relieve the symptoms of Parkinson's disease; however, dopamine does not cross the blood/brain barrier in adequate amounts. Levodopa, a precursor of dopamine, has a high affinity for an amino acid carrier system present at the blood/brain barrier. After transport across the blood/brain barrier, levodopa is decarboxylated to yield dopamine. Levodopa can be considered to be a pro-drug for dopamine.

Having explored the role of systemic circulation in the transport of drug molecules, the following sections of this chapter will now highlight the barriers encountered during the absorption of drug molecules into the systemic circulation from two of the common sites of administration – the gastrointestinal tract and skin.

7.8 GASTROINTESTINAL DRUG ABSORPTION

The most popular route of drug delivery is the oral route, which we all use routinely for the ingestion and delivery of nutrients from food and drink. Food and drinks are taken into the oral cavity, swallowed, digested by the gastrointestinal enzymes, and absorbed across the intestinal epithelium. The epithelium of the gastrointestinal tract is designed to protect the body against any hazardous chemicals that might be ingested. However, it is also designed to absorb nutrients and secrete fluids and enzymes.

The gastrointestinal (GI) tract extends from the mouth to the rectum, as shown in Fig. 7.19. Drug-delivery systems can be designed to deliver drug molecules locally to the

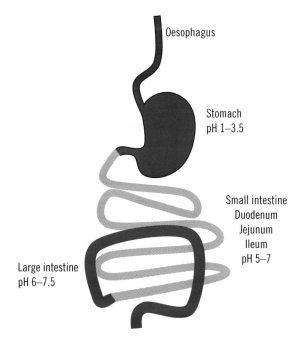

FIGURE 7.19 The gastrointestinal tract.

GI tract or systemically to the blood system through the epithelium at various points along the tract. The interior of the GI tract is lined with an inner epithelium layer, an intermediate layer of connective tissue (containing nerves, lymphatic vessels and blood vessels) and an outer layer of muscle.

When considering drug transport across a region of the gastrointestinal tract it is important to consider (1) the structure of the epithelium barrier, (2) the composition of the GI fluid and (3) the normal transition time in that specific region of the gastro-intestinal tract. In the next sections we discuss the drug absorption in the small and large intestine.

7.8.1 The small intestine

The small intestine is a major site for drug absorption after ingestion. It is a coiled narrow tube, approximately 2.5 cm in diameter and 6 m in length, and connects the stomach and the large intestine. The function of the small intestine is to breakdown food and to absorb nutrients. The transit time through the small intestine is approximately 3 hours. The small intestine can be divided into three sections; the duodenum, the jejunum and the ileum.

The duodenum is the first section of the small intestine, immediately after the stomach. It is the widest part of the small intestine and is approximately 25 cm in length. The duodenum is where the majority of food is broken down and the majority of chemical degradation takes place. The pancreas releases digestive enzymes (including trypsin, chymo-trypsin, amylase and lipase) into the duodenum through the bile duct. Bile is released into the duodenum via the bile duct from the liver and gall bladder. Bile is an alkaline liquid (due to the presence of bicarbonate ions), and neutralizes the contents exiting the stomach's acidic environment. (The pH of intestinal fluid exiting the duodenum is approximately 5.5.) Bile also contains phospholipids and bile salts that are involved in the emulsification of fats prior to digestion.

After the duodenum, the next section of the small intestine is the jejunum. The majority of adsorption of nutrients and molecules from the small intestine occurs in the jejunum. After the jejunum is the ileum, its function is to absorb the products of digestion that have not been absorbed by the jejunum. The pH of the fluid in the jejunum is between 6 and 7 and pH 7 to 7.5 in the ileum.

Structure of the small intestine epithelium

The epithelial cells of the small intestine are designed for absorption. The intestinal epithelium is composed of simple columnar epithelial cells: called intestinal absorptive cells or enterocytes. Mucus-secreting goblet cells are dispersed throughout the entero-cytes. Lymphoid tissue is also present in the small intestine epithelium in a number of ways, including lymphocytes present in the epithelium, or aggregates of lymphoid follicles called Peyer's patches.

To enable nutrient absorption from the small intestine, the epithelium surface area is approximately 200 m². Several features contribute to its large surface area: (1) the circular folds along the small intestine, (2) the finger-like projections (villi) into the centre of the small intestine channel, (3) the microvilli on the surface of the epithelial cells and (4) the

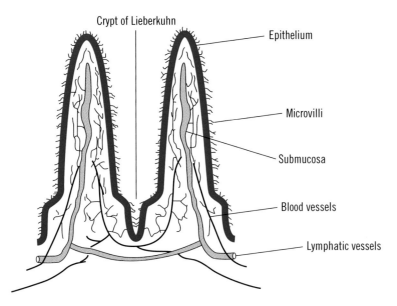

FIGURE 7.20 A diagrammatical representation of the structure of the small intestinal villi and the submucosal layer beneath.

inward folds in the epithelium at the base of the villi called the crypts of Lieberkuhn. The crypts of Lieberkuhn are glands located in the intestinal membrane that secrete enzymes. A simple schematic diagram of the villi and submucosal layer is shown in Fig. 7.20. The epithelial cells of the intestine adhere to each other due to the presence of tight junctions that restrict the paracellular transport of drug molecules. Drug absorption from the small intestine occurs primarily by transcellular transport both passive and active.

The epithelium of the interior of the small intestine is separated from the submucosal layer by a layer of connective tissue and a layer of thin smooth muscle. The smooth muscle layer is responsible for the folding structure of the epithelial layer. The submucosal layer of connective tissue beneath the epithelium contains blood vessels, lymphatic vessels and gland ducts. The capillaries in the submucosa carry blood via the portal vein to the liver. Drugs absorbed from the small intestine must pass through the liver before they can be distributed around the body via the systemic circulation. The metabolism of drug molecules in the intestine and liver before entry into the systemic circulation is called first-pass metabolism.

7.8.2 The large intestine

The large intestine, or colon, extends from the ileum to the anus and is approximately 7.3 cm in diameter and 1.5 m in length. The function of the large intestine is to absorb water, and to store and eliminate faecal waste. The role of the epithelium in the large intestine is both absorptive and protective. The epithelium of the large intestine is composed of simple columnar epithelial cells and, unlike the small intestine epithelium, does not contain villi. It does, however, contain invaginations (the intestinal glands). The large intestine has a large number of mucus-secreting goblet cells.

The first section of the large intestine is slightly more acidic than the ileum, the pH in the final section of the large intestine is pH 6–7. The transit time through the large intestine is variable and can range from 2 to 48 hours. The amount of fluid in the large intestine decreases from its junction with the small intestine to the rectum: this can have implications for the dissolution of drug from tablets and capsules, as discussed in Section 7.8.3.

The final section of the large intestine is called the rectum. It is 12–19 cm in length and has a surface area between 200 and 400 cm^2. The tissues of the rectum are highly perfused with blood vessels. The veins in the rectal area, unlike the rest of the small and large intestine, do not pass into the systemic circulation via the portal vein and liver. Instead, they pass straight into the systemic circulation. Therefore, drug molecules absorbed from the rectum do not undergo first-pass metabolism. The pH of the rectal area is approximately pH 7.5. However, its buffering capacity is poor and fluctuations in pH can occur. The amount of fluid in the rectum is low: about 3 mL of mucus.

Even though the gastrointestinal system is designed to deliver nutrients to the systemic circulation it is not always an ideal route of delivery for drug molecules. Now we will look at some of the challenges the GI tract poses for drug absorption.

7.8.3 Challenges to drug absorption in the GI tract

Due to the presence of tight junctions between the epithelial cells of the intestinal epithelium, the majority of drug transport in the GI tract is via a transcellular route. For passive transcellular drug transport, a drug molecule is required to be lipophilic. However, before a drug can be transported across the GI epithelial barrier it must first be in solution. Therefore, a balance between solubility and lipophilicity is required to achieve systemic drug absorption from the small intestine.

Solubility and lipophilicity balance

The balance between the solubility and lipophilicity of ionizable drugs can vary throughout the gastrointestinal tract due to changes in pH: the pH of GI fluids changes from being acidic in the stomach to being slightly alkaline in the large intestine. Also, the lipophilicity of ionizable drugs can alter along the gastrointestinal tract due to changes in their state of ionization. The molecular weight of a drug molecule is also an important drug property to consider as it can influence both drug solubility and its permeability of gastrointestinal epithelium.

Dissolution

The dissolution of drug molecules can be the rate-limiting step to drug absorption from the gastrointestinal tract. Dissolution – the process by which molecules in the solid state move into solution – can be influenced by the pH of the intestinal contents. The presence of bile can increase the dissolution rate of lipophilic drugs and the low volume of liquid in the lower regions of the gastrointestinal tract can limit the dissolution of drug particles and tablets. The viscosity of the mucus layer covering the epithelial cells of the GI tract may also reduce the dissolution rate for some drug molecules.

Carrier-mediated transport

Hydrophilic compounds can be transported across the intestinal epithelium by carrier-mediated transporters. For example, angiotensin-converting enzyme inhibitor (ACE inhibitor) molecules are transported by peptide transporters. However, the presence of the efflux transporter P-glycoprotein in the small intestine can inhibit the absorption of a number of drugs including digoxin.

Intestinal enzyme activity

The enzyme activity of the small intestine is high: a large number of digestive enzymes are present. There is also evidence to suggest that the activity of the enzyme cytochrome P450 3A4 is higher in the small intestine epithelia compared to the colon: the metabolism of drug molecules that act as substrates for cytochrome P450 3A4 may be reduced in the colon compared to the small intestine. The colon is a potential absorption site for orally administered protein and peptide drugs due to its absence of peptidase activity.

Bacteria

Bacteria colonize the lower regions of the gastrointestinal tract: the bacterial count in the large intestine is 10^{11}/g compared to 10^4/g in the upper small intestine. There are 400 different species of bacteria present in the large intestine and bacteria in this region are predominantly anaerobic in nature. (This isn't surprising: it is difficult for a ready supply of oxygen to reach the depths of the large intestine.) The bacteria in the lower regions of the GI tract produce enzymes that are capable of metabolizing carbohydrates and proteins: this activity can be exploited to deliver drugs to the large intestine. For example, sulfasalazine is composed of the anti-inflammatory active salicylate (5-aminosalicylic acid) joined by an azo bond an antibacterial sulfapyridine, as shown in Fig. 7.21. It is administered for the treatment of ulcerative colitis in the large intestine. The intact molecule is delivered to the large intestine where it is cleaved at the azo bond by the enzymes present in the large intestine into its active salicylate and sulfapyridine constituents.

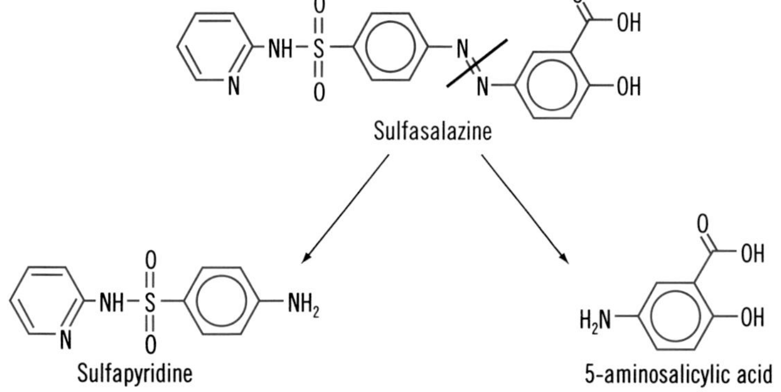

FIGURE 7.21 Molecular structure of sulfasalazine is cleaved at the azo bond by the enzymes present in the large intestine into its active salicylate and sulfapyridine constituents.

Food

The presence of food in the GI tract contributes to a number of factors that influence drug absorption. Food can alter pH, increase GI tract mobility, increase secretions of enzymes and bile and increase the viscosity of fluid in the gastrointestinal tract. Further, components of food can increase or decrease the activity of certain enzymes associated with drug metabolism. For example, grapefruit juice inhibits the activity of cytochrome p450 enzymes. Drug molecules can be displaced from the rectum prior to absorption due to the movement of faecal matter.

Having explored the GI route to drug absorption, let us now consider drug absorption transdermally (through the skin).

7.9 THE SKIN BARRIER

The skin is the largest organ of the body. It represents 10% of the body mass and covers an area of approximately 1.7 m² of the average person. It regulates heat and water loss from the body and prevents the ingress of hazardous chemicals or micro-organisms into the body. Healthy skin is designed to be a highly efficient protective barrier. Due to its protective role, the skin is not designed to be an ideal route for drug delivery. However, its large size and easy accessibility confer upon it a number of advantages for both local and systemic drug delivery. To understand how the skin can act as a route or a barrier to drug absorption, the structure and function of the skin must be understood.

7.9.1 Structure of the skin

The outside layer of skin, the epidermis, can be divided into two main layers; (1) the outer layer of non-viable epidermis (the stratum corneum) and (2) the viable epidermis. Beneath the epidermis lies the dermis, as shown in Fig. 7.22. The dermis is composed mainly of connective tissue that supports the epidermis. It contains blood vessels, lymphatic vessel and nerve endings and also supports hair follicles and sweat glands. Drug molecules administered to the skin must first pass through the non-viable and viable epidermis before they can reach the blood vessels present in the dermis.

The layer of the skin exposed to the external environment, the epidermis, is composed of stratified squamous epithelial cells, a large proportion of which are keratinocytes. Keratinocytes are epithelial cells that produce a fibrous protein called keratin that is also found in nails and hair. As keratinocytes mature, keratin is accumulated within the cell: these cells are said to be 'keratinized'. Keratinocytes play an important role in the barrier function of the skin; they are dehydrated, mechanically strong and chemically resistant. The outer layers of the epidermis are constantly being shed and replaced due to cell pro-liferation (mitosis) in the bottom layer of epithelial cells. Cells are continually moving from the bottom layer to the top layer of the epidermis and structural changes occur in the cells as they move upwards: cells become flattened, produce more keratin and lose their nucleus. The time of transit for a cell to move from the basal layer to the shedding stratum cornea is approximately 1 month.

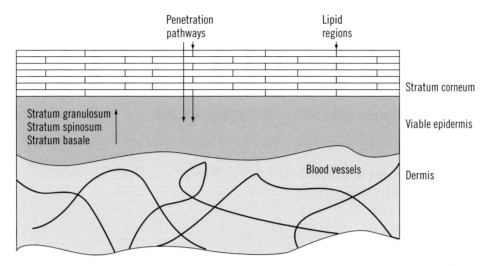

FIGURE 7.22 A schematic drawing of a skin cross-section showing the dermis and epidermis (stratum corneum and viable epidermis) indicating drug transport pathways.

Non-viable epidermis (stratum corneum)

Eventually the cells die and the top layer of dead cells is termed the non-viable epidermis or stratum corneum. The stratum corneum is 10–15 cell layers thick and approximately 10–15 μm in depth when dry and considerably thicker when wet. The non-viable cells of the stratum corneum are called corneocytes. The stratum corneum consists of 75–80% protein and 5–15% lipid. The protein is located primarily in the corneocytes. The cellular protein is highly insoluble and resistant to chemical attack. Lipids present in the stratum corneum are mainly ceramides, fatty acids, cholesterol, cholesterol sulfate and sterol/wax esters. The stratum corneum can be thought of as a brick wall with the corneocytes embedded in a cement of extracellular lipids. Together, these components provide an effective barrier to water loss and the transport of many drugs.

Viable epidermis

The viable epidermis can be subdivided into four layers. The lower layer adjacent to the dermis (the stratum basale) contains a number of cell structures; keratinocytes, melanocytes (these cells synthesis the pigment melanin), Langerhans cells (major antigen-presenting cells of the epidermis), and Merkel cells (nerve endings that facilitate the sensations of touch and pain). Mitotic division to form new cells occurs in this layer.

The layer of cells above the stratum basale, the stratum spinosum (also referred to as the prickle cell layer), is composed of keratinocytes that change in morphology from columnar to polygonal cells. The next layer, the stratum granulosum, is 1–3 cells thick and contains enzymes that begin to degrade the viable cell components. In this layer there is an increase in cell keratinization, and further morphological changes occur such as cell flattening. These cells also contain lipid-filled granules; this lipid forms the extracellular lipid that acts as a cement between the corneocytes in the stratum corneum. The stratum lucidum is the uppermost layer of the viable epidermis and is only observed in thick epidermis as seen on the palms of the hands and soles of the feet.

Now let us look at what physicochemical properties influence the transport of drug molecules through the skin.

7.9.2 Transport of drug molecules across the skin

Lipophilic drug molecules applied to the stratum corneum can partition into the extracellular lipids and the lipid cellular membranes of the corneocytes in stratum corneum. Drug molecules may pass through the stratum cornea by passive diffusion (paracellular and transcellular). The more lipophilic a molecule is, the greater its ability to partition into the stratum corneum. However, some hydrophilic molecules can permeate the skin due to their low molecular weight. The extent to which drug molecules diffuse through the stratum cornea is influenced by their diffusion coefficient: molecules of high molecular weight have lower diffusion coefficients, as explained previously in Chapter 3.

While very lipophilic molecules partition into the stratum corneum, their lipophilic nature can reduce the degree to which they partition out of the stratum corneum: they can remain stored in the outer layers of skin. The ability of a drug molecule to permeate the skin barrier can be quantified by a permeability coefficient (K_p). Table 7.4 gives details of the skin permeability coefficients of a range of compounds with varying $\log P$ (an indicator of lipophilicity) and molecular weight.

Techniques to enhance skin permeation of a drug

A number of techniques have been employed to enhance the extent to which drug molecules permeate the skin. As transport across the skin is mainly by passive diffusion, increasing the drug concentration in a delivery system applied to the skin (such as a cream, gel, ointment or patch) can increase the concentration gradient across the stratum corneum and, hence, increase the extent of permeation of some drug molecules.

TABLE 7.4 The measured skin permeability coefficient (K_p) of a group of compounds with varying $\log P$ values and molecular weights.

Molecule	K_p	$\log P$ o/w	Molecular weight (Da)
Sucrose	1.206	−3.70	342
Adenosine	1.584	−1.05	267
Aldosterone	0.586	1.08	360
Corticosterone	1.808	1.94	346
Estradiol	4.248	2.29	272
Testosterone	11.330	3.32	288

Data reproduced from Yakov Frum *et al.*, (2007), Towards a correlation between drug properties and in vitro transdermal flux variability. Int. J. Pharm., **336**, 140–147.

Penetration enhancers can be used to increase the skin permeation of a number of drugs. Compounds used as skin penetration enhancers include water, alcohols, surfactants and fatty acids.

Applying an electrical current across skin (iontophoresis) can promote the delivery of some ionizable drugs. The application of large electrical pulses (electroporation) to the skin temporarily disturbs the phospholipid bilayer of the cell membrane. The disturbance of the phospholipid bilayer allows large hydrophilic molecules to cross the skin barrier.

Microneedle devices can be used to create pores in the stratum corneum and to facilitate the delivery of large macromolecules across the skin. The depth of skin pierced by the microneedles is very small and the tips of the needles do not reach the nerves or blood vessels beneath the viable dermis. Therefore, application of microneedles to the skin does not result in bleeding or pain associated with traditional needle injections.

While two of the common routes of drug administration are the oral and transdermal routes, the barriers encountered at other routes of administration also require consideration. We cannot cover all of these extensively in this book; however, a small selection of these will be highlighted in this final section.

7.10   OTHER ROUTES OF ADMINISTRATION

The drug absorption after drug administration to the lungs, nasal cavity and oral cavity, will be discussed briefly in this section and potential barriers associated with these routes of delivery will be highlighted.

7.10.1   Drug absorption in the lungs

The primary function of the lungs is to transport oxygen from the air to the blood and carbon dioxide from the blood to the air. The lungs are designed for efficient air flow and gas exchange, and comprise airways that exhibit a branched structure, as shown in Fig. 7.23. In the alveolar region, the epithelium is designed for efficient transport of molecules between gas (air in lungs) and liquid phases (blood in the systemic circulation). The pulmonary blood/air barrier in the alveoli region is separated by three thin layers, the epithelium layer, the interstitum and the capillary epithelium. The epithelium in this region is a monolayer of squamous cells. The lungs contain a number of different epithelial cell types, as shown in Table 7.5.

The lungs have a number of features advantageous for systemic drug delivery: large surface area ($80-120m^2$), high blood perfusion, thin epithelium ($0.1-0.2$ µm) in the alveolar region and an immense capacity for solute exchange. The transport of drug molecules into the systemic circulation via the lungs is primarily by passive transcellular diffusion. Generally, small lipophilic molecules are more rapidly absorbed from the lungs than larger ones.

Inhalers and nebulizers are used to deliver drugs into the lungs, these are discussed in more detail in Section 6.7. Before a drug can act locally or be absorbed systemically across the pulmonary epithelium, it must first avoid the physiological mechanisms designed to

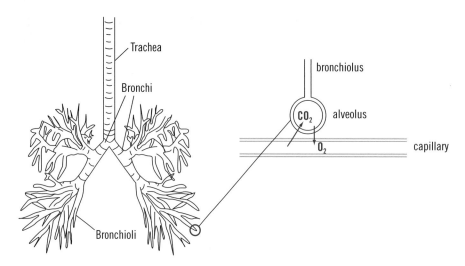

FIGURE 7.23 Schematic diagram of the pulmonary system, indicating the process of gas exchange in the alveolar region.

TABLE 7.5 The different types of epithelial cells present in lung epithelium.

Type of epithelial cell	Function
Serous	Secretes enzymes
Goblet	Secretes mucus
Clara	Secreting products, including Clara cell secretory protein (CCSP) and a component of the lung surfactant
Endocrine (Kulchitsky)	Secrete hormones
Ciliated	Move mucus
Brush cells	Uncertain
Pneumocytes type I	Offer short pathway for gas to blood
Pneumocytes type II	Secrete, renew and store lung surfactant
Alveolar macrophages	Engulf and transport particulate matter to lymph nodes and mucocillary escalator

avoid hazardous aerosols reaching the lower lungs. The first challenge is the deposition of drug particles or droplets in the lower regions of the lungs.

Deposition of drugs in the lungs

With the exception of medical gases, drug molecules are deposited in the lungs in the form of aerosols – solid particles dispersed in gas or liquid droplets dispersed in gas. These aerosols are carried through the lungs by oral breathing. However, oral breathing

TABLE 7.6 Dimensions of powder particles that reach various regions of the lungs.

Particle size	Region of disposition in the lung
> 10 μm	Particles impact on the back of throat
6 to 10 μm	Trachea and bronchi
3 to 6 μm	Bronchioli
1 to 3 μm	Alveoli

can be affected by disease states, such as cystic fibrosis or pulmonary obstruction, which can affect a drug's efficacy in a given patient.

In addition to breathing patterns, the region of the lungs where particles are deposited depends on the size and density of the particles administered. Table 7.6 gives details of the size of particles that are typically deposited in various regions of the pulmonary system. Particles are deposited both due to sedimentation and due to their collision with the walls of the airways.

Once a drug is deposited in the lungs, its can be cleared from the lungs by mucociliary and alveolar clearance.

Mucociliary and alveolar clearance

The epithelium in the upper airways is dominated by mucus-producing goblet cells and ciliated cells. Ciliated cells have tail-like projections (cilia) that extend from the surface of the epithelial cells. These cilia sweep in waves and move mucus and dirt through the airways. The mucus sitting over the ciliated epithelium is composed of two layers: the lower layer of low viscosity liquid that lubricates the cilia and the upper mucus layer that has a gel-like viscosity. Particles deposited in the upper airways can become trapped in the upper layer of mucus and some drug molecules may dissolve in the mucus. Any trapped or dissolved particles are moved upwards towards the pharynx by the movement of the mucus due to the beating movement of the cilia. From the pharynx these particles enter the gastrointestinal tract: this process is called 'mucociliary clearance'.

Alveolar macrophages are present in alveolar epithelium. Macrophages rapidly engulf particles they encounter. The process of clearing macrophages that have engulfed particles is a slow one that can take days to even weeks. This process is called 'aveolar clearance'. Macrophages can travel up to the mucus regions and be cleared by mucociliary clearance. They can also be taken up by the lymph system.

In addition to mucociliary and alveolar clearance, drug molecules in the lungs can be metabolized by pulmonary enzymes.

Metabolism

Drug molecules absorbed systemically via the lungs avoid first-pass metabolism in the liver. However, drug molecules can be metabolized by enzymes acting locally in the lungs.

Enzyme-secreting Clara cells and type II pneumocytes are responsible for most of the pulmonary enzymatic activity.

The next route of administration to be explored is the nasal cavity.

7.10.2 Drug absorption in the nasal cavity

The nasal cavity is designed to perform a number of important functions: (1) to warm and humidify incoming air, (2) to trap particles present in that air, (3) to sense smells, (4) to protect against microbes or viruses and (5) to exert an immune response in the presence of antigens. Drug molecules can be delivered to the nasal cavity using nasal sprays or nasal drops to exert local or systemic effects. Nasal drug delivery is an attractive route for drug administration because of the nose's accessibility. Also, there is a large epithelial surface area for drug absorption in the narrow horizontal passages after the nostrils, called the turbinates. The large surface area is due to the microvilli structures on the surface of these cells: each cell can contain around 300 microvilli. The epithelium in this area has a rich blood supply.

The nasal cavity offers a number of advantages for the delivery of drugs compared to oral delivery. The molecular weight of molecules that can be delivered via the nasal epithelium is much higher than those that can be delivered via the intestinal epithelium: large hydrophilic molecules can pass through the aqueous filled pores between the nasal epithelial cells. There are a number of marketed products that deliver relatively low molecular weight peptides such as calcitonin, oxytocin, desmopressin and buserelin systemically by the nasal route.

However, as in other tissues, the extent of drug absorption decreases as the molecular weight of drug molecules increases. Lipophilic molecules are absorbed by passive diffusion through the epithelium barrier. There is some evidence to suggest that amino acids can be transported by carrier transporter systems in the nasal epithelium, these carriers could be exploited for the transport of some hydrophilic drugs.

Similar to the lungs, the nose is designed with protective mechanisms that prevent particles and pathogens travelling into the lungs, these protective mechanisms can act as barriers to drug absorption.

Particle entrapment

Stiff hairs called 'vibrissae' are present in the nostrils and stop the passage of particles greater than 10 µm into the airways. The turbinates' narrow passages create turbulent air flow and this increases the entrapment of 5–10 µm airborne particles on these mucus-lined surfaces. The epithelium in this region can be ciliated or non-ciliated. Similar to the ciliated cells in the upper regions of the lungs, these cilia sweep in waves and move mucus and dirt through the nasal passages by mucociliary clearance. The mucus is swallowed into the gastrointestinal tract or ejected from the nasopharnyx region by coughing (expectoration).

7.10.3 Drug absorption in the oral cavity

Delivery of drugs systemically via the epithelium of the oral cavity has a number of advantages. First, delivery of drugs to the oral cavity is highly acceptable to patients.

TABLE 7.7 The main differences in the properties of sublingual and buccal epithelium in relation to drug transport.

	Sublingual	Buccal
Epithelium type	Non-keratinized	Keratinized
Mucosa thickness	100–200 μm	500–800 μm
Relative permeability	Higher	Lower
Relative onset of action	Faster	Slower
Relative surface area	Lower	Higher
Relative amount of saliva	Higher	Lower
Expanse of smooth muscle and immobile mucosa for retention	Lower	Higher

Second, systemic delivery of drug molecules can be achieved as the tissues of the oral cavity have a rich blood supply and the epithelium is relatively permeable compared to epithelial membranes at other sites of administration, such as the skin. Third, the membranes of the oral cavity are tolerant to potential antigens due to the virtual lack of Langerhans cells (antigen-presenting cells). Finally, drug molecules delivered systemically via the oral cavity are not subjected to the first-pass metabolism in the liver, unlike molecules delivered via the GI tract. Consequently, the risk of enzymatic degradation is reduced for drug molecules in the oral cavity compared to other regions of the gastrointestinal tract.

Delivery to the oral cavity can be classified as

(1) sublingual delivery: delivery through the membrane lining the floor of the mouth and the under side of the tongue

(2) buccal delivery: delivery through the buccal mucosa, the membrane lining the cheeks, gums, upper and lower lips; or

(3) local delivery: delivery to the surface of the oral cavity mucosa.

The sublingual region and the buccal regions of the oral cavity are used to deliver a number of drugs. Nitrolingual spray™ that contains glyceryl trinitrate for the treatment of angina is administered sublingually. By contrast, nicotine gum administered to assist in smoking cessation, is delivered via the buccal cavity. The patient is advised to chew the gum until the taste becomes strong and then to rest the gum between the gum and cheek. Table 7.7 lists the main differences in the properties of sublingual and buccal epithelium in relation to drug permeability.

Oral epithelium

Some drug molecules can reach the systemic circulation via paracellular and transcellular routes across the oral epithelium. The oral epithelium layer comprises flat cells layered on top of one another (stratified squamous epithelium cells). The oral epithelium can be divided into two cell types depending on whether the epithelium layer is keratinized or non-keratinized. Keratinized epithelium is more impermeable to drug transport compared to non-keratinized epithelium. The epithelium found on the base of the tongue, the roof of the mouth and gums is keratinized. It needs to be mechanically strong to resist damage during chewing. The mucosa lining found everywhere else in the oral cavity is non-keratinized.

Transcellular transport by passive diffusion is possible for hydrophobic molecules of low molecular weight. These molecules must travel through a number of layers of epithelial cells prior to reaching the blood vessels beneath.

Oral epithelium rarely contains tight junctions. The surface layers of oral epithelium are surrounded by an extracellular material consisting primarily of polar protein carbohydrate complexes. However, the extracellular material beneath the surface layer is lipidic. These intercellular lipids pose a major physical barrier to drug paracellular transport. Paracellular transport through these lipidic regions is dependent on the drug's ability to partition into these lipid regions, which in turn is dependent on the lipophilic nature of the drug.

The surface of the oral epithelium is covered in a layer of mucus secreted by epithelial goblet cells. The major components of mucus are glycoproteins called mucins. Mucin is negatively charged at physiological pH. Some drug substances may bind to the components of mucus, by electrostatic, hydrophobic or hydrogen bonding, which may limit their transport into the oral epithelium.

Saliva is a watery excretion secreted from three major and numerous minor salivary glands in the oral cavity. The saliva contains mucin, enzymes (mainly amylase and lysozyme), antibodies and electrolytes. Saliva is constantly being produced by the salivary glands and may wash away the drug from the oral epithelia or dilute the drug, thus reducing the concentration gradient – the driving force for drug transport by passive paracellular and transcellular transport.

We hope at the end of this chapter the basics of drug–target receptor interactions, the factors influencing the partitioning of drug molecules and the main transport mechanisms by which drug molecules are transported across biological barriers have been highlighted. The main points are summarized below.

⊙ 7.11 SUMMARY

An ideal drug molecule would cause a single specific physiological effect. Drug molecules can be selective and interact with only one specific type of target receptor. Drugs molecules that interact with a range of receptor types are non-selective and as a result cause a range of physiological effects. Drug molecules can act as agonists or antagonists. The binding of agonists to receptors results in a recognized positive physiological effect. Antagonists act by blocking or reversing the effects of agonists.

Drug receptors can be proteins embedded in cell membranes. The biological target of a drug can also be a specific enzyme. Nucleic acids are also important biological targets for drug binding. The binding of a drug to a receptor is an equilibrium process.

The partition coefficient, P, describes the extent to which a molecule will partition between two immiscible liquid phases. The partition coefficient, P, for a drug can be described in terms of the concentration of the drug in both aqueous and organic solvents. Logarithms of the partition coefficients, $\log P$ values, give a narrower range of values that are more convenient to use. Compounds with $\log P$ values lying between approximately 1.0 and 4.5 are preferred as drug molecules, as these compounds have the required balance between aqueous and lipid solubility to allow for good drug absorption and distribution.

$\log P$ can be determined by experimental measurement using the 'shake-flask' method or using high-performance liquid chromatography (HPLC). Software for calculating $\log P$ values based on molecular structure is now very sophisticated and widely used.

The partitioning of acidic, basic and zwitterionic pharmaceutical compounds is influenced by the amounts of neutral and ionized forms in solution. The effect of pH on the extent of ionization has to be taken into consideration. To account for the differences in partitioning due to ionization, the partition coefficient, P, is replaced by the distribution coefficient, D.

Most biological barriers are composed of cells. Human cells are surrounded by an exterior lipid bilayer membrane called the cell membrane. The cell membrane's bilayer structure allows the diffusion of some hydrophobic molecules through the membrane. Polar molecules can be transported through cell membranes due to the presence of carrier proteins, pores and gates.

Epithelial cells perform their barrier function by adhering to one another due to the presence of cellular junctions – tight junctions, adherent junctions, desmosomes and gap junctions. Enzymes present at the epithelium membrane pose a barrier to drug transport due to the degradation of drugs.

Drug transport across cellular barriers can be paracellular or a transcellular. Low molecular weight, lipophilic molecules can be transported passively through cells. Hydrophilic drug molecules can be actively transported through cells by carrier-mediated transport. Efflux transporters, such as P-glycoprotein, can transport drug molecules back out of cells. Large molecules can be transported into cells by endocytosis. Pore transport allows low molecular weight, hydrophilic molecules to be transported into and out of cells.

The majority of drug transport into and out of the systemic circulation occurs across capillaries. The endothelium of capillaries are only one cell deep. There are three types of capillaries – continuous, fenestrated and sinusoid. The structure of the capillaries in the CNS creates a barrier to drug transport known as the blood/brain barrier (BBB). Paracellular transport is limited due to the presence of tight junctions and pinocytosis is limited. The main route by which drugs cross the capillaries of the CNS into the brain is by passive diffusion.

The GI tract is a major site for the absorption of drug molecules. Due to the presence of tight junctions, the majority of drug transport is via a transcellular route. The lipophilicity of an ionizable drug can alter along the gastrointestinal tract due to changes in its ionization state and this can alter transcellular absorption. Hydrophilic compounds can be transported transcellularly by carrier-mediated transporters. The enzyme activity of the small intestine is high. Bacteria that colonize the lower regions of the gastrointestinal tract can be exploited to deliver drugs such as sulfasalazine.

Drug transport from the GI tract to the systemic circulation is via the portal vein and liver: the metabolism of drug molecules before entry into the systemic circulation is called first-pass metabolism. The veins in the rectal area transport the drug directly into the systemic circulation and do not pass through the portal vein and liver.

The skin is advantageous for drug delivery: it is accessible and has a large surface area. However, its protective barrier properties can restrict drug transport. Low molecular weight lipophilic molecules pass through the stratum cornea by passive diffusion (paracellular and transcellular). Penetration enhancers, iontophoresis, electroporation and microneedles can be used to enhance drug delivery across the skin.

Drug molecules are generally deposited in the lungs in the form of aerosols. The region of the lungs where particles are deposited depends on the breathing pattern and the size and density of the particles administered. Particles can be cleared from the lungs by mucociliary and alveolar clearance. Enzymatic activity in the lungs can degrade drugs.

Large hydrophilic molecules can pass through the aqueous-filled pores between the nasal epithelial cells. Lipophilic molecules are absorbed by passive diffusion through the epithelium barrier. Particles can be prevented from entering the nose by the vibrissae in the nostrils. Particles can become trapped in the mucus present in the turbinate regions and cleared by mucociliary clearance.

The oral epithelial layer can be keratinized or non-keratinized. Keratinized epithelium is more impermeable to drug transport than non-keratinsed epithelium. The surface of the oral epithelium is covered in a layer of mucus. Some drug substances may bind to the components of mucus, which may limit drug transport through the oral epithelium.

⊜ REFERENCES

Demare, S., Slater, B., Lacombe, G., Breuzin, D., Dini, C. (2007) Accurate automated log Po/w measurement by gradient-flow liquid-liquid partition chromatography, *J. Chromatography A*, **1175**, 16–25.

Frum, Y., Khan, G. M., Sefcik, J., Rouse, J., Eccleston, G. M. and Meidan, V. M. (2007), Towards a correlation between drug properties and *in vitro* transdermal flux variability. Int. J. Pharm., **336**, 140–147

Leo, A. L. (1990) Method of Calculating Partition Coefficients, from Comprehensive Medicinal Chemistry, Vol. 4, C. Hansch, P. G. G. Sammes and J. B. Taylor (eds), C. A. Ramsden (Vol. Ed.), pp 295–319

Michalets, E. L. (1998) Update: Clinically significant cytochrome P-450 drug interactions, Pharmacotherapy, **18**, 86–112

Mirrlees, M. S., Moulton, S. J., Murphy, C. T., Taylor, P. J. (1976) Direct Measurement of Octanol-Water Partition Coefficients by High-Pressure Liquid Chromatography, *J. Med. Chem.*, **19**, 615–619

⋒ FURTHER READING

Aulton M. E. (ed.) (2002). *Pharmaceutics, the science of dosage form design* (2nd edn), Churchill Livingstone, Edinburgh.

Desai, A. and Lee, M. (eds) (2007). *Gibaldi's Drug delivery systems in Pharmaceutical Care*, American Society of Health-System Pharmacists, Bethesda.

Hillery, A. M., Lloyd, A. W., Swarbrick, J. (eds). (2001). *Drug Delivery and Targeting*, Marcel Dekker, New York

Rang, H. P., Dale, M. M., Ritter, J. M. and R. J. Flower. (2007). *Pharmacology* (6th edn), Churchill Livingstone, Philadelphia

❓ EXERCISES

7.1 Figure 7.24 shows the molecular structures of the β-blockers atenolol, metoprolol and propranolol. The log P values for these compounds are 3.21, 1.88 and 0.16, but these values are not given in order. Based on the relative hydrophilicity/hydrophobicity of the structures, assign the three log P values to the three drug compounds.

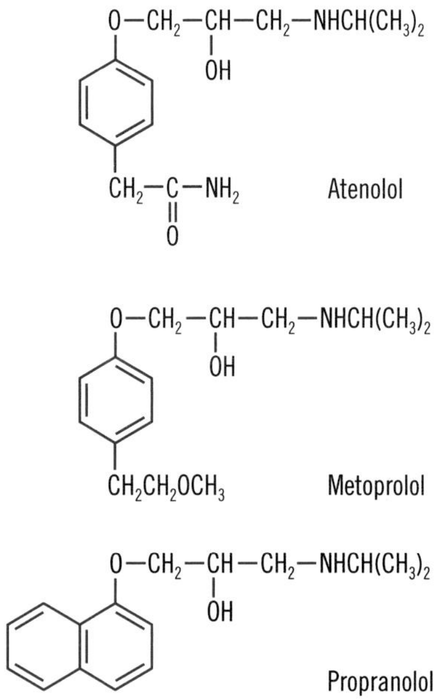

FIGURE 7.24

7.2 A new therapeutic molecule, drug Y, is identified. Its 1-octanol–water partition coefficient is determined experimentally at 25 °C. The concentration of drug X in 1-octanol is 2.7 mg mL^{-1} and in water 0.2 mg mL^{-1}. Calculate the partition coefficient and log P of drug X. Comment on the ability of drug X to move from aqueous medium into lipid rich media, such as cell membranes.

7.3   Figure 7.25 shows the variation of log D with pH for a new pharmaceutical lead compound. From the figure, state whether the lead compound is acidic or basic. Estimate its log P and its pK_a.

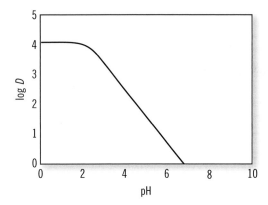

FIGURE 7.25

PHYSICOCHEMICAL ASPECTS OF PHARMACOKINETICS

08

Learning objectives

Having studied this chapter, you should be able to:

- explain what the terms pharmacodynamics and pharmacokinetics refer to
- describe how the processes of absorption, metabolism, distribution and elimination influence the level of a drug in the body
- describe the physicochemical properties that influence the processes of metabolism, distribution and elimination
- explain what pharmacokinetic parameters are used to describe drug absorption, metabolism, distribution and elimination
- calculate the pharmacokinetic parameters given in the worked examples
- describe the type of metabolism that occurs during phase I and II metabolism.

A drug must interact with its target receptor in the right amounts for it to be effective. If too little drug interacts with the target receptors, a therapeutic effect may not occur. However, if too much drug interacts with the target receptors, toxic effects may result. The duration of the therapeutic response is determined by the duration of interaction between the drug molecules and their target receptors.

When a quantity of drug is administered to a patient only a proportion of the drug will reach the target receptors, not least because only a proportion of the drug administered may be absorbed from the site of administration into the systemic circulation. The proportion of drug that reaches the target receptors can vary depending on the drug's molecular structure and its physicochemical properties. Within the systemic circulation, the drug may be degraded by enzymes in the liver, bind to proteins present in the blood, partition into tissues or be excreted by the kidneys before it reaches its specific target.

It is usually not convenient or possible to measure the amount of drug that interacts with the target receptors, instead, it is more common to measure the drug plasma concentration. Relationships between drug plasma concentration and the desired therapeutic response – for example, blood-pressure reduction or glucose blood levels – can be established. Taking samples of blood and measuring the drug plasma concentration can provide information regarding the extent of (1) drug absorption from various sites of administration, (2) drug binding to plasma proteins, (3) drug degradation by enzymes (metabolism) and (4) drug excretion by the kidneys or via the bile. It also provides information regarding the drug plasma concentrations required to achieve a desired therapeutic response, and the optimum dose of drug and frequency of dosage required to maintain the required therapeutic plasma concentrations.

Pharmacodynamics

Pharmacodynamics is the study of the relationship between drug plasma concentration and the pharmacological and toxicological responses. The relationship between pharmacological response and drug plasma concentration can be used to determine the therapeutic range of drug plasma concentrations that will yield minimal side effects. The lower concentration of this therapeutic range is known as the '**minimum effective concentration**' and the upper concentration is called the '**minimum toxic concentration**'.

Pharmacokinetics

Pharmacokinetics is the study of the time course of drug movement in the body. The amount of drug present in the body at any one time depends on the proportion of administered drug absorbed and the amount of absorbed drug eliminated from the body. Pharmacokinetics takes into account the time scale over which these events take place. It provides an insight into how the body handles a drug after administration. By determining some pharmacokinetic parameters the optimum dose of drug and dosing intervals required to achieve therapeutic drug plasma concentrations can be established.

In the previous chapter the physicochemical and biological factors that affect drug absorption from various sites of administration were explained. This chapter focuses on how the physicochemical properties of drugs can influence how a drug is handled by the body after drug absorption; we will also explore some commonly used pharmacokinetic parameters.

8.1 ABSORPTION, DISTRIBUTION, METABOLISM AND ELIMINATION

The fate of a drug in the body after administration can be described by a series of processes. The first process is the **absorption** of the drug into the tissue at the site of administration and, from there, into the systemic circulation. Absorption can be considered to be an 'input' process as it *increases* the level of drug reaching the target receptors. Immediately

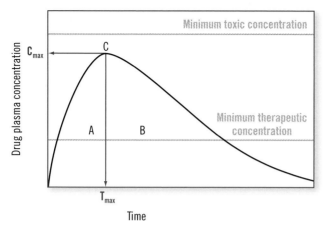

A   Absorption > Excretion and Metabolism
B   Absorption < Excretion and Metabolism
C   Absorption = Excretion and Metabolism

FIGURE 8.1 Plot of drug plasma concentration against time after non-intravenous drug administration, indicating the C_{max} and T_{max}.

after drug absorption, the second process of drug **distribution** occurs during which the drug is distributed throughout the body.

During absorption and distribution, the drug will encounter enzymes present within cells and extracellular fluid. These enzymes may transform drug molecules, thereby inactivating them. This process is called **metabolism**. **Elimination** is the process of drug removal from the body by **excretion**. A drug can be excreted via the kidneys, bile and lungs. In some circumstances, elimination can refer to metabolism as well as excretion because metabolism may break down a drug into an inactive form, therefore essentially eliminating it as an active species. Metabolism and elimination can be described as 'output' processes; both processes *decrease* the level of drug reaching the target receptors.

Drug plasma concentration will vary according to the balance between drug input and output processes. If the input process of absorption is greater than the output processes of metabolism and excretion, then the drug plasma concentration will increase. However, if the input process is less than the output processes, then the plasma drug concentration will decrease. If the input and output processes are equal, then the drug plasma concentration will remain constant, as shown in Fig. 8.1.

8.2 BIOAVAILABILITY

The fraction of drug administered that reaches the systemic circulation is called the **bioavailability fraction** (*F*). A drug administered intravenously is assumed to be 100% present in the systemic circulation and would therefore have a bioavailability fraction of 1.

The amount of drug that reaches the systemic circulation by other routes of administration is determined by measuring the drug plasma concentration after administration at regular time points and plotting the drug plasma concentration against time, as shown in Fig. 8.1. The area under the plot of drug plasma concentration against time is a measure of the amount of drug that reaches the systemic circulation. The area under this plot is commonly referred to as the area under the curve (AUC) and can be calculated using the trapezoidal rule, which is explained in Box 8.1. The bioavailability fraction can be calculated by comparing the AUC for a drug administered by a non-intravenous route with the AUC for the same drug after intravenous administration (AUC_{iv}), using eqn 8.1.

$$F = \frac{AUC}{AUC_{iv}}.$$

(8.1)

BOX 8.1 Calculation of the area under the curve (AUC) using the trapezoidal rule

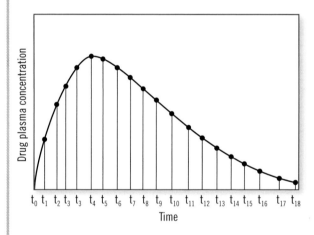

FIGURE 8.2 Plot of drug plasma concentration against time after non-intravenous drug administration split into trapezoids.

The most common way to calculate the area under the curve (AUC) is to use the trapezoidal rule. The plasma profile is divided up into trapezoids, each representing a set time interval, as shown in Fig. 8.2. The area for each trapezoid is calculated using the following equation.

$$\text{Area 1} = \frac{C_1 + C_2}{2} \times (t_2 - t_1) \quad \text{Area 2} = \frac{C_2 + C_3}{2} \times (t_3 - t_2) \text{ etc.}$$

The area under the curve after the last concentration value (AUC_{last}) is extrapolated to infinity using the following equation.

$$\text{Area}_{last} = \frac{C_{last}}{k},$$

where C_{last} is the last concentration value and k is the rate constant of the slope.

Equation 8.1 assumes that the amount (dose) of drug delivered by both routes of administration is equivalent. Where the amounts of drug differ, then the equation used to determine the bioavailability fraction must be rewritten to take this into account

$$F = \frac{D_{iv} \cdot AUC}{D \cdot AUC_{iv}}.$$ (8.2)

D_{iv} refers to the dose of drug administered intravenously and D refers to the dose of drug administered by the non-intravenous route of administration – for which the bioavailability fraction is being calculated.

▷ **EXAMPLE 8.1**

Digoxin was administered to patients via both an intravenous injection and an oral tablet. Each patient was administered a single intravenous injection containing 0.2 mg digoxin and a single tablet containing 0.5 mg digoxin. After plotting digoxin plasma concentration against time, the average AUC for the intravenous dose was calculated to be 35.1 ng mL^{-1} h and the AUC for the oral dose was calculated to be 64.0 ng mL^{-1} h. What is the bioavailability fraction (F) of the oral tablet?

Equation 8.2 above allows the bioavailability fraction (F) to be calculated if the dose administered and the AUC are known for the injection and tablet dosage forms. The bioavailability fraction can be calculated as shown below.

$$F = \frac{D_{iv} \cdot AUC}{D \cdot AUC_{iv}} \rightarrow F = \frac{0.2 \text{ mg} \times 64.0 \text{ ng mL}^{-1} \text{ h}}{0.5 \text{ mg} \times 35.1 \text{ ng mL}^{-1} \text{ h}} = 0.73.$$

Absorption rate

The bioavailability fraction (F) gives information regarding the level of drug absorption into the system circulation but it does not give information regarding the time period over which this drug absorption occurs. The extent of drug absorption over a time period is referred to as the drug absorption rate. A drug's absorption rate can be obtained from a plot of drug plasma concentration against time, and can be expressed as the time required to reach maximum drug concentration (C_{max}). This value is called the T_{max}, as indicated on Fig. 8.1.

The bioavailability fraction is influenced by drug absorption, and as explained in Chapter 7, the mechanism and extent of drug absorption is influenced by a drug's molecular weight, its lipophilicity and also by specific functional groups that can influence carrier-mediated transport. Drug bioavailability is also influenced by drug distribution, metabolism and excretion. In the next section we will explore drug distribution in the body.

8.3 DISTRIBUTION

Distribution is the process by which a drug present in the systemic circulation moves from the systemic circulation into other biological tissues and fluids. Drug plasma concentration will be initially high, driving drug partitioning into surrounding tissues and fluids. The drug will first partition from the blood into well-perfused tissues (such as the lungs) and later into poorly perfused tissues (such as adipose tissue and muscle). Drug distribution into tissues can be non-specific (a drug may pervade all tissues rather than 'homing in' on a specific biological target). As a result, a drug can be present in tissues other than those where the target receptors are present. This can result in the need for high drug doses to be administered to achieve sufficient drug concentrations at the target receptors. (The non-specificity can mean that a drug is effectively 'diluted' throughout the body, so more drug is needed to increase the concentration at the required site of action.)

A drug can also bind to non-target receptors and cause unwanted side effects, such as the hair loss and nausea observed after the administration of some drugs for the treatment of cancer. However, drug-delivery systems can be designed to target the drug to specific receptors and to minimize unwanted side effects.

The distribution of a drug in the body can vary depending on the physicochemical properties of the drug. A drug's physicochemical properties can influence drug–protein binding, transport through biological membranes and drug partitioning into tissues. These physicochemical properties include molecular size, log P or log D, aqueous solubility and pK_a. Some drugs will be distributed to tissues throughout the body and others will be highly concentrated in certain tissues and fluids. A parameter called the '**volume of distribution**' (V_d) gives an indication of the extent of drug distribution throughout the body. A large volume of distribution indicates that the drug is either distributed widely throughout the body or highly concentrated in certain tissues. A small volume of distribution indicates that the drug is primarily present in the systemic circulation.

Volume of distribution (V_d)

In order to explain what the volume of distribution of a drug refers to, envisage that a dose of 50 mg drug X is injected intravenously into a patient. Immediately after administration blood samples are taken and the drug plasma concentration at time zero (C_p^o) is found to be 5.0 mg/L. The drug's volume of distribution (V_d) can be calculated from the initial dose of drug and the drug plasma concentration after administration, using

$$V_d = \frac{\text{dose}}{C_p^o}. \tag{8.3}$$

The calculated volume of distribution for drug X is found to be 10 L. Compare this with a second drug, drug Y, administered intravenously at the same dose, 50 mg. For drug Y, the drug plasma concentration immediately after administration is determined to be 0.5 mg/L. The volume of distribution for the second drug Y is calculated (using eqn 8.3) to be 100 L.

The difference between the volume of distribution of drugs X and Y may be due to differences in their physicochemical properties. For example, if drug X has a hydrophilic nature, drug X may bind to plasma proteins such that it is retained in the systemic circulation. As a result, we would see a higher concentration of drug in the plasma. Its hydrophilic nature would result in the poor distribution of drug X into the lipophilic tissues. By contrast, drug Y may have a less hydrophilic and more lipophilic nature and therefore be readily distributed from the systemic circulation into the surrounding tissues; as a result, a lower concentration of drug Y would be measured in the plasma.

Lipophilic drugs such as amiodarone, imipramine and digoxin have high volumes of distribution, 5000 L, 2100 L and 420 L, respectively, they are distributed in the tissues throughout the body. Warfarin, which is tightly bound to the plasma protein albumin, has a volume of distribution close to that of the blood volume, 8 L. The binding of drugs to plasma proteins is explained in greater detail later in this section.

By carrying out pharmacodynamic studies, the drug plasma concentration required to achieve a specific therapeutic response can be determined. If the volume of distribution of a drug is known, the dose of drug required to achieve the therapeutic drug plasma concentration can be calculated using eqn 8.3.

▷ **EXAMPLE 8.2**

A phenytoin plasma concentration of 10–20 mg L^{-1} is required to achieve control of epileptic seizures. Phenytoin has a volume of distribution of 0.7 L kg^{-1}. What intravenous dose of phenytoin is required to achieve plasma levels of 20 mg L^{-1}?

The dose required can be determined using eqn 8.3 as shown below,

$$V_d = \frac{dose}{C_p^o},$$

therefore the equation can be rearranged to calculate the dose, dose $= C_p^o . V_d$ for this example dose $= (20 \text{ mg } L^{-1} \cdot 0.7 \text{ L } kg^{-1}) = 14 \text{ mg } kg^{-1}$.

The volume of distribution of a drug is influenced by drug binding to plasma proteins because the binding of drugs to plasma proteins decreases their distribution from the systemic circulation into the surrounding tissues. Drugs can also bind to other components of the blood, such as red blood cells. If we measure drug plasma levels only, we may not account for a drug bound to, or present in, non-plasma blood components. The pharmacokinetic parameter '**blood plasma ratio**' can provide information regarding the proportion of drug in the blood but not accounted for in plasma.

Blood plasma ratio

In order to determine the concentration of drug present in the body, we typically take a blood sample. Blood is not homogeneous and can be considered as a dispersion containing blood cells. Drug can be present in blood in several forms: as free drug (unbound) in solution, as drug bound to plasma proteins, as drug bound to blood cells, and as drug bound to membrane receptors. Due to the difficulty of measuring drug concentration in whole blood (C_{blood}), it is normal to determine drug concentration in plasma (C_{plasma}). Lipophilic molecules, such as cyclosporine, can penetrate red blood cells. Therefore, the amount of drug present in the whole blood (C_{blood}) for some drugs will be greater than or equal to the amount present in the plasma (C_{plasma}).

The equation used to calculate the blood plasma ratio (λ) of a drug is

$$\lambda = \frac{C_{blood}}{C_{plasma}}. \tag{8.4}$$

A blood plasma ratio of greater than one would indicate that the drug is relatively concentrated in components of the whole blood other than plasma. A blood plasma ratio of less than one would indicate that the drug is mainly present in the plasma.

▷ **EXAMPLE 8.3**

After eight patients were administered with a dose of tacrolimus, the average blood drug concentration was 7.92 ng mL^{-1} and the average plasma concentration was 0.40 ng mL^{-1}. What would be the blood plasma ratio and explain what the blood plasma ratio determined means in terms of drug distribution in the blood?

Equation 8.4 above allows the 'blood plasma ratio' (λ) of a drug to be calculated if the concentrations of the drug in the blood and plasma are known. Therefore, the blood plasma ratio can be calculated as shown below.

$$\lambda = \frac{C_{\text{blood}}}{C_{\text{plasma}}} \rightarrow \lambda = \frac{7.92 \text{ ng mL}^{-1}}{0.40 \text{ ng mL}^{-1}} = 19.80.$$

The blood plasma ratio calculated is considerably greater than one. This would indicate that the drug is concentrated in components of the blood other than the plasma. Tacrolimus is known to bind strongly to red blood cells.

A drug present in blood may be free (in plasma) or bound to plasma proteins and other blood components. In order to interact with target receptors and be pharmacologically active, drug molecules must be free – that is, they must not be bound to another entity. The concentration of free drug or unbound drug in plasma (C_{unbound}) is an important indicator of the amount of pharmacologically active drug in the body. If drug molecules bind to proteins, they essentially become sequestered in a non-active form, and the concentration of free drug present in plasma is reduced.

Protein binding

Plasma proteins that bind to drugs include albumin and α_1-acid glycoprotein. There are two binding sites on albumin that bind acidic drugs (site I and site II). Warfarin and phenylbutazone bind to site I. Diazepam and ibuprofen bind to site II. By contrast, α_1-acid glycoprotein has one binding site that binds basic drugs such as lignocaine.

Drug binding to plasma proteins is reversible: an equilibrium exists between unbound drug and bound drug in the plasma, as shown by

$$\text{unbound drug} + \text{free protein} \rightleftharpoons \text{drug protein complex.} \tag{8.5}$$

As the concentration of unbound drug in plasma (C_{free}) is reduced, the concentration of bound drug (C_{bound}) is also reduced. While the amount of unbound and protein bound drug in plasma changes with time, the fraction of total drug, unbound, remains constant. The '**fraction of unbound drug**' (f_u) can be expressed as shown by eqn 8.6. $C_{\text{total plasma}}$ is the total concentration of drug in plasma.

$$f_u = \frac{C_{\text{free}}}{C_{\text{total plasma}}} = \frac{C_{\text{free}}}{C_{\text{free}} + C_{\text{bound}}}. \tag{8.6}$$

Determining the fraction of unbound drug present in the plasma enables the fraction of pharmacologically active drug present to be determined.

Drug–protein binding is a competitive interaction and a drug can be displaced from its binding sites on proteins by the presence of endogenous molecules and other drugs in the plasma. Conditions such as hypoalbuminaemia (abnormally low levels of albumin in the plasma) can reduce the fraction of bound drug and therefore increase the fraction of pharmacologically active unbound drug.

 EXAMPLE 8.4

If the fraction of unbound phenytoin in plasma (f_u) is 0.1, what total concentration of phenytoin in plasma would be required to achieve an unbound phenytoin plasma concentration of 5.2 ng mL^{-1}?

Equation 8.5 relates the fraction of unbound drug (f_u) to the unbound drug plasma concentration (C_{free}) and total drug plasma concentration ($C_{total\ plasma}$). If the fraction of unbound drug and the required unbound drug plasma concentration are known, then the required total phenytoin plasma concentration can be calculated as shown below.

$$f_u = \frac{C_{free}}{C_{total\ plasma}},\ \text{if we fill in the values we know } 0.1 = \frac{5.2\ \text{ng mL}^{-1}}{C_{total\ plasma}},\ \text{we can then}$$

rearrange the equation to calculate the total plasma concentration

$$C_{total\ plasma} = \frac{5.2\ \text{ng mL}^{-1}}{0.1} = 52\ \text{ng mL}^{-1}.$$

8.4 DRUG METABOLISM

Drug compounds are **xenobiotics** – that is, they are 'alien' chemical entities that are not part of the endogenous biochemistry of the body. The biochemical system of the body is not inert towards xenobiotics. Rather, the enzymes and co-factors that act as the catalysts and reagents of the biochemical systems of the body act on xenobiotics as well as on endogenous biochemicals. The result of the action of the body's enzymes and co-factors on drugs and other xenobiotics is the formation of new products known as metabolites. The metabolites formed from a drug will have different chemical structures from the parent drug, and so will also have different biological and physical properties.

The general trend of the metabolic processes is to produce metabolites that are more polar and more hydrophilic than the starting substrates. The key benefit of these outcomes is that they aid excretion. More hydrophilic compounds have greater water solubility and hence greater tendency to be excreted. Hence, these metabolic processes promote the removal of the xenobiotics from the body through excretion.

It is also the case that some drugs, known as 'pro-drugs' are deliberately designed so as to undergo metabolism to generate an active form *in vivo*. For these reasons, drug metabolism and the physicochemical properties of the metabolites must be taken into consideration. Drug metabolism is a specialized topic. What follows is not intended to cover drug metabolism in depth. Instead, texts giving more detailed coverage of this topic are given under Further Reading at the end of the chapter.

Drug metabolism is generally considered to be divided into two phases. In **Phase I metabolism**, drug compounds undergo reactions such as hydrolysis, oxidation and/or reduction. In **Phase II metabolism**, drugs or their metabolites are coupled to certain biochemical species to form new metabolites known as conjugates. Both the processes of Phase I and Phase II metabolism directly affect the physicochemical properties of the drug substrate. The processes of drug metabolism could occur in any organ or tissue of the body, but are especially associated with the liver.

8.4.1 Phase I metabolism

The main processes that occur during Phase I metabolism of a drug are the hydrolysis, oxidation or reduction of the drug molecules. For metabolic hydrolysis to occur, the drug molecule must possess a functional group that is susceptible to hydrolysis, the cleavage of a covalent bond by the action of water. The main functional groups that have a tendency to undergo metabolic hydrolysis are, in particular, esters, and to a lesser extent, amides. The enzymes that catalyse these reactions include hydrolases, esterases, lipases, amidases, peptidases and proteases. Examples of metabolic hydrolyses are shown in Fig. 8.3.

In the first of the two examples shown in Fig. 8.3, the ester functionality of enalapril is hydrolysed to give a carboxylic acid, enalaprilate. This is an example of a metabolic conversion from a pro-drug to an active drug, as enalaprilate is the active form of the drug. In fact, it is an inhibitor of angiotensin-converting enzyme. Ester groups are often used in the design of pro-drugs, as metabolic hydrolysis of esters is usually an efficient process.

In the second example, the central amide bond on lidocaine is hydrolysed to give an amine and a carboxylic acid. The metabolic hydrolysis of esters and amides opens up these bonds to give carboxylic acids and alcohols or amines. These products are generally more polar than the substrate esters and amides. They are also lower in molecular weight.

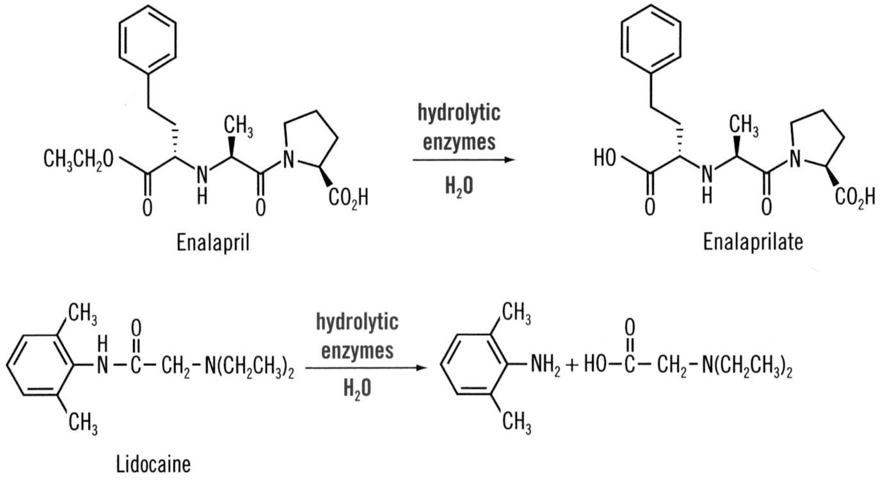

FIGURE 8.3 Metabolic hydrolysis of enalapril and lidocaine.

Therefore, they will usually be more water soluble than the parent esters or amides, and their distribution in the body may be significantly different. Another indication of the increase in polarity as a consequence of hydrolysis is a general decrease in log P value (see Section 7.2).

The body possesses many highly complex enzyme systems that carry out oxidation and reduction processes. One major group of oxidoreductase enzymes are the cytochrome P450 oxidoreductases (CYPs). The monoamine oxygenases (MAOs) are another class.

The action of oxidoreductases generally requires molecular oxygen and co-factors such as nicotinamide adenine dinucleotide (NADH). In the presence of oxygen, these enzymes generally introduce an oxygen atom onto the substrate structure. Hence, the metabolite formed may often contain an extra oxygen atom so that, for example, a hydrocarbon may be converted into an alcohol. However, the initially formed metabolite may also be unstable, and may degrade with cleavage into two products. This type of process often occurs at carbons joined to heteroatoms such as nitrogen or oxygen. Hence, this type of metabolic oxidation is often observed as loss of an alkyl group from a nitrogen or oxygen atom.

Another type of metabolic oxidation features the addition of an oxygen atom to a tertiary amine or a sulfide to give amine-N-oxides and sulfoxides, respectively. Examples of these types of metabolic oxidation are shown in Fig. 8.4.

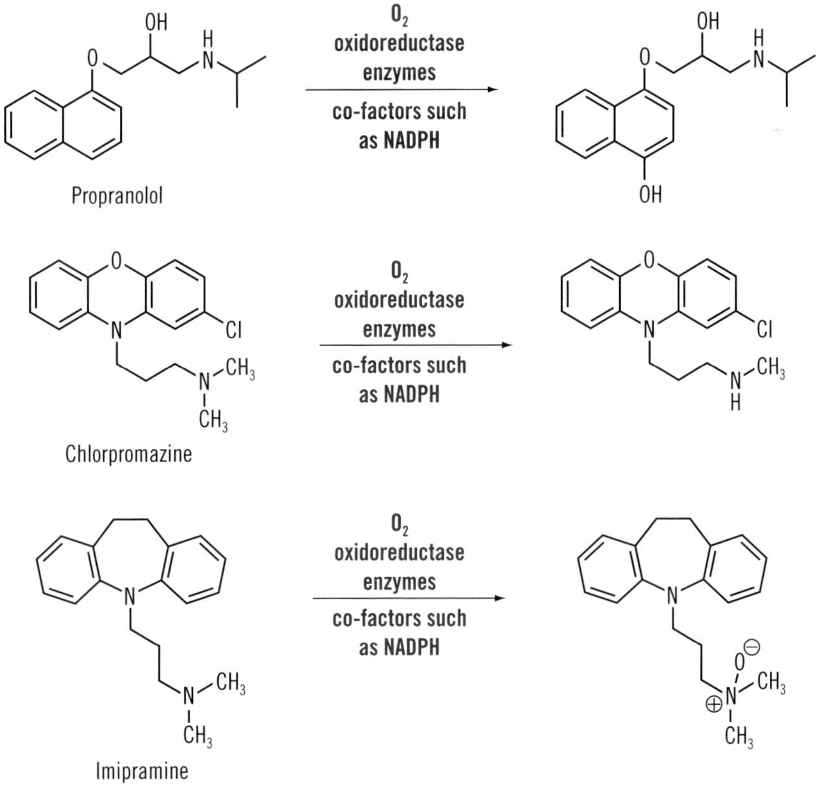

FIGURE 8.4 Metabolic oxidation of propranolol, chlorpromazine and imipramine.

In the first example shown in Fig. 8.4, the β-blocker propranolol is oxidized to the phenolic metabolite shown. This process replaces a C–H bond with a C–OH bond. It therefore introduces a new polar functional group into the structure, and so increases its hydrophilicity and decreases its lipophilicity.

In the second example, metabolic oxidation results in the removal of an N-methyl group from chlorpromazine. The result is that an N–CH$_3$ bond is replaced by an N–H bond. This change increases the hydrogen-bond-donating capacity of the compound, and so again increases its hydrophilicity.

In the last example, an oxygen atom is introduced onto the nitrogen of a tertiary amino group of imipramine. The resulting metabolite is an amine-N-oxide. This is a polar functional group, so again the polarity of the product metabolite is increased compared to the substrate. In general, therefore, Phase I oxidations tend to produce metabolites with increased water solubility and decreased lipid solubility, with correspondingly lower log P values.

Phase I reductions are carried out by the same or similar enzyme systems as the oxidations, usually at low oxygen levels. The reduction processes are less significant in terms of their impact on physicochemical properties, and so will not be considered further.

8.4.2 Phase II metabolism

In Phase II metabolism, compounds are coupled to certain specific biochemicals to produce conjugates. For this to happen, the substrate must have a free protic functional group, that is, an –OH, –NH, –SH or similar group. The substrates for Phase II metabolism are either xenobiotics with free protic functional groups, or the Phase I metabolites of xenobiotics. As can be seen from the propranolol and chlorpromazine examples in Fig. 8.4, Phase I processes often result in the introduction of protic groups. Hence, these metabolites can become substrates for Phase II processes.

A second requirement for Phase II metabolites is that reactive 'high-energy' forms of the conjugating species are available. These 'high-energy' forms are usually strongly electrophilic compounds. The enzymes that catalyse the conjugation reactions are generally known as transferases. Figure 8.5 shows some examples of the more common Phase II processes.

The first example shown in Fig. 8.5 is a conjugation to glucuronic acid. This is the most important of the Phase II processes in terms of frequency of occurrence. The high-energy form of glucuronic acid is uridine diphosphate glucuronic acid (UDP-glucuronic acid). The catalysing enzyme is uridine diphosphate glucuronyl transferase (UDP-glucuronyl transferase), and the product conjugates are known as glucuronides. Addition of the glucuronic acid group significantly increases both the molecular weight and the polarity of the substrate. The increase in molecular weight can reduce the rate of excretion of these molecular conjugates in the urine and prolong the lifetime of the dose by **biliary recycling**. Biliary recycling is explained in Section 8.5.1.

The second example shown in Fig. 8.5 is the sulfation of oestrone. The group transferred by conjugation in this case is a sulfate group. The high-energy form of sulfate for these processes is 3′-phosphoadenosine-5′-phosphosulfate, and the catalysing

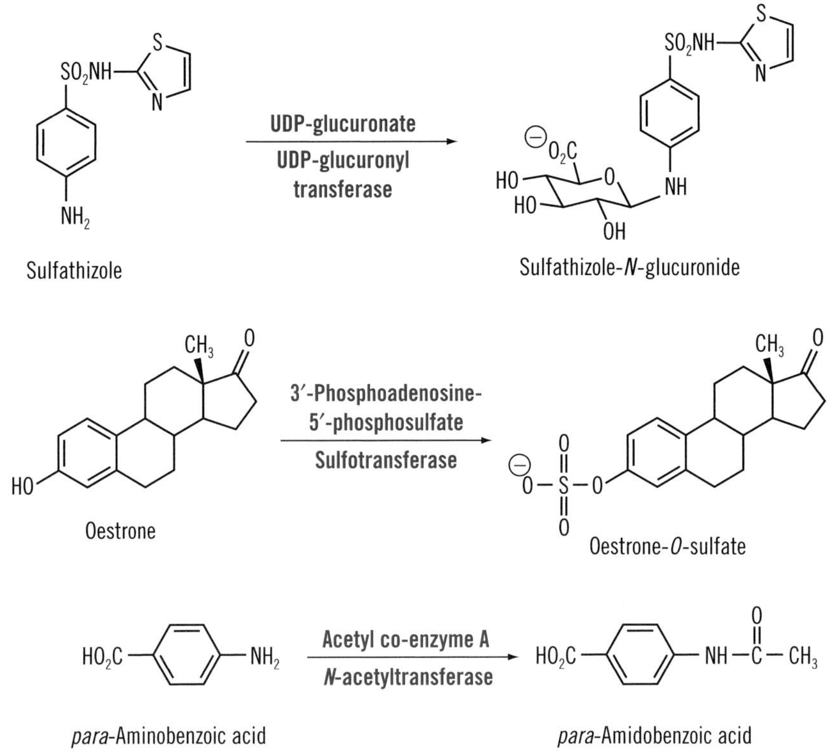

FIGURE 8.5 Examples of Phase II metabolic processes.

enzyme is sulfotransferase. The sulfate group is highly polar as it remains negatively charged even at relatively low pH. The effect of this process is to convert a relatively lipophilic compound, oestrone, into a hydrophilic one, oestrone-O-sulfate.

The third example in Fig. 8.5 is the acetylation of *para*-aminobenzoic acid. The high-energy form in this case is acetyl co-enzyme A, and the catalysing enzyme is *N*-acetyltransferase. In this case, both the substrate and the product are relatively polar compounds.

Other species that are conjugated to xenobiotics by Phase II metabolic processes include glucose, amino acids, and methyl groups. The processes follow the pattern shown in Fig. 8.5, in which the xenobiotic substrate reacts with a high-energy form of the conjugating group, and the process is catalysed by a transferase enzyme. An exception to the pattern shown in Fig. 8.5 is conjugation by glutathione. In these cases, it is the substrates that are the high-energy forms.

In the body, a drug substance with the appropriate functional group can be exposed to both Phase I and Phase II metabolism. The two phases may operate in series, that is, the Phase I metabolites of a drug may become the substrates for Phase II processes. The metabolic pathways for phenacetin, shown in Fig. 8.6, illustrate how Phase I and Phase II processes can combine.

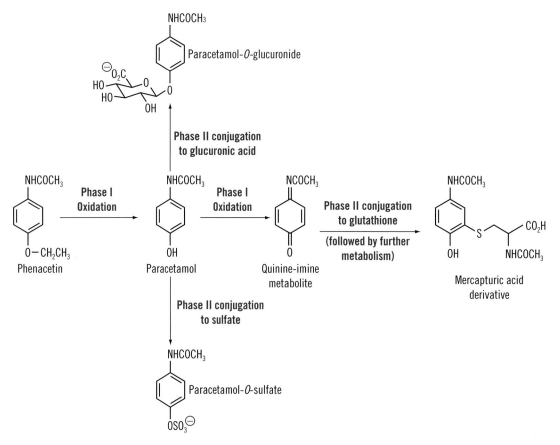

FIGURE 8.6 Metabolic pathways for phenacetin.

A Phase I oxidative process results in cleavage of the $O–CH_2CH_3$ bond in phenacetin. The product metabolite happens to have the structure of paracetamol. The −OH group in paracetamol is available to undergo Phase II conjugations, to give the glucuronide and sulfate metabolites. However, it can also undergo further Phase I oxidation, giving a reactive quinone-imine metabolite, which is toxic. This metabolite can be 'trapped and removed' by reaction with glutathione, ultimately giving a mercapturic acid derivative, which is excreted.

The general trend of the processes shown in Figs 8.3–8.6 is to produce metabolites that are more polar and more hydrophilic that the starting substrates. As stated earlier, the key benefit of these outcomes is that they aid excretion. More hydrophilic compounds have greater water solubility and hence greater tendency to be excreted. Now let us move on to understand the excretion process.

8.5 EXCRETION

Drugs can be excreted from the body by a number of routes: via the kidneys, the bile, and the lungs. Metabolism transforms compounds into molecules that can be readily excreted from the body. Compounds are required to have certain physicochemical properties to be excreted by the kidneys: they must be of a low molecular weight and polar at physiological pH. In general, Phase I and Phase II metabolism tends to produce metabolites that are more polar, hence more water soluble and sometimes of a lower molecular weight than the initial drug molecule. In this section we see how this aids excretion.

8.5.1 Excretion processes

The most common route of drug excretion is via the kidneys. Drugs can also be excreted via the bile and lungs. The excretion of a drug from the kidneys involves a number of stages. The first stage is 'glomerular filtration'.

Glomerular filtration

Due to the pressure of blood in the arteries, some of the blood passes into the kidney through the glomerulus. The glomerulus is a clump of capillaries, which have a number of openings (fenestrations) in their walls. The glomerulus acts as a filter for blood. The openings in the capillaries are large enough to allow unbound drug molecules, < 60 000 Da, to pass out of the capillaries but small enough to retain protein-bound drugs and blood cells. The fluid that passes through the walls of capillaries passes into the proximal tubule, as shown in Fig. 8.7.

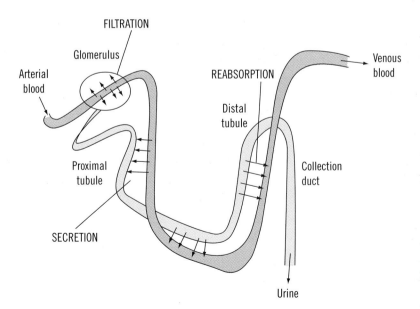

FIGURE 8.7 Schematic diagram illustrating the renal excretion process.

Tubular secretion

During the second stage '**tubular secretion**', weak acids and weak bases are actively transported (against the concentration gradient) from the capillaries surrounding the proximal tubule into the proximal tubule by carrier-mediated transporters.

These carrier transporters are saturable and the interaction of drugs with these carrier transporters is competitive. If two drugs are present then both can interact with a transporter such that the drug with the higher binding activity may inhibit the binding of the drug with the lower binding activity. For example, both penicillin and probenecid compete for the acid transporter. The presence of probenecid in the systemic circulation can reduce the excretion of penicillin via the kidneys. A second example is cimetidine and procainamide that both compete for the basic transporter.

Tubular reabsorption

Water reabsorption back into the systemic circulation from the proximal tubule reduces the volume of liquid in the renal tubule from approximately 180 L to 1–2 L each day. As water is reabsorbed from the proximal tubule, the concentration of the solution in the proximal tubule is increased. This increase in concentration creates a concentration gradient between the solution within the tubule and the capillaries, which acts as a driving force for the reabsorption of lipophilic molecules back into the systemic circulation by diffusion from the distal tubules. This process of reabsorption is called '**tubular reabsorption**'. Ionized and polar drugs cannot be reabsorbed and are excreted in the urine. Whether a drug is ionized or un-ionized in the distal tubule depends on the pH of the urine and the pK_a of the drug. (The pH of the urine can vary widely – between 4.5 and 8.0.)

Biliary excretion

Bile is secreted by the liver via the bile duct into the gastrointestinal tract and is eliminated from the body in the faeces. Some drugs and their metabolites are excreted by the liver into bile. Specifically, ionized, polar and lipophilic molecules with molecular weights greater than around 400 Da are excreted in this way. However, lower molecular weight molecules are reabsorbed into the systemic circulation before being excreted from the bile duct. Conjugation to glucuronic acid increases the molecular weight of a compound by 192 Da. Hence, conjugation of a compound with molecular weight between 200 and 400 Da to glucuronic acid can alter its excretion from the urinary to the biliary route. An example of such a compound is the antibiotic chloramphenicol, Fig. 8.8.

Biliary recycling

Chloramphenicol (and its Phase I metabolites) have molecular weights less than 400 Da and so would normally be excreted in the urine. Phase II conjugation to glucuronic acid gives metabolites of molecular weight greater than 400 Da. These are excreted via the bile duct into the gastrointestinal tract. However, while in the gastrointestinal tract, a significant portion of the glucuronide metabolites are hydrolysed by β-glucuronidase enzymes. This process reverses the glucuronidation process, releasing unconjugated

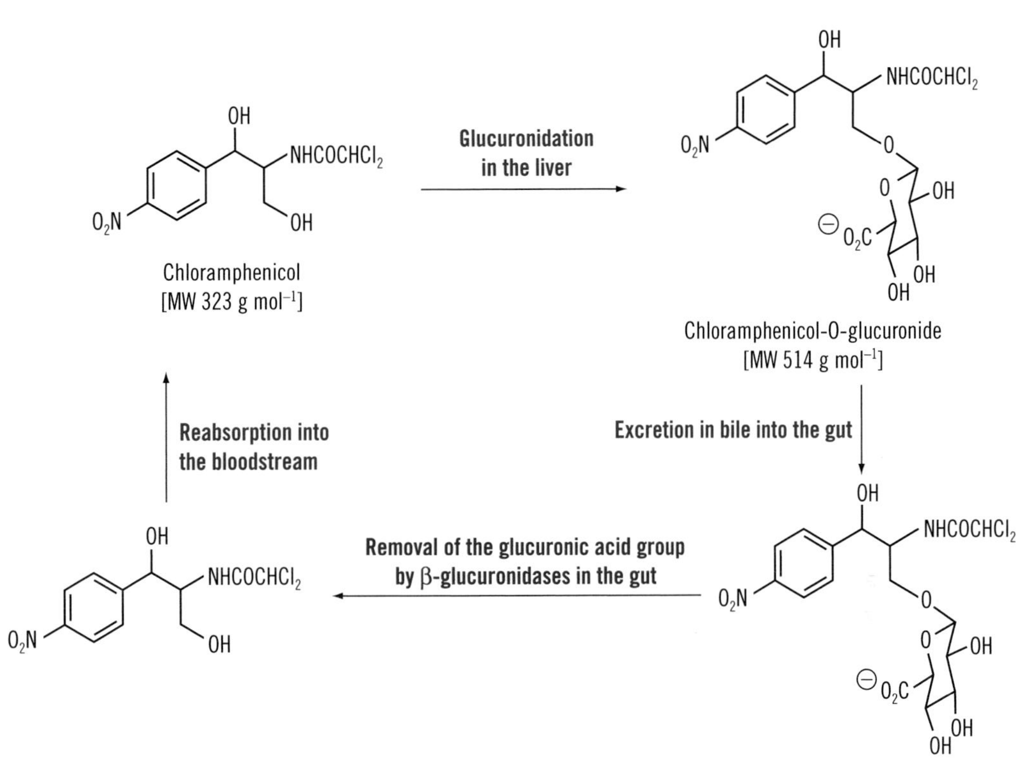

FIGURE 8.8 Biliary recycling of chloramphenicol.

chloramphenicol. A portion of the released chloramphenicol is then reabsorbed from the gastrointestinal tract into the blood stream (Fig. 8.8).

This process of glucuronidation, excretion into the bile, degradation by β-glucuronidases, and reabsorption into the bloodstream is known as '**biliary recycling**'. It can be significant for drugs of molecular weight in the range $200-400$ g mol^{-1}. The effect of biliary recycling is to reduce the rate of excretion of the drug and to prolong the lifetime of the dose. This can be significant for drugs such as steroid-based contraceptives, as biliary recycling enhances the duration of action of the drug and reduces the frequency of administration.

Pulmonary excretion

Gaseous and volatile compounds can be excreted from the lungs. Indeed, the breathalyzer test exploits excretion from the lungs in order to quantify the amount of ethanol in the blood stream. A proportion of gaseous anaesthetics is also excreted in expired air.

8.5.2 Pharmacokinetic aspects of elimination

A number of pharmacokinetic parameters can be calculated to describe the time course over which drug molecules are eliminated from the body. It is important to understand the rate of drug elimination: in order to maintain drug plasma concentrations within

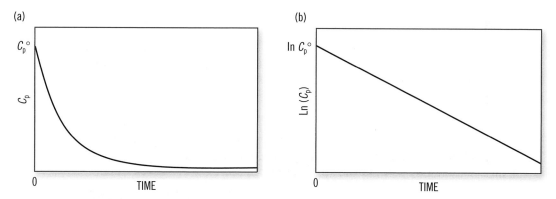

FIGURE 8.9 (a) Plot of drug plasma concentration (C_p) against time, and (b) plot of the natural logarithm of drug plasma concentration (ln C_p) against time. C_p^o refers to drug plasma concentration immediately after intravenous administration ($t = 0$).

the therapeutic range, the level of drug input into the body must equal the level of drug elimination. If the drug is administered at the same rate as the drug is removed, then the plasma concentration levels can be maintained at a state referred to as a **steady-state plasma** concentration.

Elimination rate constant

The **elimination rate constant**, k_{el}, represents the fraction of intact drug eliminated from the body per unit time, and takes into account the processes of excretion and metabolism. If the drug plasma concentration (C_p) is plotted against time, the relationship is not linear but exponential, as shown in Fig. 8.9(a) and described mathematically by

$$C_p = C_p^o \cdot e^{-k_{el} \cdot t}, \tag{8.7}$$

where C_p^o is the plasma concentration at time, $t = 0$.

If the drug plasma concentrations are converted to the natural log of their values then a straight line relationship is obtained. The elimination constant is the slope of this line, as shown in Fig. 8.9(b) and described mathematically by eqn 8.8. The steeper the slope of the line is, the faster the rate of elimination of the drug from the body

$$\ln C_p = \ln C_p^o - k_{el} \cdot t. \tag{8.8}$$

Clearance

The pharmacokinetic parameter '**clearance**' can be used to describe elimination. Clearance (Cl) refers to the volume of blood cleared of drug per unit time, L min^{-1} or L h^{-1}. The volume of blood in a patient will vary with patient size. Therefore, clearance is often expressed as L min^{-1} kg^{-1} or L h^{-1} kg^{-1}. Clearance can describe total clearance (Cl_{tot}) or partial clearance such as clearance from the kidneys, renal clearance (Cl_R), or from the liver, hepatic clearance (Cl_H).

EXAMPLE 8.5

A drug X has a volume of distribution of 60 L. The clearance of the drug is 6 L min⁻¹. This means that 6 L of plasma is cleared of drug every minute. The dose of drug administered intravenously is 60 mg. The initial plasma concentration of a drug immediately after administration will be 1 mg L⁻¹ (60 mg/60 L).

After 1 minute, 6 L of plasma will be cleared of drug: as 1 L contains 1 mg drug, 6 L will contain 6 mg, so 6 mg of drug will be removed, as shown in Table 8.1. The amount of drug remaining in the plasma is reduced to 54 mg after 1 minute, and the new plasma concentration is 54 mg/60 L (0.9 mg L⁻¹).

In the next minute a further 6 L will be cleared of drug. This time, 5.4 mg of drug will be cleared and the amount remaining will be 48.6 mg and so on as time proceeds the amount of drug remaining will decrease as the drug is cleared.

If the amount of drug remaining is plotted against time the relationship is exponential, as shown in Fig. 8.10(a). The amount of drug eliminated per unit time is not constant but decreasing with time. If the amount remaining is converted to its natural log, then a straight line relationship is obtained. The slope of this line is the elimination constant, as shown in Fig. 8.10(b).

TABLE 8.1 This table shows the determination of the amount of drug X (volume of distribution 60 L and clearance 6 L min⁻¹) that is removed from and remaining in the plasma at time intervals after administration.

Time (min)	Volume of distribution (L)	Drug present (mg)	Concentration (mg/L)	Volume cleared (L)	Drug cleared (mg)	Drug remaining (mg)
0	60	60.0	1.0	6.0	6.00	54.00
1	60	54.0	0.9	6.0	5.40	48.60
2	60	48.6	0.8	6.0	4.86	43.74
3	60	43.7	0.7	6.0	4.37	39.37
4	60	39.4	0.7	6.0	3.94	35.43
5	60	35.4	0.6	6.0	3.54	31.89
6	60	31.9	0.5	6.0	3.19	28.70
7	60	28.7	0.5	6.0	2.87	25.83
8	60	25.8	0.4	6.0	2.58	23.25
9	60	23.2	0.4	6.0	2.32	20.92
10	60	20.9	0.3	6.0	2.09	18.83

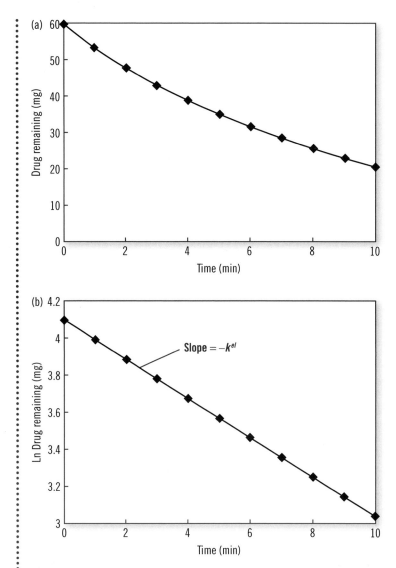

FIGURE 8.10 (a) Plot of drug remaining against time, and (b) plot of the natural logarithm of a drug remaining against time.

Extraction ratio

The '**extraction ratio**' is a measure of the fraction of the drug that is irreversibly removed from the body during a single pass through an organ, such as the kidneys or liver. For example, the hepatic extraction ratio (E_H) is a measure of the amount of drug removed from the blood during one pass through the liver compared to the total amount of drug in the blood, as shown in eqn 8.9. It can be calculated if the clearance from that organ is known. It is also dependent on the blood flow (Q) through the organ.

$$E_H = \frac{Cl_H}{Q_H}.$$

(8.9)

EXAMPLE 8.6

The blood flow through the liver is 100 L h^{-1}. The clearance of the drug from the liver is 10 L h^{-1}. This means that 10% of the drug passing through the liver is extracted by the liver. The extraction ratio is 0.1.

A drug with a high extraction ratio by an organ is extensively metabolized or eliminated by that organ. Immediately after a drug is injected into the systemic circulation elimination begins due to a combination of metabolism and excretion. The plasma concentration decreases rapidly initially and more gradually at later time points, as shown in Fig. 8.9.

Half-life

The **half-life** ($t_{1/2}$) of a drug in the body after administration is the time required to reduce the drug plasma concentration by half. The half-life is useful for estimating the clearance of the drug from the system. The half-life of a drug can be determined from the elimination constant (k_{el}), The time required to reduce the initial drug plasma level by half can be determined using eqn 8.8. $C_p^o/2$ can be substituted for the plasma concentration C_p at time t and $t_{1/2}$ can be substituted for t. Equation 8.8 can be rewritten as shown by

$$k_{el} \cdot t_{1/2} = \ln C_p^o - \ln \left(\frac{C_p^o}{2} \right). \tag{8.10}$$

This equation can be rearranged and the $t_{1/2}$ can be calculated as shown in

$$t_{1/2} = \frac{0.693}{k_{el}}. \tag{8.11}$$

The elimination rate constant can be determined from the slope of the line obtained when the drug remaining is plotted against time, as shown in Example 8.5.

 EXAMPLE 8.7

The initial plasma concentration of theophylline after intravenous injection to a patient is 6 μg mL^{-1}. The theophylline plasma concentration is measured at regular intervals and after 2 hours the plasma concentration is reduced to 5.6 μg mL^{-1}. Calculate the elimination constant (k_{el}) and the half-life ($t_{1/2}$) of theophylline in this patient.

In order to calculate the half-life ($t_{1/2}$) we need to first calculate the elimination constant (k_{el}). The elimination rate constant can be determined from the slope of the line obtained when the drug remaining is plotted against time. As two points on this line are known, the slope of the line can be determined, as

$$\text{slope} = \frac{\ln C_1 - \ln C_2}{t_1 - t_2} \rightarrow \text{slope} = \frac{\ln 6 - \ln 5.6}{2 \text{ h}} = 0.05 \text{ h}^{-1}$$

The elimination constant (k_{el}) is calculated to be 0.05 h^{-1} and the half-life can be determined using eqn 8.11 as

$$t_{1/2} = \frac{0.693}{k_{el}} \rightarrow t_{1/2} = \frac{0.693}{0.05 \text{ h}^{-1}} = 13.86 \text{ h}.$$

The half-life of drugs in the body is normally between 1 and 24 hours. Drugs with half-lives of greater than 24 hours can stay in the body for a long time. While this may be desirable for prolonged action it can be undesirable from a safety point of view. If the patient takes an overdose of a drug or a drug causes toxic effects in a patient, it will take a long time for a drug with a long half-half to clear the patient's system.

⊡ 8.6 SUMMARY

Only a proportion of the drug administered may be absorbed from the site of administration into the systemic circulation. Drug plasma concentration can provide information regarding drug absorption, plasma proteins, metabolism and drug excretion.

Pharmacodynamics is the study of the relationship between drug plasma concentration and pharmacological and toxicological responses. Pharmacokinetics is the study of the time course of drug movement in the body.

Immediately after drug absorption the drug is distributed throughout the body. During absorption and distribution the drug will encounter enzymes that may metabolize drug molecules. Elimination is the process of drug removal from the body by excretion.

The bioavailability fraction (F) gives information regarding the extent of drug absorption into the system circulation after non-intravenous administration. The rate of absorption can be expressed as the time required to reach the maximum drug concentration (C_{max}) and is referred to as the T_{max}.

A large volume of distribution indicates that the drug is distributed widely throughout the body or highly concentrated in certain tissues. A small volume of distribution indicates that the drug is primarily present in the systemic circulation.

Proteins present in plasma can bind a number of drug molecules preventing their distribution from the systemic circulation into other tissues. Drug binding to plasma proteins is reversible. Equilibrium exists between free drug and bound drug in the plasma.

The action of the body's enzymes and co-factors on drugs and other xenobiotics is the formation of new products known as metabolites. In Phase I metabolism, drug compounds undergo reactions such as hydrolysis, oxidation and/or reduction. In Phase II metabolism, drugs or their metabolites are coupled to certain biochemical species to form new metabolites known as conjugates. Both the processes of Phase I and Phase II metabolism directly affect the physicochemical properties of the drug substrate and facilitate excretion.

Drugs can be excreted from the body via the kidneys, the bile and the lungs. Compounds are required to have certain physicochemical properties to be excreted by the kidneys. Bile is secreted by the liver via the bile duct into the gastrointestinal tract and eliminated from the body in the faeces. Some drugs and their metabolites are excreted by the liver into bile. Gaseous and volatile compounds can be excreted from the lungs.

The elimination rate constant, k_{el}, represents the fraction of intact drug eliminated from the body per unit time. Clearance (Cl) refers to the volume of blood cleared of drug per unit time. The extraction ratio is a measure of the fraction of drug that is irreversibly removed from the body during a single pass through an organ. The half-life ($t_{1/2}$) is the time required to reduce the drug plasma concentration by half.

📖 FURTHER READING

Coleman, M. D. (2005) *Human Drug Metabolism, An Introduction.* Chichester: John Wiley & Sons.

Gibson, G. G., Skett, P. (1994) *Introduction to Drug Metabolism* (2nd edn). London: Blackie Academic & Professional.

Gilaldi, M. and Perrier, D. (2007) *Pharmacokinetics* (2nd edn). New York: Marcel Dekker.

Patrick, G. L. (2005) *An Introduction to Medicinal Chemistry* (3rd edn). Oxford: Oxford University Press.

Rang, H. P., Dale, M. M., Ritter, J. M., Flower, R. J. (2007) *Pharmacology* (6th edn). Edinburgh: Churchill Livingstone.

❓ EXERCISES

8.1 The drug timolol is administered to patients in the form of an intravenous injection and an eye drop. Each patient is administered a single intravenous injection containing 0.2 mg timolol and eye drops containing 0.3 mg timolol. After plotting timolol plasma concentration against time, the average AUC for the intravenous dose is calculated to be 6.47 ng mL^{-1} h and the AUC for the eye drops is calculated to be 4.78 ng mL^{-1} h. Calculate the bioavailability fraction of the eye drop.

8.2 A digoxin plasma concentration of 1.562 µg L^{-1} is required to treat heart failure. Digoxin has a volume of distribution of 7.2 L kg^{-1}. What intravenous dose of digoxin is required to achieve plasma levels of 1.562 µg L^{-1} in a 60 kg patient?

8.3 After administration of a dose of atorvastatin to 10 patients, the average blood drug concentration determined was 4.56 ng mL^{-1} and the average plasma concentration determined was 6.77 ng mL^{-1}. Calculate the blood plasma ratio and then explain what this ratio means in terms of drug distribution in the blood.

8.4 If the fraction of unbound phenytoin in plasma (f_u) is 0.07 and the total concentration of phenytoin in plasma is 82 ng mL^{-1} what would be the resultant unbound phenytoin plasma concentration?

8.5 Calculate the extraction ratio of a drug from the liver if the blood flow through the liver is 125 L/h. The clearance of drug from the liver is 25 L/h.

8.6 The initial plasma concentration of theophylline after intravenous injection to a patient is 4.5 µg mL^{-1}. The theophylline plasma concentration is measured at regular intervals and after 2 h the plasma concentration is reduced to 3.2 µg mL^{-1}. Calculate the elimination constant (k_{el}) and the half-life ($t_{1/2}$) of theophylline in this patient.

A.1

COMPREHENSIVE LIST OF ROUTES OF ADMINISTRATION

Term	Description	Example of API	Indication
Otic	to the ear	Gentamicin	Infection of the ear
Ophthalmic	to the eye		
Conjunctival	to the conjunctiva	Cromolyn	Conjunctivitis
Subconjunctival	within the cornea	Triacinolone	Scleritis
Intracorneal	beneath the conjunctiva	Amphothericin B	Fungal keratitis
Intraocular	into the eye		
Retrobular	behind the eyeball	Lignocaine	Anaesthesia prior to surgery
Intravitreal	into the vitreous humour	Triamcinolone	Macular oedema
Nasal	to the nose	Betamethasone	Inflammation of the nose
Nasogastric	to the stomach via nose	Cimetidine	Gastric ulceration
Oropharyngeal	to the mouth and pharynx	Hexetidine	Infections of oropharynx
Oral	to or by way of the mouth	Paracetamol	General analgesia
Peroral	through or by way of the mouth		
Buccal	toward the cheek	Prochlorperazine	Vertigo
Peridontal	around the teeth	Doxycycline	Peridontal infection
Sublingual	beneath the tongue	Glyceryl trinitrate	Angina pectoris
Enteral	directly into the intestines		
Rectal	to the rectum	Meloxicam	Rheumatoid arthritis
Intravesical	within the bladder	Pharmorubicin	Superficial bladder carcinoma
Topical	to the outer surface of the body		
Cutaneous	to the skin	Benzoyl peroxide	Acne vulgaris
Subcutaneous	beneath the skin	Insulin	Diabetes mellitus
Intradermal	into the dermis	Triamcinolone	Inflammation of dermal lesions
Transdermal	through the dermal layer to systemic circulation	Fentanyl	Analgesia

Term	Description	Example of API	Indication
Implantation	by implanting	Estradiol	Hormone replacement therapy
Intravascular	within a vessel		
Intra-arterial	within an artery	Floxuridine	Colon, kidney or stomach cancer
Intravenous	within a vein	Gentamicin	Systemic infections
Intracoronary	within the coronary arteries	Sirolimus	Ischaemic disease in coronary arteries
Intramuscular	within a muscle	Lorazepam	Acute anxiety states
Transmucosal	across the mucosa	Fentanyl	Break through pain in cancer patients
Intracardiac	within the heart	Isoproterenol	Cardiac arrest
Intrapericardial	within the pericardium	Mechlorethamine	Malignant effusions
Intraspinal	within the vertebral column		
Epidural	upon or over the dura mater	Morphine	Analgesia
Intradiscal	within a disc	Bupivacaine	Degenerative disc analgesia
Intrathecal	within the cerebrospinal fluid	Pethidine	Anaesthesia
Perineural	surrounding a nerve	Ropivacaine	Acute pain management
Intratumour	within a tumour	Carmustine	Brain tumour
Intra-articular	within a joint	Methylprednisolone	Rheumatoid arthritis
Intratendinous	within a tendon	Methylprednisolone	Tendinitis
Intraductal	within the duct of a gland	Doxorubicin	Breast cancer
Endocervical	within the canal of the cervix uteri	Prostaglandin E2	Induction of labour
Vaginal	into the vagina	Clotrimazole	Candidal vaginitis
Intrauterine	within the uterus	Progestin	Contraceptive
Intracorporus	within the corporus cavernosa	Phentolamine	Erectile dysfunction
Intrapulmonary	within the lungs or its bronchi	Salbutamol	Asthma
Endotracheal	directly into the trachea	Adrenaline	Cardiac arrest
Transtracheal	through the wall of the trachea	Oxygen	Chronic obstructive airways disease
Intrapleural	within the pleura	Bleomycin	Malignant pleural effusion

A.2
BRIEF REVIEW OF LOGARITHMS

Logarithms (or 'logs') are widely used in the pharmaceutical sciences, especially to make it easier to deal with very large or very small numbers.

If

$$y = a^x,$$

then we define

$$\log_a y = x.$$

We say that the logarithm of y to the base a is x.

We cannot state the logarithm of a number without specifying the base. The most commonly used base is 10 (logarithms to the base 10 are called common logarithms). Logarithms to the base e (approximately 2.718), known as natural logarithms, are also often encountered. These are given the symbol 'ln'.

As logarithms are exponents (in a^x, x is called the exponent), they follow the laws of exponents. The most important of these laws are listed here.

$$a^x a^y = a^{x+y}$$

$$a^x/a^y = a^{x-y}$$

$$(a^x)^y = a^{xy}.$$

Based on these, the main rules for manipulating logarithms are the following.

$$\log_a x + \log_a y = \log_a(x + y)$$

$$\log_a x - \log_a y = \log_a(x/y)$$

$$n \log_a x = \log_a(x^n).$$

Special cases to note are

$$\log_a 1 = 0,$$

$$\log_a a = 1,$$

and

$$\log_a(a^n) = n.$$

A.3

ACTIVITIES OF IONS IN SOLUTION

Many pharmaceuticals are used in salt forms that behave as electrolytes in aqueous solutions. (Electrolytes are solutions in which the solutes are ionic and that can conduct electric current.) It is also very common that the pharmaceutically active forms of acidic or basic compounds are the ionized forms. We need to be able to relate the energy of an ion to its concentration. The energy we need to consider is the chemical potential, μ. Equation 3.20 (see Chapter 3, Section 3.1.3) can be used to describe the chemical potential of an ion in solution as follows

$$\mu_i = \mu_i^\circ + RT \ln a_i,$$

in which a_i is the activity of an ion of type i. From eqn 3.21 ($a_i = \gamma_i m_i$) in which m_i is the concentration of ion i, and γ_i is the activity coefficient of that ion at that concentration. The activity coefficient, γ, allows us to convert the concentration of the ion (m), which is something we can measure, into the activity (a) of the ion, which we cannot measure directly. This is important, because it is the activity of an ion that determines its contribution to the overall energy of the system.

Combining these equations, the expression for the chemical potential of ion i may be written as

$$\mu_i = \mu_i^\circ + RT \ln m_i + RT \ln \gamma_i. \tag{A.1}$$

Equation A.1 tells us that the chemical potential of an ion in solution is made up of three contributions. These are the chemical potential of the ion in the standard state (μ_i°), a contribution from the concentration of the ion ($RT \ln m_i$), and a contribution from the activity coefficient of the ion ($RT \ln \gamma_i$).

For a solution in which no ion–ion interactions are occurring, the last term in eqn A.1 is equal to zero. This only occurs for so-called ideal solutions. The term $RT \ln \gamma_i$ therefore allows for the difference in chemical potential arising from departures from non-ideality – that is, from ion–ion interactions that exist in real pharmaceutical solutions.

The relationship between the activity coefficient of an ion, γ, and the overall composition of an ionic solution is a complex one. However, it is important pharmaceutically, as many pharmaceutical environments, such as biological fluids, are ionic solutions. Debye and Hückel showed that, for dilute ionic solutions,

$$\log_{10} \gamma_i = -A z_i^2 \sqrt{I}, \tag{A.2}$$

in which A is a constant, z_i is the valence of ion i, and I is a quantity known as the ionic strength of the solution. (Note that eqn A.2 uses the logarithm to the base ten of the activity coefficient rather than the natural logarithm.)

The ionic strength I was introduced by Lewis and Randall to deal with a complication that affects many ionic solutions, including biological and pharmaceutical ones. The complication is that ions with multiple charges, such as phosphate, have greater effects on activity coefficients than singly charged ions such as chloride. The ionic strength I is defined as

$$I = \frac{1}{2}\sum m_i z_i^2. \tag{A.3}$$

Equation A.3 has terms for both the concentration and the charge (in the form of valency z_i) of the ionic species present in a solution. Equations A.2 and A.3 describe the relationship between the activity of a single species of ion and the ionic strength of the medium. An electrolyte $C_x A_y$ will dissociate in solution as follows

$$C_x A_y \rightleftharpoons xC^{z+} + yA^{z-}$$

to give x cations of valence z^+ and y anions of valence z^-. (For a strong electrolyte, the equilibrium will lie entirely to the right-hand side.) The activities of the anions and cations are a_+ and a_-, respectively. The corresponding activity coefficients are γ_+ and γ_-. The following relationship, known as the Debye–Hückel limiting law can be derived

$$\log \gamma_\pm = -A z_+ z_- \sqrt{I}, \tag{A.4}$$

in which the term $\gamma_\pm$ is known as the mean ionic activity coefficient. The constant A is solvent dependent. For example, for water at 25 °C, A has the value 0.509 $mol^{-1/2}\,kg^{1/2}$. The Debye–Hückel limiting law is valid only for very dilute solutions, that is of concentration < 0.02 mol kg^{-1}. However, at higher concentrations, the following extended form of the Debye–Hückel equation may be used

$$\log \gamma_\pm = -\frac{A z_+ z_- \sqrt{I}}{1 + B\alpha\sqrt{I}},$$

in which B is a solvent-dependent constant and α is the mean effective ionic diameter.

▷ **WORKED EXAMPLE**

Calculate (i) the ionic strength and (ii) the mean ionic activity coefficient of a 0.5×10^{-3} mol kg^{-1} solution of histamine dihydrochloride (Fig. A.1) in water at 25 °C.

Equation A.3 is used to calculate the ionic strength. From the structure of histamine dihydrochloride shown in Fig. A.1, dissociation gives two moles of chloride ions for every mole of histamine ions. The valency of the chloride ions, z_-, is equal to one; while that of the histamine ion, z_+, is equal to two. Ionic strength I is therefore equal to

$$\tfrac{1}{2}[(0.5 \times 10^{-3} \times 2 \times 1^2) + (0.5 \times 10^{-3} \times 1 \times 2^2) = 1.5 \times 10^{-3} \text{ mol kg}^{-1}.$$

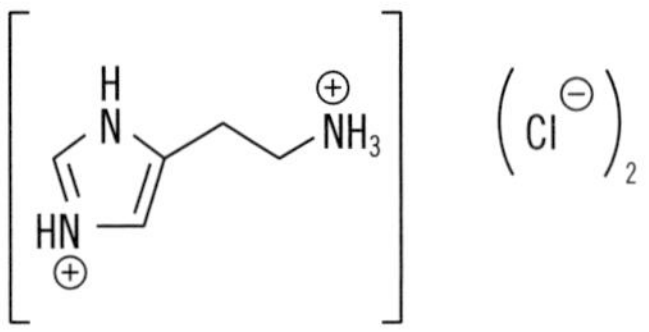

FIGURE A.1 Structure of histamine dihydrochloride.

The mean ionic activity coefficient can then be obtained from eqn A.4 using the value of constant A for water at 25 °C given above.

$$\log \gamma_{\pm} = -(0.509)(2)(1)\sqrt{1.5 \times 10^{-3}} = -0.0394,$$

hence, $\gamma_{\pm} = 0.9614$.

The mean ionic activity coefficient for an ionic substance gives us some idea of the extent to which that substance contributes to the energy of a system when it is in solution, that is, it gives us the *activity* of the ions. For example, the above value for histamine dihydrochloride tells us that the histamine and chloride ion contribute slightly less than their concentration values alone would suggest.

A.4

OSMOTIC PRESSURE

It is possible to have two solution phases separated by a membrane that allows passage of solvent molecules, but not of solute molecules. Such a membrane is known as a semipermeable membrane. Many biological membranes are semipermeable. As discussed in Section 3.2.1, a substance will transfer spontaneously from a phase in which it has higher chemical potential to a phase in which it has lower chemical potential. Solvent molecules therefore cross semipermeable membranes from the solution in which the chemical potential of the solvent is higher to that in which it is lower. This process is known as osmosis.

Figure A.2(a) illustrates an initial situation in which solutions A and B, having solvent molecules of differing chemical potentials, are separated by a semipermeable membrane. If the chemical potential of the solvent in solution A is μ_A and that of the solvent in solution B is μ_B, and if $\mu_A > \mu_B$, solvent molecules will flow across the membrane in the direction of solution B. This process will continue until it is counterbalanced by the increased pressure exerted by the atmosphere on solution B relative to solution A. This situation is illustrated in Fig. A.2(b). The difference in pressures exerted upon the two solutions necessary to prevent the spontaneous flow of solvent molecules across the membrane is equal to the osmotic pressure, Π. If two solutions separated by a semipermeable membrane are of equal osmotic pressure, no flow of solvent across the membrane occurs, and the solutions are said to be isotonic.

The following equation, relating osmotic pressure, Π, with volume V, temperature T and number of moles of solute n_2, was determined empirically by van't Hoff.

$$\Pi = \frac{n_2}{V} RT. \tag{A.5}$$

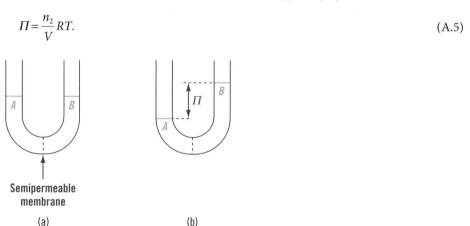

FIGURE A.2 (a) Schematic illustration of a U-tube containing two solutions A and B, separated by a semipermeable membrane, in which the chemical potentials of the solvent are μ_A and μ_B, respectively. (b) If $\mu_A > \mu_B$, solvent molecules cross the semipermeable membrane from solution A to solution B, until the difference in pressure exerted on solutions A and B is equal to the osmotic pressure, Π. Brown, T. E.; Lemay, H. E.; Bursten, B. E.; Burdge, J. R., Brunauer, L., Chemistry: The Central Science, 9th edn, © 2003, p. 507. Reprinted by permission of Pearson Education, Inc., Upper Saddle River, NJ.

Van't Hoff's equation tells us that osmotic pressure is directly proportional to solute concentration (n_2/V) and temperature. The equation holds only for dilute solutions that exhibit behaviour approaching ideality. To allow for this drawback, a more sophisticated version of the equation can be derived. First, the concentration term (n_2) is replaced. If the number of moles of the solute, n_2, is replaced by m_2/M_2, in which m_2 is the mass of the solute and M_2 is the molecular weight of the solute, eqn A.5 can be given as

$$\Pi = \frac{m_2 RT}{V M_2}. \tag{A.6}$$

We then need to introduce terms that allow for deviations from ideality, especially for ionic solutions. Van't Hoff introduced a correction factor, i. In the case of ideal solutions, the correction factor approaches the number of molecular or ionic species, v, expected to be formed when a substance dissolves. For example, for a neutral molecular substance such as glucose, $i = v = 1$. For an ionic substance such as NaCl, if behaving ideally, $i = v = 2$. However, a manifestation of non-ideal behaviour is that often, $i < v$. For example, $i = 1.9$ for 0.05 M solutions of NaCl at 25 °C. In applying eqns A.5 and A.6, allowance has to be made for ionization of solutes, as separation into ions increases the number of components in the solution. The right-hand side of the equations are therefore multiplied by v in the case of ionic solutes. However, to allow for derivations from ideality, a factor φ, equal to i/v, is also included, to give

$$\Pi = \frac{v\varphi m_2 RT}{V M_2}. \tag{A.7}$$

Factor φ is known as the practical osmotic coefficient. As ideal behaviour is approached, i approaches v, and φ approaches one. Equation A.7 is a more sophisticated version of eqn A.5 that attempts to allow for deviations from ideality.

The concentration of a solution can be described in terms of its osmolarity or osmolality, which is defined as the number of osmotically active particles per litre of solvent, or kilogram of solvent, respectively. This is equal to the molar or molal concentration, multiplied by $v\varphi$. Hence, for an ideal un-ionized substance, the osmolarity equals the molar concentration.

▶ **WORKED EXAMPLE**

A physiological saline solution contains 0.16 mol L^{-1} NaCl. The practical osmotic coefficient of such a solution is 0.928 [A. T. Florence and D. Attwood, *Physicochemical Principles of Pharmacy*, 3rd edn, 1998, Palgrave, reproduced with permission of Palgrave Macmillan]. The molecular weight of NaCl is 58.5 amu. Calculate the osmotic pressure, Π, of 500 mL of such a solution at 25 °C.

Use eqn A.7. For NaCl, $v = 2$. $R = 0.08314$ L bar K^{-1} mol^{-1}. 25 °C = 298 K. The mass of NaCl, m_2, in 500 mL equals $0.08 \times 58.5 = 4.68$ g. Hence

$$\Pi = \frac{2 \times 0.928 \times 4.68 \text{ g} \times 0.08314 \text{ L bar K}^{-1} \text{ mol}^{-1} \times 298 \text{ K}}{0.5 \text{ L} \times 58.5 \text{ g mol}^{-1}} = \frac{215.20 \text{ L g bar mol}^{-1}}{29.25 \text{ L g mol}^{-1}}$$

$$= 7.36 \text{ bar.}$$

A.5

SOLUTIONS TO 'END OF CHAPTER' EXERCISES

2.1 Molality: 0.071 mol kg^{-1}; mole fraction: 1.3×10^{-3}.

2.2 Equilibrium defining of propranolol – see Fig. A.3 Propranolol pK_b = 4.5 at 25 °C.

2.3 Percentage ibuprofen ionized at pH 3.5 = 11.2%.

2.4 Precipitation at pH 8.92 and lower.

3.1 $K_{eq} = [NO_2]^2 / [NO]^2[O_2] = 33.6$.

3.2 64 J mol^{-1} K^{-1}.

3.3 1.5 kJ mol^{-1}.

3.4 Two components (H$_2$O and CuSO$_4$), because the composition of any phase can be described using the equilibrium CuSO$_4$ + x H$_2$O $\rightleftharpoons$ CuSO$_4$.xH$_2$O.

3.5 The partial pressure of methanol is 4.86×10^4 N m^{-2}. The partial pressure of ethanol is 1.8×10^4 N m^{-2}.

3.6 1.61×10^{-5}.

4.1 The Miller index for the plane shown is 200.

4.2 The following process occur: CuSO$_4$.5H$_2$O $\rightarrow$ CuSO$_4$.3H$_2$O $\rightarrow$ CuSO$_4$.H$_2$O $\rightarrow$ anhydrous CuSO$_4$.

5.1 Particles at the upper end of the range will sediment fastest. Stokes' equation (5.2) indicates that the larger the particle radius the faster the sedimentation rate.

5.2 188 Pa.

5.3 3115 Da.

5.4 Surfactant B.

7.1 Atenolol 0.16, metoprolol 1.88, propranolol 3.21.

7.2 1.13. Due to its positive log P value, the drug will have a higher solubility in lipid-based media compared to aqueous media. Therefore, the drug will partition from aqueous medium into lipid-rich media such as cell membranes.

7.3 The new lead compound is acidic. Its log P is 4.0 and its pK_a is ~ 2.0.

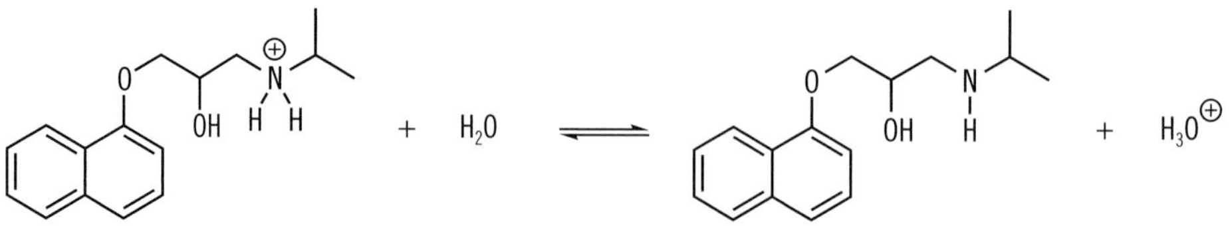

FIGURE A.3 Dissociation of the conjugate acid of propranolol.

8.1 0.49.

8.2 675 µg.

8.3 0.67. This value < 1 would indicate that the drug is mainly located in the plasma.

8.4 5.74 ng mL^{-1}.

8.5 0.2.

8.6 k_{el} is 0.17 h^{-1}, $t_{1/2}$ is 4.065 h.

GLOSSARY

Adipose tissue: Fat-based tissue

Albumin: Water-soluble proteins found in blood serum, milk, egg white and many other animal and plant tissues

Anaerobic bacteria: Bacteria with respiration that does not make use of oxygen

Analgesic: Pain-relieving drug

Apical surface: The surface of a layer of cells to which there is nothing attached

Astrocytes: Star-shaped cells found in the central nervous systems

Bacterial count: Number of viable bacteria

Biopharmaceutical class II drugs: Biopharmaceutical Classification System (BCS), was proposed by Amidon *et al.* in 1995. It classifies drugs into four different groups, depending on their solubility and permeability. Class II drugs have low aqueous solubility and high permeability

Bronchodilator: Substance that causes relaxation of bronchial muscle

Common ion effect: Effect on an equilibrium involving ionic species that occurs when additional quantities of one of the ionic species are introduced into the system. The equilibrium adjusts so as to reduce the concentration of that particular ionic species (the 'common ion')

Deliquescence: a solid becoming liquid, usually due to absorption of moisture

Dielectric constant: A measure of the ability of a substance (such as a solvent) to stabilize ions in solution. Solvents with higher dielectric constants have greater ability to stabilize and solvate ions

Dipole–dipole interactions: Attractive forces between molecules containing localized or overall dipole moments

Dipole moment: Measure of the polarization caused by unsymmetrical charge distribution

Dispersion forces: see 'London dispersion forces'

Effervescence: Release of bubbles

Efficacy: Power to produce effects or intended results

Freeze-drying: Removal of water by sublimation

Glycerides: Esters of glycerol and fatty acids

Heterocycles: Cyclic molecular segments containing at least one carbon atom and at least one non-carbon atom in the cycle

High-performance liquid chromatography (HPLC): Method of separation of molecular substances, normally for purposes of quantitative analysis. Involves differential migration of compounds partitioned between a mobile liquid phase and a solid stationary phase. The stationary phase is packed into a steel column and the mobile phase pumped through the column under pressure

Ischaemic: Deficient supply of blood to a body part, can be caused by an obstruction

London dispersion forces: Weak attractive forces between molecules that form as molecules approach each other, in which the electrons of one approaching molecule are attracted to the nuclei of the other

Macrophages: Large cells present in blood, lymph, and connective tissues that remove waste products, harmful microorganisms, and foreign material from the blood stream by engulfing the material within the cell

Milling: Processes for physically reducing the size of solid particles

Pharmacology: The science of drug action

Prophylaxis: A treatment that prevents a disease or stops the disease from spreading

Receptor: Biochemical entity that is the molecular 'point of contact' for a biologically active molecule. In current use, the term usually refers to a protein occurring (at least in part) at the surface of a cell

Steroid: Member of a class of biochemical entities characterized by a core structure consisting of four fused cycloalkanes, three of which are normally six-membered and one of which is normally five-membered

Van der Waals forces: Combination of London dispersion forces and dipole–dipole interactions

INDEX